Martin Jacobson

U.S. Department of Agriculture (Retired)
Beltsville, Maryland

Glossary of Plant-Derived Insect Deterrents

CRC Press, Inc.
Boca Raton, Florida

Library of Congress Cataloging in Publication Data

Glossary of plant-derived insect deterrents / Martin Jacobsen.
   p. cm.
Includes bibliographical references.
ISBN 0-8493-3278-8
1. Botanical insecticides—Dictionaries. 2. Insect baits and
repellents—Dictionaries. I. Title. II. Title: Plant-derived insect
deterrents
SB951.54.J33   1990
632'.951 — dc20   89-25310

Direct all inquiries to CRC Press, Inc., 2000 Corporate Blvd., N.W., Boca Raton, Florida, 33431.

© 1990 by CRC Press, Inc.

International Standard Book Number 0-8493-3278-8

Library of Congress Card Number 89-25310
Printed in the United States

THE AUTHOR

Martin Jacobson received his B.S. degree in chemistry from the City University of New York in 1940. He accepted an offer as a chemist with the Industrial Hygiene Division of the National Institutes of Health in Bethesda, Maryland. In 1942 he transferred to the Bureau of Entomology and Plant Quarantine of the U.S. Department of Agriculture's Agricultural Research Center, Beltsville, Maryland, as a research chemist to isolate, identify, and synthesize phytochemical pesticides, insect hormones, and insect sex pheromones. During this period, he pursued evening graduate studies in chemistry and microbiology at George Washington University, Washington, D.C. He also served as a part-time Research Associate in Chemistry at that university during the period 1944 to 1948.

From 1964 to 1972, Mr. Jacobson was an Investigations Leader with the Entomology Research Division at Beltsville, Chief of the Biologically Active Natural Products Laboratory from 1973 to 1985, and Research Leader (Plant Investigations) with the Insect Chemical Ecology Laboratory until his retirement from Federal Service in 1986. He is currently an agricultural consultant in private practice in Silver Spring, Maryland.

During his long career with the USDA, Mr. Jacobson spent several weeks in 1971 as a Visiting Scientist teaching a graduate course on insect pheromones and hormones in the Department of Chemistry, University of Idaho, Moscow, Idaho. He was invited to speak at and to organize numerous symposia at national and international scientific meetings in the United States, Europe, Asia, and Africa, in the field of pesticides and sex pheromones occurring naturally in plants and insects, respectively. His awards include the Hillebrand Prize of the Chemical Society of Washington in 1971, USDA Certificates of Merit and cash awards for research in 1965, 1967, and 1968, the McGregory Lecture Award in Chemistry at Colgate University (Syracuse, New York), two bronze medals for excellence in research at the 3rd International Congress of Pesticide Chemistry, Helsinki, Finland in 1974, USDA Director's Award on Natural Products research in 1981, and an Inventor's Incentive Award for commercialization of a boll weevil deterrent in 1983.

Mr. Jacobson has been a member of the American Chemical Society, Entomological Society of America, Chemical Society of Washington, Pesticide Science Society of Washington, American Association for the Advancement of Science, New York Academy of Sciences, and a Fellow of the Washington Academy of Sciences. He is the author or coauthor of more than 300 scientific reports in numerous journals, the author of four books (*Insect Sex Attractants*, Wiley, 1965; *Insect Sex Pheromones*, Academic Press, 1972; *Insecticides From Plants. A Review of the Literature, 1941-1953*, USDA Handbook 154, 1958; *Insecticides From Plants/A Review of the Literature, 1954-1971*, USDA Handbook 461, 1975), and editor of the book *Naturally Occurring Insecticides*, Marcel Dekker, 1971. He also holds six U.S. Patents on naturally occurring insecticides.

ACKNOWLEDGMENTS

I acknowledge, with thanks, the kind assistance of Dr. Douglas C. Ferguson and Dr. Robert D. Gordon, of the U.S. Department of Agriculture (U.S. National Museum), Washington, D.C., for supplying the author names of several of the insect scientific names.

To my daughter, Barbara, with love.

TABLE OF CONTENTS

I. Introduction

I. INTRODUCTION

Crop damage and loss due to feeding by insect larvae and adults is estimated to total billions of dollars each year in the U.S. alone, with comparable losses in many other developed countries. The loss is even more staggering in the developing countries, mainly in Asia and Africa. Many of the synthetic insecticides previously used for insect control have been banned or their use seriously curtailed because of concern about health and environmental effects.

There is ample evidence to show that the plant kingdom is a vast storehouse of chemical substances manufactured and used by these plants for defense from attack by insects, bacteria, and viruses. A wealth of literature has accumulated in reports,[1-4] scientific journals,[5-20] and books,[21-23] especially during the past 20 years on the subject of naturally occurring insect feeding deterrents and growth inhibitors. However, feeding by pests and other destructive insects is not limited to crop feeders, Considerable destruction of homes and household items is caused by feeding by such pest insects as termites, cockroaches, ants, and carpet beetles. A large group of blood-feeding insects, including mosquitoes, fleas, bedbugs, ticks, and biting flies cause much discomfort, and bites and stings by wasps, hornets, and enraged bees can make life unpleasant for both man and animals. Plant-derived deterrents to feeding by all of these insect species will be covered in this book, which is expected to serve a very useful purpose in view of the fact that the heretofore published information is completely scattered through the scientific literature. Review articles such as those by Beck,[5] Rosenthal,[19] Painter,[21] McIndoo,[24] Heal et al.,[25] and Jacobson[26,27] treat the older literature satisfactorily, but coverage of the more recent research is rather fragmentary.[28,29]

The mechanism by which insects detect feeding deterrents is chemosensory in nature, involving impingement of odor molecules from the plants upon special deterrent receptors usually found on the insect antennae.[5,15,30] The deterrent may be effective from a long distance, so that the insect will not immediately land on the plant, or from only a few inches. Alternatively, the insect may land on the plant but will not feed, or the insect may bite into the plant but refuse to feed further.[31-33] There is, however, another mechanism of insect olfaction involving bright coloration or color designs in or on the plant leaves serving to ward off insect attack (or in some cases attract).

This handbook brings together in one place sources which entomologists, chemists, botanists, insect ecologists, physiologists, and pharmacologists may consult for information to supplement that reported herein on the subject of plant-derived feeding deterrents. Almost 1500 plant species from 175 families are treated. It is my intention to report not only those plants found to be effective as feeding deterrents, but also which were tested and found to be ineffective. In addition, results of toxicity, repellency, sterility, or growth disruption are reported for those plants found to be effective as deterrents, as well as the identities, where known, of those chemical compounds identified as being responsible for the activity.

As the reader will see, there is a considerable amount of research that has been done

3

in this field, as well as follow-up research that is still in progress. A number of the plant species show promise for practical use as insect control agents. However, one must avoid becoming complacent about these insect feeding deterrents. Professor Metcalf and his students[34] at the University of Illinois, reporting on their research with the cucurbitacins, have shown that these intensely bitter compounds occurring in plants of the family Cucurbitaceae (squash, melon) that arose to repel herbivores and protect the plants from attack have become specific kairomone *feeding stimulants* for the destructive cucumber beetles of the genus *Diabrotica*. A striking example of counteradaptation by an insect to an insect feeding deterrent is that reported by Carroll and Hoffman[35] of the State University of New York at Stony Brook, The squash beetle, *Epilachna borealis* (Fabricius) biting into a squash leaf causes the plant to mobilize deterrent substances to the damaged region within 40 min. However, this wily beetle has learned to adapt to this situation by cutting a circular trench in the leaf before leisurely feeding on the encircled leaf material. Trenching behavior requires approximately 10 min. The beetle's strategy effectively prevents the mobilization of deterrent substances to the feeding area. Question: Who is more cunning, man or beast, in our struggle against harmful insects?

Payne[33] has presented a review of insect interactions with their host trees, Metcalf[36] has reviewed insect-plant co-evolution, and Dowd et al.[37] have reviewed the detoxification of plant toxins by insects. Duffy[38] has reviewed very well the sequestration of plant products by insects. Excellent discussions of the chemical aspects of deterrency are those by Grisebach and Ebel,[39] Meinwald et al.,[9] Rembold and Winter,[40] and Picman.[41]

II. Methods for Antifeedent Bioassay

II. METHODS FOR ANTIFEEDANT BIOASSAY

A. Crop Pests

In general, the test methods of choice or no-choice with vegetable and fruit insects involve incorporation of the plant extract or test compound(s) into the diet at definite concentration, exposure to the insect, and comparison of the percentage feeding over a period of time with the test material and with an untreated control. For field trials, the test materials are sprayed or dusted on the crop. Using various modifications of these methods, the occurrence of deterrency or acceptance was experimentally determined for *Spodoptera frugiperda*,[40] *S. littoralis*,[41,42] *Oncopeltus fasciatus*,[40] *Ostrinia nubilalis*,[43] *Plutella xylostella*,[44] *Plathypena scabra*,[45] *Argyrotaenia velutinana*,[46] *Cydia pomonella*,[46] *Conotrachelus nenuphar*,[47] *Carpophilus hemipterus*,[47] *Amyelois transitella*,[48] *Acalymma vittatum*,[49-51] *Diabrotica undecimpunctata howardi*,[52] and *Nilaparvata lugens*.[53]

Using an electronic device, Saxena and Khan[54,55] monitored the feeding behavior of *Nephotettix virescens* on rice plants sprayed with neem oil (*Melia azadirachta*) as compared with a control (plants sprayed with acetone alone).

Antonius et al.[56] were able to determine, by the use of electrophysiological responses, the primary sites of feeding deterrence on *Spodoptera litura* by various plant extracts. Electrophysiological and behavioral analyses were used by Dethier[57] to compare the gustatory sensitivity of *Danaus plexippus* and of *Manduca sexta* with that of *Phormia regina*. Similar methods were used by Norris[58] to measure the deterrency of *Scolytus multistriatus* by various plant-derived quinones.

Adler and Uebel[59] measured the feeding deterrence of a commercial preparation (Margosan-O) of neem seed kernels to adult *Dissosteira carolina*, *Diapheromera femorata*, and *Gryllus pennsylvanicus*. Application of the extract was made to pieces of corn leaves for *D. carolina* and to white oak leaf squares for *D. femorata* and *G. pennsylvanicus*.

B. Household Pests

Olfactory repellency of *Blattella germanica* by various plant essential oils was determined by a test tube method and beaker method.[60] Of 92 such substances, several mint oils were highly repellent to the roaches. Oils of lavender, citronella, rosemary, and spearmint, which are strong mosquito repellents, had little or no effect on the cockroach. The choice method of Goodhue and Tissol[62] using glass dishes was employed to test for repellency to four species of cockroaches.[63]

C. Blood-Feeding Pests

For several years, the Entomology and Plant Quarantine station of the U.S. Department of Agriculture at Gainesville, Florida used human subjects to test thousands of synthetic and natural substances as possible repellents for mosquitoes in the laboratory. These subjects, all of whom were staff members of the facility, smeared a bare arm with a solution of the test-substance and then inserted the treated arm through

7

a cloth sleeve into a wire screen cage containing several hundred adult mosquitoes. Rabbits, which later replaced the human subjects, were shaved on a patch of skin to which the test material was then applied. The effectiveness of the test materials was determined by the number of mosquito strikes made to the test site.[64]

III. Biological Test Results

III. BIOLOGICAL TEST RESULTS

The biological test results obtained with plant extracts or their components are given. The scientific and common names of the plant, in alphabetical order, are supplied within each plant family. The plant part used, test method utilized, and the test insects are discussed. For those plants shown to be effective the substance(s) responsible (of known) and their method of isolation are given, together with any available reported toxicity test results and pharmacological or toxicological properties reported for animal life.

A. Cryptogams

1. Algae

Dictyota dichotoma **Lamour,** brown alga: Volatile components obtained by steam distillation of the fresh plants, impregnated on castor bean leaf disks at 1000 ppm, strongly deterred 4th instar larvae of *Spodoptera littoralis*.[65] The component responsible for this activity was identified as 1-(1,3,4,5,6,7-hexahydro-4-hydroxy-3,8-dimethyl-5-azulenyl)ethanone. The crude distillate possessed insecticidal activity against *Musca domestica* and *Sitophilus oryzae*.

Sargassum tortile **C. A. Ag.,** brown alga: The insect growth inhibitor, crinitol (3,7,11,15-tetramethyl-2,6,10,14-hexadecatetraen-1,9-diol), isolated from a methanol extract of this alga inhibited the growth of *Pectinophora gossypiella* larvae after incorporation into the diet at 500 ppm.[66] It also completely inhibited the growth of *Escherichia coli* microorganisms.

2. Lichens

Letharia vulpina **(L.) Rue,** wolf lichen, waino: Vulpinic acid and atranorin (aβ-orcinoldepside) isolated from this lichen inhibited feeding by larvae of *Spodoptera ornithogalli* exposed to broccoli leaf disks at <0.6 and <0.03%, respectively, in Petri dishes in the laboratory.[67]

3. Fungi

Lactarius rufus **Fr.:** Aqueous extracts of this fungus contain lactorufin A, lactorufin B, isolactorufin, and a mixture of monohydroxy lactones.[68] Compound A and the mixture were strongly deterrent to feeding by adults of *Sitophilus granarius* and by larvae of *Tribolius confusus* and *Trogoderma granarium* in laboratory choice and nochoice tests. Lactorufin B was fairly deterrent and isolactorufin was only slightly deterrent.

Plagiochila fruticosa **Schust.,** liverwort
Plagiochila hattoriana **Blomquist,** liverwort
Plagiochila ovalifolia **Mitt.,** liverwort
Plagiochila yokogurensis **Inoue,** liverwort: Of four *ent*-seaquiterpene hemiacetals and two pungent methoxylagiochilines isolated from these liverworts,[69] only plagiochiline A was very strongly deterrent to feeding by *Spodoptera exempta* at 1 to 10 ng/cm^2.[70]

4. Ferns

Pteridium aquilinum **(L.) Kuhn,** bracken fern, eagle fern: Fresh bracken and crude extracts thereof deterred feeding or settling of nine species of insects in no-choice laboratory bioassays.[71] The insect species used were *Liocoris tripustulatus*, *Brachypterus glaber*, *Phyllobius pyri*, *P. argentatus*, *Pieris brassicae*, *Chilo partellus*, and *Schistocerca gregaria*. The resistance of these ferns was present all year but it was highest in May and June. The most active compounds isolated were several sesquiterpene pterosins, such as pterosin F.[72]

B. Phanerogams

FAMILY ACANTHACEAE

Justicia adhatoda **L.,** adhatoda, malabar nut tree: A mixture of 8 parts of the powdered leaves with 92 parts of gram seed exposed to adult *Callosobruchus chinensis* did not prevent feeding,[73] and 0.05 and 1% admixtures of the leaves with groundnut kernels was likewise ineffective against *Sitotroga cerealella* larvae.[74,75] However, Chellapa and Chelliah[76] reported that 1% powdered leaves in rice was very effective against *S. cerealella* and fairly effective against *Rhyzoper-*

11

tha dominica larvae. Five alkaloids (vasicine, vasicinol, deoxyvasicine, vasicinone, and deoxyvasicinone) isolated by Saxena et al.[77] from an ethanol extract of the leaves were tested at 0.05 and 0.1% in Petri dishes against *Acaulophora foveicollis* and *Epilachna vigintioctopunctata* beetles. All compounds were effective in reducing feeding, with deoxyvasicine the most effective against the former and vasicine most effective against the latter species. Vasicinol induced severe antifertility effects in *Dysdercus koenigii* and *Tribolium castaneum* when applied at 0.1 and 0.3%, respectively, to the insect's drinking water, due to blockage of oocytes in the oviduct. An ethanol extract of the powdered leaves applied as a 1% spray to filter papers on which adult *T. castaneum* were confined for 24 h killed 75% in 2 d and 100% in 4 d. The extract is said to be harmless to man.[78] The spray was not toxic to *Musca domestica* and to *Aedes aegypti* mosquitos.[79]

Justicia gendarussa **Burm. f.:** In India the natives scattered the leaves among their clothes to preserve them from insects.[80] A 5% aqueous extract of the leaves killed 100% of the larvae of *Spodoptera litura* and *Euproctis fraterna*. Powdered roots, leaves, and stems dusted upon *Callosobruchus chinensis* killed 75 to 80% in 5 d.[81]

Justicia procumbens **Linnaeus:** A methanol extract of the leaves incorporated into an artificial diet was a moderate repellent for larvae of *Bombyx mori*. Those larvae that fed showed growth retardation over a 6-d period.[82] Two active compunds were isolated from the extract and identified as justicidin A and B, which gave 100 and 90% mortality, respectively, of feeding 4th instars at 20 ppm.[83]

FAMILY ACERACEAE

Acer negundo **L.,** boxelder, ash-leaved maple
Acer pennsylvanicum **L.,** striped maple: Aqueous extracts of the leaves of these species did not deter feeding by *Lymantria dispar* larvae.[84]
Acer platanoides **L.,** Norway maple

Acer rudrum **L.,** red maple: Aqueous extracts of the leaves of these species deterred feeding by *Lymantria dispar* larvae,[85] but the wood failed to deter feeding by *Reticulitermes flavipes* termites and did not prevent attack by the brown-rot fungus, *Lenzites trabea*.[86] The wood was also susceptible to feeding by *Cryptotermes brevis* termites.[86] However, *Coptotermes formosanus* termites refused to feed on the wood in both force-feeding and choice-feeding tests.[87] Red maple was not infested during an outbreak of *Malacosoma disstria* in southern Louisiana.[88]
Acer saccharinum **L.,** sugar maple: The wood is susceptible to attack by *Coptotermes formosanus* termites.[87]

FAMILY ADIANTACEAE

Adiantum pedatum **L.:** A diet supplemented with the dried leaves deterred feeding by *Eurema hecabe mandarina* larvae.[90]

FAMILY AIZOACEAE

Dianthus cercidifolia **Maxim.:** Leaves of this plant were not accepted for feeding by larvae of *Spodoptera littoralis*.[91]
Mesembryanthemum chilense **Nolina:** An ether extract of the combined stems, leaves, and flowers placed on cantaloupe leaves was highly deterrent to feeding by adult *Acalymma vittatum*.[89]

FAMILY ALTINGIACEAE

Liquidambar formosana **Hance:** An ether extract of the fruits placed on cantaloupe leaves was highly deterrent to feeding by adult *Acalymma vittatum*.[89]
Liquidambar styraciflua **L.,** sweetgum, redgum: The wood is susceptible to attack by *Coptotermes formosanus* termites.[87]

FAMILY AMARANTHACEAE

Achyranthes japonica **Nakai:** Leaves of this plant were normally accepted for feeding by larvae of *Spodoptera littoralis*.[91]
Achyranthes caudatus **L.,** love-lies-bleeding:

An acetone solution containing 45 mg of the stem extract failed to deter feeding by *Leptinotarsa decemlineata* or *Mythimna unipuncta*.[92]

FAMILY AMARYLLIDACEAE

Agave americana L., century plant: Wallpaper in India is impregnated with the expressed juice of this plant to protect it from white ants.[93] An infusion of the leaves is used as an insecticide in the Philippine Islands.[94] An aqueous extract of the leaves was toxic to *Periplaneta americana* when injected into the blood stream, but *Blattella germanica* and *Oncopeltus fasciatus* were not affected by immersion in the extract.[25]

Allium ampeloprasum var. *porrum* (**L.**) Regel, leek, elephant garlic: In Belgium, an infusion prepared by immersing small pieces of the plant in water for 1 week was said to repel flies (unidentified)[95] and *Delia antiqua*.[96]

Allium cepa L., onion: Gram seeds (100 parts) mixed with 153 parts of a petroleum ether extract of onion did not repel *Callosobruchus chinensis*.[73]

Allium sativum L., garlic: Gram seeds mixed with 1 to 3 parts of a petroleum ether extract of garlic repelled *Callosobruchus chinensis* for at least 135 d.[73] However, garlic bulbs stored with grain in closed containers did not protect the grain from attack by *Sitophilus oryzae*.[97] Exposure of *Epilachna varivestis* larvae to a methanol extract of garlic bulbs strongly deterred feeding and caused larval and pupal mortality. The adults that developed from exposed pupae showed deformed wings.[98] Garlic is attacked by *Delia antiqua*.[96] Many sources attest to the value of garlic for repelling and killing mosquitos. The odor of garlic stunned mosquitos in 5 to 10 min and killed them in 5 h.[100] Greenstock and Larrea[101] reported that the essential oil distilled from garlic repelled not only mosquitos but also killed a large variety of crop pests in the field; this was played up in several newspapers[102,103] and a journal.[104] Laboratory tests showed that 5% extracts of five

kinds of garlic quickly killed *Culex pipiens* larvae.[105,106] Steam-distilled garlic oil and a crude methanol extract of garlic killed 3rd instar larvae of *Spodoptera litura* and 2nd and 4th instar larvae of *Euproctis* sp.[106] Crude garlic juice was highly potent as a contact toxicant. Crude extracts of garlic at 1.25, 2.5, and 5% concentrations in laboratory dishes killed *Syrphus carollae* plant lice and larval and pupal forms of *Chrysopa carnea* and *Coccinella septempunctata*.[107] Garlic oil was highly repellent to adult *Blattella germanica*.[61] Application of an acetone extract of the whole garlic plant to the tergum of thoracic and abdominal segments of *Dysdercus cingulatus* larvae and *Spodoptera litura* larvae failed to cause malformations.[108] Larvae of *Pericallia ricini* allowed to feed on cabbage leaves smeared with the extract showed no effect. Applications of a petroleum ether extract of garlic rhizomes to abdominal tergites of 5th instar *D. cingulatus* resulted in death.[109] The mosquito larvicidal principles of garlic have been isolated and identified as diallyl disulfide and diallyl trisulfide, which are fatal to *Culex* mosquitos at 5 ppm.[110]

Crinum asiaticum L.
Crinum bulbospermum Milne-Redhead
Hippeastrum hybridum Herb.
Hymenocallis littoralis Salisb.
Zephyranthes grandiflora Lindl.: Leaf pieces of each of these five plants offered to 24-h-old starved *Schistocerca gregaria* were not fed upon.[111,112] Acetone extracts of *Z. grandiflora* leaf cell sap completely inhibited feeding by the locusts at 0.5%[99] Lycorine, an alkaloid isolated from the bulbs of *H. littoralis* (350 mg from 3 kg), sprayed on cabbage leaves inhibited feeding by the locusts.[113] It is highly likely that lycorine or a closely related alkaloid is responsible for the activity of all five plant species.

Hypoxis obtusa Busch: The rhizome is protected from mold and insect attack in East Africa. A new glycoside, designated "hypoxide", isolated from the rhizome may be responsible for this activity.[114]

Lycoris radiata Herb: A diet supplemented with the dried leaves detered feeding by larvae of *Eurema hecabe mandarina*.[90]

FAMILY ANACARDIACEAE

Anacardium excelsum (Bert. & Balb.) Skiels, espave: The wood is resistant to termites.[86,115] An aqueous extract of the stem bark was slightly toxic to *Periplaneta americana* and *Oncopeltus fasciatus* but not to *Blattella germanica.*[25]

Anacardium occidentale L., cashew, scajou: Cashew nut shell oil is used in India to preserve floors, timbers, and books from termite attack. The tree yields a gum that is useful in bookbinding because of its insect-repellent properties.[80,116,117] Oil extracted from the husk of the nuts prevented termite attack on treated wood for only a short time, but a 1% solution of anacardic acid, obtained from the oil, prevented attack for more than 3 months.[118,119] An extract of the pericarp of the nuts gave complete protection against *Amphicerus cornutus* for 7 months when brushed on timber,[120] and laboratory tests demonstrated the toxicity of the nut oil to the stored product insects *Oryzaephilus surinamensis* and *Ahasverus advena.*[121] A mixture of the shell oil and kerosene was tested in India against the mosquito *Armigeres obturbans;* it caused nearly 100% mortality of the larvae and pupae within 2 h.[121,122] These effects are due to cardol and anacardic acid, both of which are present in the oil. An acetone extract of the stems applied topically to the last immature instar stage of *Dysdercus cingulatus* caused juvenomimetic activity.[123]

Cotinus cocgyarta Scop., smoke tree: Fustic crystals obtained from this tree were not repellent to termites when tested at considerable dilution.[118]

Mangifera indica L., Indian mango: The powdered plant is used in India as a fumigant against mosquitos.[116] Extracts of the leaves had no effect on feeding by *Bombyx mori* larvae.[82] An ether extract of the fruit did not deter feeding by *Lymantria dispar* larvae.[124] The wood is susceptible to termite attack.[124] The gum resin, mixed with lime juice or oil, was used as a cure for scabies, and the powdered flowers were used for fummigating against mosquitos.[125]

Metopium toxiferum (L.) Krug & Urban, poisonwood: An ether extract of all parts of the plant failed to deter feeding by *Lymantria dispar* larvae.[84] The wood is susceptible to termite attack.[86]

Pistacia chinensis Bunge: An ether extract of all parts of the plant deterred feeding by *Lymantria dispar* larvae.[84]

Rhus canadensis Marsh: Extracts of the whole plant did not repel *Popillia japonica* adults.[126]

Rhus viminalis Vahl.: The wood is reputed to be indestructible and is not subject to insect attack.[116]

Schinopsis quebracho-colorado (Schlecht. Barkl. & T. Meyer), red quebracho: A susceptible wood treated with an extract of red quebracho wood remained susceptible to termite attack.[118]

Schinus molle L., California pepper tree

Tapirira guianensis Aubl.: Ether extracts of the stem bark of these two species were moderately deterrent to feeding by adult *Acalymma vittatum.*[89]

FAMILY ANNONACEAE

Annona cherimola Mill., cherimoya: The seeds were reported to contain a poisonous substance, irritating to the eyes, that is used as an insecticide in West Africa to destroy human parasites.[121]

Annona muricata L., sour sop: The roots and leaves are said to be parasiticidal and the seed to be insecticidal especially to *Pediculus humanus capitis.*[116]

Annona glabra L., pond-apple, alligator apple: An aqueous extract of the seeds deterred feeding by *Attagenus megatoma.*[125]

Annona purpurea Moc. & Sesse: An ether extract of the twigs strongly deterred feeding by adult *Acalymma vittatum* at 0.5% and showed moderate deterrency at 0.1%.[89]

Annona reticulata L., custard apple: An aqueous extract of the seeds was deterrent and toxic to *Aphis fabae* and *Oryzaephilus surinamensis.*[127] A mixture of the leaves with sorghum millet and cowpeas gave complete protection from *Callosobruchus maculatus* for 3 months.[129,130] The powdered seed was

toxic to *Spodoptera eridania* and *Acyrthosiphon pisum* but not to *Udea rubigalis, Oncopeltus fasciatus,* and *Tetranychus urticae*.[128] The plant is used in India on domestic animals and the seed is used in the Philippines to kill *Pediculus humanus capitis*.[116]
A n n o n a s q u a m o s a L ., c u s t a r d apple,.sweetsop: An ether extract of the seeds deterred feeding by *Oncopeltus fasciatus*.[25] The powdered seeds are used in the Philippines as an insecticide and against *Pediculus humanus capitis*.[94] Hexane extracts of the leaves and seeds strongly deterred feeding by Nialparvata lugens, *Nephotettix virescens, Dicladisopa armigera,*[131] and *Callosobruchus chinensis*.[132] Numerous species of insects have been reported to be susceptible to either the powdered seeds or an extract of these,[26] but an ether extract of the seeds was much less effective than DDT as a toxicant for *Tribolium castaneum*.[132] Oil extracted from the seeds was highly effective in reducing the survival of *Nephotettix virescens,* and its transmission of the rice tungro virus.[133-137] The oil has also been reported[138] to be used by farmers for protecting rice from leafhoppers and planthoppers.[139]

***Annona* spp.** (Chemistry): Although the specific compounds responsible for the pesticidal effects of species of *Annona* have not been identified, a considerable number of compounds have recently been isolated and characterized which may prove to be implicated. These are described here. The 47 volatile flavor components of *A. atemoya* were all monoterpenes or sesquiterpenes, the most prolific being α-pinene, β-pinene, germacrene D, and bicyclogermacrene.[140] Bullatantriol. a sesquiterpene, was isolated from the leaves of *A. bullata* Rich.[141] Annonacin, a polyketide with antimicrobial properties and cytotoxic to KB cells (mouse leukemia) in vitro, was isolated from the stem bark of *A. densicoma*.[142] An alkaloid, annonelliptine, was isolated from the leaves and stems of *A. elliptica* R. E. Fries.[143] Annonaquinone A, isolated from the stem bark of *A. montana,*[144] is also cytotoxic (KB). Several kaurane dit-

erpenes were isolated from the stem bark of *A. reticulata*.[145] Three bis-indole alkaloids, annonidines A, C, and D were isolated from *Annidium mannii* Engl. & Diels, and synthesized.[146] Coniothalenol, a tetrahydrofurano-2-pyrone, was isolated from an ethanol extract of the stem bark of *Coniothalamus giganteus* Hook fil. & Thomas.[147]

***Asimina triloba* (L.) Dunal,** pawpaw: Pawpaw trees planted to serve as a mosquito repellent were found to have no value for this purpose.[148]

***Canangas odorata* (Lam.) Hook f. & Thoms.,** ylang-ylang: Ylang-ylang oil, whose chief constituent is geraniol, was moderately repellent to worker bees, *Apis florea,* at 0.0625% and strongly repellent at 0.3 to 0.5%.[149] Adult *Callosobruchus chinensis* kept for 48 h in a mixture of mung seed and the powdered roots, leaves, or pericarp of this plant showed little mortality, but mung mixed with caraway seeds caused significant mortality.[150]

***Dennettia tripetala* G. Baker:** Oil obtained by steam distillation of the edible fruits conferred effective protection as a seed preservative for up to 14 weeks in laboratory tests.[151] Topical application of the oil to nymphs and adults of *Periplancta americana* and to adult *Zonocerus variegatus* was highly toxic to these insects.[152]

***Desmas elegans* (Thw.) Safford:** *Callosobruchus chinensis* kept for 48 h in contact with a powdered mixture of mung seed and *Desmas* roots or leaves showed no harmful effects.[150]

***Eriosanthum acuminatus* (Thw.) Airy-Shaw:** *Callosobruchus chinensis* kept for 48 h in contact with a powdered mixture of the flowers or seeds of this plant showed little or no deterrence or harmful effects.[150]

***Monodora tenuifolia* Benth.:** Larvae of *Acre aponina* exposed to the dry fruits or the essential oil of the fruits were not deterred from feeding.[151] Topical application of the oil to the larvae of this insect and to adults of *Dysdercus suturellus, Ootheca mutabilis,* and *Riptortus densipes* was not toxic to these insects.[152]

FAMILY APIACEAE

***Anethum graveolens* L.**, dill: Dill seed oil failed to repel *Blattella germanica* in dish tests,[61] but larvae of the ticks *Ixodes redikorzevi, Haemaphysalis punctata, Rhipicephalus rossicus,* and *Dermacentor marginatus* exposed to the powdered leaves died in 30, 35, 30, and 45 min, respectively.[153] Exposure to the fresh whole leaf caused death in 80, 112, 80, and 130 min, respectively.

***Angelica archangelica* L.**, garden angelica: Angelica root oil failed to repel *Blattella germanica* in dish tests.[61] The powdered root and its extracts were not repellent to *Popillia japonica,*[126] *Tineola bisselliella,*[154] *Cochliomyia hominivorax,*[155] *Cuclotogaster heterographus,*[155] *Ctemocephalides,*[156] *Cimex lectularius,*[157] and roaches (unidentified).[157] An ether extract of the seed did not inhibit feeding by adult *Acalymma vittatum* at 0.1% but it was effective at 0.5%.[89]

***Angelica silvestris* L.:** An aqueous extract of the roots at 63 mg/ml deterred feeding by *Mythimna unipuncta,*[92] but constituents of the seeds (oxypeucedanin, bisabolangelone, imperatorin, and umbelliprenin) failed to deter feeding by this insect or *Leptinotarsa decemlineata.*[158] However, isobergaptene (another constituent) deterred feeding by larvae of *Spodoptera litura.*[158] Topical application of 0.01 mg of bisabolangelone to pupae of *Tribolium confusum* and *Trogoderma granarium* caused malformation and reduced fecundity, but application to the larvae or adults did not.[159]

***Apium graveolens* L. var. Dulce (Mill.) Pers.**, celery: Celery seed oil was mildly repellent to *Blattella germanica.*[61]

***Apium sellowianum* H. Wolff.:** Squash leaves that had been dipped in a 0.5% solution of the petroleum ether extract of the seeds completely deterred feeding by adult *Acalymma vittatum* over a 24-h period; 0.1% deterred feeding for 3 h but not for 6 or 24 h.[52]

***Carum bulbocastanum* (L.) Koch.:** This plant was used in India to protect clothes and skins from the ravages of insects.[125]

***Carum carvi* L.**, caraway: (-)-Carvone, a constituent of caraway seeds, was incorporated at 1% into a wheat bran diet and offered to *Spodoptera littoralis* larvae. The average larval weight on this diet was very low, only 2.5% of the larvae pupated, and no adults emerged.[160] A piece of cloth impregnated with carvacrol, a constituent of the oil, was tested against *Pediculus humanus humanus;* all died within 12 h.[161] Oil of caraway was sometimes slightly attractive but usually quite repellent to *Blatta orientalis.*[162] It failed to repel *Blattella germanica.*[61] Although Novak[163] reported that extracts of the plant were not toxic to *Aedes* mosquito larvae, Hartzell and Wilcoxon[164] found that acetone extracts of the seed killed 90% of the larvae tested. Hartzell[165] later reported that acetone and water extracts of the seed were ineffective against mosquito larvae.

***Conium maculatum* L.**, poison hemlock: Aqueous extracts of the whole plant were not repellent to *Popillia japonica.*[126] A German patent describes the addition of coniine (an alkaloid constituent) to a drenching solution to hides to serve as a mothproofing agent.[166] An acetone extract of the leaves was ineffective against mosquito larvae.[167] The powdered fruit was ineffective against larvae of *Musca domestica.*[168]

***Coriandrum sativum* L.**, coriander: Oil of coriander was reported to be one of the best repellents for *Cochliomyia hominivorax.*[155] It was also repellent to *Blattella germanica.*[61] However, the oil impregnated on filter papers stimulated feeding by *Papilio ajia.*[169] When applied as 2% emulsion sprays, the oil killed 51 to 80% of red spiders and *Aphis gossypi* within 24 hr.[170]

***Cuminum cyminum* L.**, cumin: Oil of cumin exhibited good repellency against *Cochiiomyia hominivorax* for 1 or 2 d,[155] and it was repellent to *Blatella germanica* and *Blatta orientalis.*[162]

***Daucus carota* L.**, carrot, Queen-Anne's lace: Extracts of the whole plant did not repel *Popillia japonica*[126] and the seed oil of subspecies *sativus* did not repel *Blattella germanica.*[61] Acetone and aqueous extracts of the seeds

were not effective against mosquito larvae,[165] and an aqueous extract of the roots was nontoxic to *Blattella germanica, Periplaneta americana,* and *Oncopeltus fasciatus.*[25] A glycosidal bitter principle isolated from dried carrot greens showed no anthelmintic action.[171] *Trans*-2-nonenal isolated from carrot root is responsible for the toxicity to larvae of the plant's parasite, *Psila rosae.*[172]
Falcaria vulgaris **Bernh.**: The insect juvenilizing agent, germacrene-D, was isolated from the essential oil of the fresh plant.[173]
Ferula assa-foetida **L.**, asafetida: The oil was ineffective against *Lygus lineolaria,*[174] but it strongly repelled the cornfield ant.[175] Grain stored with asafetida in closed receptacles was not protected against *Sitophilus oryzae.*[97] Asafetida did not reduce white ant attacks and attacks on sugar cane in India, but large doses slightly reduced attacks on wheat.[176] An alcoholic solution of asafetida failed to repel *Cochliomyiahominivorax.*[155] An acetone extract of the gum was toxic to *Diaphania hyalinata* but not to *Spodoptera eridania* and *Herpetogramma bipunctalis.*[128] The dust was also ineffective against the larvae of *Ostrinia nubilalis, Musca domestica,* and *Cydia pomonella.*[26]
Foeniculum vulgare **Mill.**, fennel, finocchio: Fennel oil was not repellent to *Blattella germanica*[61] but it exhibited good repellent action against *Cochliomyia hominivorax* for 1 or 2 d.[155] Extracts were not repellent to *Popillia japonica* adults.[126]
Heracleum spondylium **L.**, Imperatorin, isolated from the seeds, deterred feeding at 8 mg/ml by *Mythimna unipuncta* larvae.[92]
Imperatoria ostruthium **L.**, master wort: Extracts of the whole plant were not repellent to *Popillia japonica* adults.[126]
Laser trilobum **(L.) Borkh.**: Trilobolide, a sesquiterpene lactone isolated from this plant, failed to deter feeding by adult *Sitophilus granarius* and *Tribolium confusum* and by the larvae of *T. confusum* and *Trogoderma granarium.*[177]
Pastinaca sativa **L.**, parsnip: An acetone extract of the stems deterred feeding by larvae

of *Mythimna unipuncta.*[92] Extracts of parsnip were not repellent to *Popillia japonica.*[126] The edible portions of parsnip contain myristicin, which is toxic to *Drosophila melanogaster, Musca domestica, Epilachna varivestis, Acyrthosiphon pisum,* and *Aedes* mosquito larvae.[178,179]
Petroselinum crispum **(Mill.) Nym. ex A. W. Hill,** parsley: An ether extract of the whole plant did not deter feeding by adult *Diabrotica undecimpunctata howardi* and *Acalymma vittatum.*[52] Oil of parsley was moderately repellent to *Blatta orientalis.*[162] An aqueous extract of the seeds was toxic to *Periplaneta americana* but not to *Blattella germanica* and *Oncopeltus fasciatus.*[25] Apiol, obtained from parsley seed oil, markedly increased the toxicity of a pyrethrum spray to *Musca domestica.*[180]
Pimpinella anisum **L.**, anise, star anise: Anise oil did not repel *Blattella germanica* adults.[61] Adult *Tribolium castaneum* maintained on anise seed supplemented with yeast failed to oviposit or even survive for 20 d.[181] Exposure of *Anopheles* and *Aedes* mosquito larvae to anise oil resulted in complete mortality in 24 and 1 h, respectively.[182] Anethole and anisaldehyde, constituents of the oil, increased the toxicity to *Musca domestica* when applied together with pyrethrum.[183] Extracts of the plant prepared with organic solvents were toxic to *Drosophila melanogaster* and aqueous extracts were toxic to *Aedes* mosquito larvae.[183] In laboratory tests, anise oil was strongly repellent to colonies of *Lasius alienus,*[184,185] *Cochliomyia hominivorax,*[155] and gnats.[166]
Pimpinella saxifraga **L.**, pimpinella: Extracts of the dry rhizomes and roots were more or less effective repellents for *Popillia japonica* adults.[126]
Sium suave **Walt.**, water parsnip: Extracts of the whole plant were not repellent to adult *Popillia japonica.*[126]
Trachyspermum ammi **(L.) Sprague ex Turrill:** The seed oil, of which *p*-cymene is a major constituent, is highly repellent to *Apis florea* bees.[149]

FAMILY APOCYNACEAE

Acokanthera spectabilis **Hook.:** Friedelin (a diterpene), several cardenolide glycosides, sterols, and alkaloids were isolated from the leaves of this plant. Castor leaf disks containing 0.5% friedelin were exposed to 3rd instar larvae of *Spodoptera littoralis* in Petri dishes; no feeding occurred, and field test confirmed the deterrency of friedelin. The other compounds isolated were inactive.[186]

Alstonia compensis **Engl.:** Shavings of the wood were neither repellent nor resistant to the termite, *Reticulitermes lucifugus*.[187]

Alstonia scholaris (**L.**) **R. Br.,** white cheesewood: Ether extracts of the stems and of the leaves did not deter adult *Acalymma vittatum* from feeding when exposed to leaf disks impregnated with 0.1 or 0.5% of the extract.[89] The wood is susceptible to termite attack.[188]

Apocynum androsemifolium **L.:** Consecutive extraction with petroleum ether and ethanol of the stems, leaves, flowers, and roots, followed by exposure of the extracts to *Popillia japonica*, indicated no repellency to these adults.[126] However, the extracts were slightly toxic to *Aedes aegypti* larvae.[189]

Apocynum cannabinum **L.,** dogbane, Indian hemp. hemp dogbane: An aqueous extract of the whole plant did not repel *Popillia japonica* adults, but the juice expressed from the plant showed some deterrency to feeding by *Melanotus communis* in a choice test.[190] Acetone extracts of the whole plant deterred feeding by adult *Acalymma vittatum* at 0.5% but not at 0.1%.[89]

Carissa carandas **L.,** caranda: This plant was used in India to repel flies and, when pounded with lime juice and camphor, as a remedy for itch.[191]

Catharanthus roseus (**L.**) **G. Don,** Madagascar periwinkle, bright-eyes: Methanolic and aqueous extracts of the fresh leaves were highly deterrent to feeding by larvae of *Spodoptera littoralis*. Extracts of dry leaves were less effective. An alkaloid mixture isolated from the leaves was highly inhibitory at 0.125%, and one of the alkaloids

(vinblastine) was strongly deterrent at 0.04%. The larvae lost weight and failed to pupate.[192] Topical application of a petroleum ether extract of the whole plant, in acetone solution, to the abdominal tergites of 5th instar *Dysdercus cingulatus* did not result in juvenilization,[109] but application of the alkaloid portion of the leaf and root extract to unmated adults induced sterility.[193]

Melodinus khasianus **Hkf.:** Third instar larvae of *Spilosoma obliqua*, exposed to castor leaves that had been dipped in an acetone extract of the entire plant, were deterred from feeding to a moderate extent.[194]

Nerium oleander **L.,** common oleander: An acetone extract of the whole plant strongly deterred feeding by *Attagenus megatoma*, and an ethanol extract of the foliage deterred feeding by *Lymantria dispar* larvae.[84] Larvae of *Plutella xylostella* would not feed on this plant,[195] and a petroleum ether extract of the dried leaves tested at 0.5 to 5% mixed with cowpea seeds was highly effective in preventing feeding by *Callosobruchus chinensis*.[196] A concentration of 0.1% was less effective but still gave some protection. The leaf and bark have been used as an insecticide.[116] The powdered leaves and stems and a decoction thereof had no effect on aphids *(Macrosiphum* sp.).[197] The leaves were ineffective against *Spodoptera eridania* and only slightly toxic to *Diaphania hyalinata* and *Ostrinia nubilalis*. Combined petroleum ether and ethyl ether extracts, as well as combined alcohol and chloroform extracts were ineffective against *Musca domestica* but somewhat effective against *Cydia pomonella*.[26] Aqueous extracts of the leaves and of the combined branches and leaves were very toxic to *Periplaneta americana* and *Attagenus megatoma* larvae, but nontoxic to *Blattella germanica* and the larvae of *Tineola bisselliella* and *Aedes* and *Anopheles* mosquitos. Alcohol extracts were ineffective against all of these species (not tested against *Periplaneta americana)*. Aqueous extracts of the roots and of the flowers were toxic to *Attagenus megatoma* larvae.[25]

Rauvolfia caffra **Sond.:** An ether extract of

the woody stems and stem bark deterred feeding by adult *Acalymma vittatum* only slightly.[89]

Strophanthus kombe Oliver, strophanthus: Extracts of the whole plant did not repel adult *Popillia japonica.*[126] An acetone extract of the seeds was not toxic to mosquito larvae.[167]

Tabernaemontana chippii (Stapf.) Pichon: Tabernaemontana coriacea Link ex R. & S.: Ether extracts of *E. chippii* stem bark and of the combined woody stems and bark, of the leaves, and of the twigs of *T. coriacea* almost completely inhibited feeding by adult *Acalymma vittatum.*[89]

Thevetia peruviana (Pers.) K. Schum., yellow oleander, lucky-nut: Neriifolin, a known cardiotonic glycoside isolated from the seeds, was active as a feeding deterrent and as a toxicant to *Acalymma vittatum* in laboratory and greenhouse tests.[52,198] Application through the roots of cantaloupe plants indicated a systemic action. Protection from feeding by *Popillia japonica* on soybeans was obtained for 5d.[198] Neriifolin was obtained by the method of McLaughlin et al.[199] Neriifolin retarded the growth and caused considerable mortality of *Cydia pomonella* larvae when fed a diet containing 3 parts of neriifolin/million, and caused reduced oviposition and egg hatch; it also acted as a contact toxicant (complete mortality) at dosages above 50 mg/ml.[198] *Blattella germanica* and *Oncopeltus fasciatus* were not affected after immersion in an aqueous extract of the branchlets and leaves. An alcoholic extract was also nontoxic.[25] Topical application of a petroleum ether extract of the whole plant to the abdominal tergites of *Dysdercus cingulatus* nymphs did not cause juvenilization of the adults.[109] An alcohol extract of the whole plant deterred feeding by *Attagenus megatoma.*[25] Grain seed mixed with 1 or 2 parts of the powdered drupes per 100 parts of grain protected the grain from damage by *Callosobruchus chinensis.*[73] A methanol extract of the defatted leaves showed slight synergism when mixed with pyrethrum in tests against *Tribolium castaneum.*[200] An alcohol extract of the leaves

tested as a spray was not toxic to *Musca domestica* and *Aedes aegyptii* mosquitos,[79] but it was toxic to *Ostrinia nubilalis* larvae.[46] The insecticidal constituents of the plant are the glucoside, thevetin, nereifolin, and 2'-acetylnereifolin, and an unidentified constituent isolated by aqueous extraction of all parts of the plant except the leaves and fruit pulp.[199,201]

Vinca rosea L.: Alcohol extracts of the leaves were excellent feeding deterrents for *Schistocerca gregaria,* and the alkaloidal fraction gave 82.7% protection.[99] The alkaloids leurocristine and vincaleukoblastine, two of the more than 60 alkaloids reported to be present in this plant, have been used in cancer therapy.[202]

FAMILY AQUIFOLIACEAE

Ilex aquifolium L., English holly: Larvae of *Lymantria dispar* fed little or not at all in no-choice Styropor tests with dilute ethanol extracts of the leaf powder.[203]

Ilex cassine L., dahoon, dahoon holly: An ether extract of all plant parts failed to deter feeding by *Lymantria dispar* larvae.[84]

Ilex denticulata Wall.: Castor leaf bits dipped in the acetone solution of an acetone extract of the leaves failed to deter feeding by *Spilosoma obliqua* larvae.[194]

Ilex nitida (Vahl) Maxim., hueso prieto: The wood is susceptible to attack by *Cryptotermes brevis* termites.[86]

Ilex opaca Ait., American holly: Ether and methanol extracts of the leaves failed to deter feeding by adult *Acalymma vittatum* beetles.[52] First instar *Spodoptera frugiperda* larvae fed on a diet containing 15,000 ppm of an ether extract of the combined leaves and twigs, but they fed very little on a diet containing the same concentration of a methanol extract prepared from the ether-extracted plant. No toxic effects were seen on the larvae.[204] Juice expressed from the fresh leaves was impregnated on potato pieces and presented to late instar larvae of *Melanotus communis,* which fed well.[190] However, extracts of the fresh leaves were more or less repellent to *Popillia* *japonica* adults.[126] The wood is susceptible to attack by *Cryptotermes brevis* termites.[86]

Ilex paraguariensis St. Hil., maté, Paraguay tea: Extracts of the dry leaves showed some repellency to *Popillia japonica* adults.[126] Acetone and aqueous extracts of the leaves were nontoxic to mosquito larvae.[165]

Ilex verticillata (L.) A. Gray, common winterberry: Extracts of the whole plant were not repellent to *Popillia japonica* adults.[126]

Ilex wightiana Wall.: Castor bean bits dipped in an acetone solution of the acetone extract of the leaves gave some protection from feeding by *Spilosoma obliqua* larvae.[194]

FAMILY ARACEAE

Acorus calamus L., calamus, sweetflag: The plant was used by the natives in India chiefly for protecting woolen and flannel clothing from insects.[205] The aromatic rhizome was held in high esteem as an insectifuge, especially for fleas and moths. An infusion of the roots sprinkled in infested places also drove away vermin. The rhizome was used to keep moths from woolen goods and fleas from rooms.[191] In Malaya and Java the roots were dried, powdered, and scattered around fruit trees to protect them from ants,[206] or they were mixed with rice paddy before storage.[207] Paper is rendered insectproof by adding a decoction of sweetflag to the pulp during manufacture. Fabrics are also rendered insectproof with this preparation.[208] Distilled Indian calamus root oil was highly repellent to *Musca domestica* in a Peet-Grady chamber; flies buzzed around but would not land.[209] Insect-repellent formulations consisting of oil of calamus and oil of *Curcuma* (1:1), and oil of calamus with synthetic pine oil compared favorably with dimethyl phthalate in their repellency to mosquitos.[210] Good kills of *Sitotroga cerealella* and *Corcyra cephalonica* were obtained by treating paddy with the powdered calamus rhizome tied in a piece of cloth and placed in the center of the storage bin.[211] Admixture of 1% of the rhizome pieces with the paddy protected the grain for at least 3 months.[75] Good control of *Callosobruchus chinensis, Sito-*

philus oryzae, C. cephalonica, and *Trogoderma granarium* was obtained on pulses and cereals treated with the root oil or the rhizome fragments.[73,212] An ether extract of the dried rhizomes was outstanding in preventing feeding by *Athalia proxima.*[213] The root oil did not repel *Blattella germanica.*[61] Vapors of the oil affected the egg hatch of *Dysdercus koenigii;* the nymphs that hatched out did not molt and died within 24 h.[212] An acetone extract of the roots applied topically to the abdominal tergites of *D. cingulatus* nymphs at 7 and 10 μl/nymph caused complete mortality in the nymphal stage; dosages of 4 μl/nymph caused 15% mortality and 32.5% mortality at molting.[108,109] The oil killed more than 85% of *Sitophilus oryzae* exposed to the vapors at 3 to 4% concentration for 24 h.[214] Grain containing 2 lb powdered rhizome/100 lb was protected from insect attack over a period of 1 year; insects repelled were plant lice, coconut beetle grubs, *Tineola biselliella,* fowl lice, and *Cimex lectularius* (but not their eggs).[215,216] Petroleum ether extracts of the rhizomes as well as a steam-volatile fraction were toxic to *Musca nebulo* and, especially, to *Culex fatigans.* The toxic action appeared to be due to the presence of *trans*-asarone.[217] Female *Callosobruchus chinensis* and female *Trogoderma granarium* exposed to the vapors of the root oil became sterile.[218,219] Powdered sweetflag rhizomes at 1% in stored wheat completely protected the grain from *Corcyra cephalonica* for at least 2 months. The 1st instar larvae were killed before the next instar could be reached, which resulted in a reduction of adult emergence.[220] A distillate of the rhizome oil tested as foliar sprays (choice and no-choice) at 0.5 and 1% deterred feeding by *Spodoptera litura* larvae.[221] The essential oil, which contains 82% of asarone, is very effective in controlling various species of stored grain insects. The dose of asarone necessary to obtain LD_{50} was 2.75% for *Tribolium castaneum,* 1.35% for *Callosobruchus chinensis,* 1.5% for *Rhyzopertha dominica,* 4.5% for *Trogoderma granarium,* and 4.5% for *Corcyra cephalonica.*[222] A method was devised for quantitatively estimating the asarone content in the oil by column chromatography.[223] A pilot plant method for isolating asarone from the oil has been published.[224] Patra and Mitra[225] isolated acoradin, 2,4,5-trimethoxybenzaldehyde, galangin, and sitosterol from the rhizomes; their possible roles in the biological activity have not been reported. Despite the fact that calamus root oil has deterrent, repellent, and insecticidal properties against many species of insects, it has been shown to be highly *attractive* to both male and female *Dacus dorsalis, D. cucurbitae,* and *Ceratitis capitata* fruit flies. The active compounds were isolated and identified as acoragermacrone, *cis*-asarone, and asarylaldehyde.[226] It is interesting to note that, whereas *cis*-asarone is highly attractive to the flies, *trans*-asarone is highly repellent to many species of insects. b-Asarone was attractive to male *Dacus dorsalis,* acoragermacrone was attractive to female *D. cucurbitae,* and asarylaldehyde was attractive to male and female *C. capitata.* Of the three positional isomers of asarylaldehyde, only the 3,4,5- isomer was attractive.[226]

Arisaema tortuosum (Wall.) Schott.: In India, this plant was known to have insecticidal and repellent properties for insects.[227]

Arisaema tripholum (L.) Schott., jack-in-the-pulpit: Extracts of this plant were not repellent to *Popillia japonica* adults.[126]

Peltandra virginica (L.) Schott. & Endl., arrow-arum: Extracts of this plant deterred feeding by *Tineola bisselliella.*[25]

Symplocarpus foetidus (L.) Nutt., skunk cabbage: Ethereal and methanolic extracts of the whole plant did not deter feeding by adult *Acalymma vittatum.*[52] Consecutively prepared extracts of the combined leaves and stems with ether and ethanol were deterrent to feeding by *Spodoptera frugiperda* larvae at 15,000 ppm.[204]

FAMILY ARALIACEAE

Aralia hispida Vent., bristly aralia
Aralia racemosa L., American spikenard: Extracts of these plants were not repellent to *Popillia japonica* adults.[126]
Aralia elata Seem., Japanese angelica

Aralia humilis Cav., angelica tree: Acetone extracts of these plants deterred feeding by *Attagenus megatoma.*[25]

Aralia spinosa L., devil's walkingstick, Hercules-club: An ethanol extract of the leaves deterred feeding by *Lymantria dispar* larvae, and an acetone extract of the leaves repelled *Attagenus megatoma.*[25] An aqueous extract of the branchlets and leaves was toxic to *Periplaneta americana* but not to *Blattella germanica* and *Oncopeltus fasciatus.* An extract of the fruits was not toxic to any of these insects. An acetone extract of the roots showed some toxicity to *Attagenus megatoma* larvae but not to *Blattella germanica, Oncopeltus fasciatus,* and the larvae of *Tineola biselliella* or *Aedes* and *Anopheles* mosquitos.[25] Petroleum ether, ethyl ether, and chloroform extracts of the bark, as well as petroleum ether and ethyl extracts of the twigs and stems, were all ineffective against *Musca domestica.*[26]

Evodiopanax innovans (Sieb. & Zucc.) Nakai
Gilibertia trifida Makino: The leaves of these plants are normally accepted for feeding by *Spodoptera littoralis* larvae.[91]

Hedera helix L., English ivy: The juice pressed from the fresh leaves and shoots significantly restricted damage to corn seeds by *Melanotus communis* and *Diabrotica undecimpunctata howardi.*[190,228] Larvae of *Plutella xylostella* refused to feed on this plant.[195] The leaf of this plant has been used from remote antiquity to destroy vermin on the body. In India a decoction of the leaves is applied locally to destroy *Pediculus humanus capitis.*[116] Filter paper impregnated with a 3% solution of saponin C from the leaves, or of its aglycone a-hederin, were toxic to *Reticulitermes flavipes* termites and inhibited feeding by this insect; 0.05 and 0.5% solutions were not effective.[229] An alcohol extract of the bark, used as a spray, was nontoxic to *Musca domestica* and to *Aedes aegypti* mosquitos.[79] An aqueous extract of the bark was highly toxic to *Periplaneta americana* when injected into the bloodstream, but *Blattella germanica* were not affected by immersion in the extract. Petroleum ether and alcohol extracts showed some toxicity to *Attagenus megatoma* larvae but not to *Blattella germanica* or *Oncopeluts fasciatus.*[25] *Kalopanax septemlobus* Koidz., hari-giri: The heartwood of this tree is naturally resistant to the subterranean termite, *Coptotermes formosanus.* A saponin extracted from the wood seems to be responsible for this effect; it is composed of oleanolic acid, glucose, and arabinose in a 1:2:2 molar ratio.[230]

FAMILY ARAUCARIACEAE

Agathis robusta (C. Moore ex G. Muell.) *Agathis vitiensis* (Seem) Benth & Hoof f ex Drake: Ether extract of the stem bark and twigs of *A. robusta* and the twigs of *A. vitiensis* gave little or no protection from feeding by *Acalymma vittatum.*[89]

FAMILY ARECACEAE

Areca catechu L., betel palm, areca-nut: Extracts of the plant were not repellent to *Popillia japonica* adults.[126] Arecoline, an alkaloid obtained from the seed oil, prevented the production of viable eggs of *Boophilus microplus.*[231] The alkaloid occurs in the nut to the extent of 0.07 to 0.1%. A 0.6 and 0.4% solution of arecoline hydrochloride killed 100 and 80% respectively, of bean aphids; 0.75 and 0.5% solutions of arecoline killed 100 and 90% respectively.[232]

Cocos nucifera L., coconut: Coconut oil mixed at 1% with stored paddy rice strongly repelled *Rhyzopertha dominica* and *Sitotroga cerealella,* but it adversely affected germination of the grain.[233] Mixtures of the oil with 0.05 ml/100 g of green gram caused complete mortatlity of *Callosobruchus chinensis* larvae and adults and prevented oviposition by this insect.[234] Infestation of bambara groundnut with *C. maculatus* was considerably reduced by mixing with 3 ml oil/kg seed.[235] Mixutes of 1 and 5 ml of the oil/kg of stored bean seed gave good control of *Zabrotes subfasciatus;* at 5 or 10 ml/kg, complete control was achieved for more than 75 d. Crude oil was more effective than the puri-

fied oil.[236] *C. chinensis* could be eliminated from green gram by mixing the seeds with 0.5% of the coconut oil; germination of the seed was not affected.[237] Coconut oil impregnated on filter paper was highly repellent to *Schistocerca gregaria* and *Melanoplus sanguinipes.*[238] An acetone extract of the dried stems of the coconut tree applied topically to allatectomized *Dysdercus cingulatus* caused juvenomimetic activity.1[23]

Elaeis guineensis Jacq., African palm: A mixture of the oil with bean seeds at 1 ml/kg provided good control of *Zabrotes subfasciatus.* Complete control was obtained for 75 d at 5 or 10 ml/kg. Crude oil gave better protection than pure oil; it increased adult mortality and reduced oviposition, egg hatch, and the number of adult progeny. Germination of the seed was not affected.[236] The triglyceride content of the oil appears to be responsible for the biological activity.[239]

FAMILY ARISTOLOCHIACEAE

Aristolochia gigantea H.: An ether extract of the flowers failed to repel *Musca domestica* in laboratory tests.[209]

Aristolochia serpentaria L., Virginia snakeroot: An acetone extract of the whole plant deterred feeding by *Attagenus megatoma* larvae,[25] but other organic extracts of the plant did not repel *Popillia japonica* adults.[126]

Aristolochia taliscana Hook & Arn.: Ether and ethanol extracts of the combined stems, twigs, and leaves deterred feeding by adult *Acalymma vittatum* at 0.1 and 0.5%.[89]

Asarum canadense L.,Canada snakeroot, wild ginger: Although this plant contains asarone, it neither repelled or attracted *Cochliomyia hominivorax*[155] and acetone extracts did not repel *Popillia japonica* adults.[126] Methods are reported for synthesizing cis- and trans-asarones and (E)-1-(3-methyl-1-butenyl)-2,4,5-trimethoxybenzene, a structural analog of precocene.[240]

FAMILY ASCLEPIADACEAE

Ascleptias curassavica L., bloodflower milkweed, false ipecac: The Indians of southern Mexico were reported to have swept the floors and walls of their huts with this plant to repel fleas.[241,242] Alcohol extracts of the stems and roots were ineffective against *Aphis fabae*, but an extract of the flowers had a slight insecticidal effect.[232] An aqueous extract of the stems and leaves was very toxic to *Periplaneta americana* when injected into the blood stream, but *Blattella germanica* and *Oncopeltus fasciatus* were not affected by immersion in the extract.[25] Feeding known doses of cardenolides from this plant to 4th instar larvae of *Danaus plexippus* led to more efficient larval tissue incorporation at low doses than at high doses.[243]

Asclepias eriocarpa **Benth.**, woolypod milkweed

Asclepias kansana **Vail.**

Asclepias labriformis **M. E. Jones,** milkweed: Acetone extracts of these three species of milkweed deterred feeding by *Attagenus mrgatoma* larvae.[25]

Asclepias syriaca **L.**, common milkweed: An aqueous extract of the leaves deterred feeding by late instar larvae of *Melanotus communis*.[190] A minimum concentration of 0.05 to 0.1 g of the aqueous extract incorporated in the basic diet was required to produce adult *Melanoplus femurrubrum* and *M. sanguinipes*.[244] A methanol extract of the whole plant at 1000 ppm was not toxic to larvae of *Aedes aegypti* mosquitos.[245] A 1:1 misture of methylene chloride and acetone was used to extract the flower heads of this plant. Topical application of 0.5 ml of an acetone solution of the extract to *Tenebrio molitor* pupae resulted in normal adults.[246] An aqueous extract of the leaves, but not of the stems, was highly toxic to 4th instar *Aedes aegypit* larvae.[189] An acetone extract of the whole plant did not repel *Popillea japonica* adults.[126]

Asclepias tuberosa **L.**, butterfly milkweed: An aqueous extract of the leaves, tested on a potato slice, was highly deterrent to feedingy by late instar larvae of *Melanotus communis*.[190,247] Seed corn treated with the extract showed slight phytotoxicity, delayed germination, and depressed growth rate.[247] An

aqueous homogenate of the leaves and shoots applied to corn seeds strongly deterred feeding by *Diabrotica undecimpunctata howardi* in choice and no-choice test.[228] Topical application of an acetone solution of a methylene chloride-acetone extract (1:1) of the flower heads to nymphal *Tenebrio molitor* abdomens did not induce juvenilization in the resulting adults.[245]

Calotropis gigantea **(L.) R. Br.**, crown plant: An acetone extract of the whole plant deterred feeding by *Sitophilus oryzae*.[25] A methanol extract of the fruits and root bark deterred feeding by nymphus and adults of Schistocerca greparia after being painted or sprayed on maize leaves.[248] An extract of the leaves, flowers, and latex stimulated feeding. Acetone and water extracts of the latex were moderatley deterrent to the locusts.[99] Deterrency is probably due mainly to the action of the alkaloidal fraction, although an aqueous extract also showed some inhibitory activity.[249] This plant was used as an insect deterrent in India. A 5% alcoholic extract of the stems killed 55% of *Plutella xylostella* larvae.[81] Scme of the plant parts were toxic as dusts or extracts to *Musca domestica*, mosquito larvae, and several species of leaf-eating larvae.[250] A petroleum ether extract of the whole plant applied topically to the abdominal tergites of *Dysdercus cingulatus* nymphs did not affect development to normal adults.[109] Extracts of the flowers tested against *Sitophilus oryzae* on filter papers caused 92% mortality.[251,252]

Gymnema sylvestre **(Retz.) Schult.**, miracle fruit: Gymnemic acids, which are triterpene saponins present in the leaves of this vine, are feednig deterrents for larvae of *Spodoptera eridania*. These compounds probably do not act by suppressing the sweetness of sugars.[253]

Tylophora asthmatica **Wight & Arn:** The alkaloids, tylophorine (A), tylophorinine (B), and pergulatinine (C), were isolated from the leaves and incorporated into semisynthetic cellulose diets. Each of these alkaloids, at 0.01% inhibited feeding by *Spodoptera litura* larvae; activities were of the order A > B > C.[254]

FAMILY ASTERACEAE

Achillea ageratifolia (Sibth. & Sm.) Boiss.:
The sesquiterpene lactone, ageratriol, tested on wafers was moderately deterrent to feeding by larvae and adults of *Tribolium confusum*, larvae of *Trogoderma granarium*, and adult *Sitophilus granarius*.[255]

Achillea millefolium L., common yarrowL An acetone extract of the stems at 33 mg/ml reduced feeding by *Mythimna unipuncta* arvae but the activity against *Leptinotarsa decemlineata* larvae was only slight.[92] Extracts were not repellent to *Popillia japonica* adults.[126] The powdered plant was nontoxic to *Epilachna varivestis* larvae.[256] An acetone extract of the whole plant was not toxic to mosquito larvae.[167] However, a methanol extract of the plant was toxic to 24-hr-old *Aedes triseriatus* larvae; the action was due to *N*-(2-methylpropyl)-(E-E)-2,4-decadienamide and the *N*-2-methylpropyl amides of decanoic and *(E)*-2-decenoic acids.[257] An aqueous extract of the tops and lower parts of the plant was nontoxic to *Blattella germanica*, *Periplaneta americana*, and *Oncopeltus fasciatus*.[25] Exposure to the ticks *Ixodes redikurzevi*, *Haemaphysalis punctata*, *Rhipicephalus rossicus*, and *Dermacentor marginatus* to the powdered flowers caused death in 30, 38, 30, and 60 min, respectively.[153]

Achillea sibirica Ross.: A benzene extract of the whole plant was moderately effective in deterring feeding by larvae of *Spodoptera litura*.[258]

Adenostyles alliariae (Conan) Kern: The sesquiterpene lactones, neoadenostylone and deacyladenostylone, isolated from this plant were tested by dipping wafers in their ethanol solutions at 10 mg/ml. These wafers were moderately deterred to adults of *Sitophilus granarius* and both larvae and adults of *Tribolium confusum*.[259]

Ageratum conyzoides L.
Ageratum houstonianum Mill., ageratum, bedding plant: The hormones ageratochromene and 6-demethoxyageratochromene were isolated from *A. conyzoides* (whole plant).[260]

Treatment of adult *Leptocarsia chinensis* with 1000 ppm of ageratochromene accelerated oviposition, egg hatch, and molting. Topical application of 10 mg of this compound per square centimeter to *Sitophilus oryzae* resulted in complete mortality, and exposure of *Thlaspida japonica* eggs to 7 mg/cm^2 caused degeneration of 78% of these eggs. Precocene II (6,7-dimethoxy-2,2-dimethylchromene), a powerful juvenilizing hormone for insects, is obtaned from *A. houstonianum*.[261,262] This compound, added to the diet of *Rhodnius prolixus*, had a strong antifeedant effect at 48 mg/ml.[263] Application to larvae and pupae of *Corcyra cephalonica* induced serious abnormalities in the histomorphology of the female reproductive organs,[262] but application to the adult female *Pyrrhocoris apcerus* and *Dysdercus cingulatus* did not inhibit ovarian development.[264] Exposure of 2nd instar larvae of *Oncopeltus fasciatus* and 4th instar *Melanoplus* at 1 mg/cm^2 and 100 mg (topical) gave total precocious metamorphosis.[265] Topical application of 1 mg/cm^2 of precocene II to unfed immature *Dermacentor variabilis* was lethal to larvae and nymphs; many larvae exposed to lower doses failed to feed.[266]

Ambrosia artemisiifolia L., common ragweed: Extracts of the whole plant did not repel *Popillia japonica* adults.[126]

Anacyclus pyrethrum dc, pellitory: An acetone extract of the whole plant was not repellent to *Popillia japonica* adults.· The powdered root was toxic to armyworms and *Acyrthosiphon pisum* but not to *Udea rubigalis*, *Tetranychus urticae*, or *Ostrinia nubilalis*.[26] A petroleum ether extract of the roots was toxic to *Musca domestica*, but ethyl ether and ethanol extracts of the petroleum ether-extracted root were not toxic to this insect. The alcohol extract was worthless against *Ctenocephalodes felis*, *Amblyomma americanum*, chiggers, *Pediculus humanus humanus*, and as a body louse ovicide, but it showed some toxicity to mosquito larvae.[26] Pellitorine *(N*-isobutyl-*E,E*-2,4-decadienamide), a compound isolated from the petroleum ether extract of the roots, was toxic to *Musca domestica*[267] and *Tenebrio molitor*.[268] Several acetylenic compounds have also been isolated from the roots.[269] Pellitorine has been synthesized by several methods.[270-272] Anacyclin, another compound isolated from the roots[267] and prepared synthetically,[273] becomes potent to *T. molitor* upon catalytic semihydrogenation of its two acetylenic linkages; it structure has been confirmed by total synthesis.[274]

Ambrosia psilostachya dc, western ragweed: Extracts of the whole plant were not repellent to *Melanoplus differentialis* and *M. sanguinipes*.[275]

Anthemis arvensis L., corn chamomile: Extracts of the whole plant were not repellent to *Popillia japonica* adults.[126] Consecutively prepared either and methanol extracts of the combined flowers, leaves, and stems tested at 15,000 ppm did not deter feeding by *Spodopter frugiperda* larvae.[204]

Arctium minus Bernh., common burdock

Arnica alpina (L.) Olin. & Landau

Arnica montana L., arnica: Extracts of these plants did not repel *Popillia japonica* adults.[126]

Artemisia absinthium L., wornwood, absinthe: Althogh the early literature contains many references to the use of the leaves and shoots of this plant against weevils in granaries,[13] alcohol, acetone, or benzene extracts showed little or no effect against stored product insects.[232] However, the plant is used by Italian farmers to protect grain in storehouses from attack by *Tinea granella, Sitotroga cerealella,* and *Calendra granaria*.[276] Wornwood oil from the leaves failed to repel *Blattella germanics*,[61] but tick larvae exposed to the powdered seeds died quickly.[153]

Artemisia annua L., annual wormwood: An ether extract of the stems at 0.5% was highly deterrent to feeding by *Acalymma vittatum* adults.[89]

Artemisia capillaris Thumb.: 1-Phenyl-2,4-pentadiyne and capillene, two phenylacetylenes isolated from the growing buds of this plant, deterred feeding by larvae of *Pieris rapae crucivora* when tested by the leaf disk method.[277] Capillin, capillarin, methyleu-

genol, arcurcemene, and bornyl acetate were subsequently tested and found to be deterrent to feeding.[278,279]

Artemisia ludoviciana Nutt., western mugwort

Artemisia rigida Willd., fringed sagebrush: Larvae of *Melanoplus differentialis* fed and developed normally on the leaves of these plants.[275]

Artemisia pauciflora Web., Levant wornseed: The oil of this plant was moderately repellent to *Blattella orientalis*.[162] Santonin, which is obtained from this species, killed only 10% of the mosquito larvae exposed.[164]

Artemisia pontica L., Pontica epoxide, isolated from this plant, was both repellent and toxic to *Aedes atropalpus* mosquito larvae.[280]

Artemisia tridentata Nutt., big sagebrush: An aqueous extract of the branch ends, tested in a two-choice bioassay, deterred feeding by larvae of *Leptinotarsa decemlineata*. Fractionation of an ethanol extract of the branches yielded deacetoxymatricarin, which was biologically active.[281] An acetone extract of the stems was nontoxic to mosquito larvae. A crystalline compound isolated from a petroleum ether extract of the whole plant was also non-toxic.[26] Aqueous extracts of the upper portions, leaves, and flowers were nontoxic to *Blattella germanica, Periplaneta americana,* and *Oncopeltus fasciatus*.[25]

Aster novae-angliae L., New England aster: Extracts of the fresh leaves and flowers were repellent to *Popillia japonica* adults.[126]

Aster paniculatus L., An extract of the whole plant did not repel *Popillia japonica* adults.[126]

Aster tartaricus L., A methanol extract of the leaves did not deter feeding by *Bombyx mori* larvae.[82]

Asteriscus maritimus (L.) Less.: A petroleum ether extract of the aerial portions was a good feeding deterrent for *Sitophilus granarius, Tribolium confusum,* and *Trogoderma granarium*. A lactone isolated from this extract was also effective.[282]

Atractylis gummifera L., add-add: The natives of Morocco reportedly use the root as an insecticide by burning it on red cinders.[232]

Atractylis ovata Thunb.: This plant is used iun China for fumigating grain stores.[283]

Baccharis halimifolia **L.**, An ethanol extract of the foliage did not deter feeding by *Lymantria dispar* larvae.[84]

Baccharis megapotamica **(Spreng):** A hexane extract of the aerial portion at 0.5% was somewhat deterrent to feeding by *Acalymma vittatum* adults, but 0.1% was ineffective.[52] Several potent antileukemic (against P-388) tricothecenes have been isolated from the twigs and leaves.[284] A number of antileukemic and antiviral *ent*-clerodanes have been isolated and identified from *B. tricuneata* (L. f.) Pers.[285]

Bidens pilosa **L.**, hairy beggarticks: An aqueous extract of the whole plant deterred feding by *Attagenus megatoma* larvae.[25] Phenylheptatriyne, a polyacetylene isolated from the leaves and stems, deterred feeding by *Euxoa messoria* larvae.[286] An aqueous extract of the entire plant was toxic, and an extract of the tops, fruits, and flowers was very toxic, to *Periplaneta americana* but not to Blattella germanica and Oncopeltus fasicatus. Alcoholic, petroleum ether, and chloroform extracts were not toxic to *B. germanica, O. fasciatus, Tribolium confusum, Tinea bisselliella,* and *Aedes* mosquito larvae.[25]

Blumea aurita **DC,** pleadura: This plant has been suggested as a repellent for insects in the Gold Coast of Africa, and as a possible source of insect powder.[287]

Blumea eriantha **DC:** The leaves of this plant were totally rejected as food by larvae of *Earias vitella*. The leaf oil reduced oviposition and increased the preoviposition period significantly.[288]

Blumea lacera **(Burm. f.) DC:** The plant is used as an insect repellent in India, especially for fleas. In Tanganyika the leaves were spread on the floor of sleeping places to provide a hiding place for bedbugs during the day, which are then swept up in the morning and burned.[116] The essential oil extracted from this plant and from *B. malcomii,* when mixed with pyrethrins, acted as a synergist in sprays against *Musca domestica*.[289]

Brachylaena hutchinsii **Hutchison,** muhuhu: The wood of this tree is very durable, being resistant to both termites and borers.[119]

Calendula arvensis **L.**
Calendula suffruticosa **Vahl.:** The oils extracted from the seeds of these species with pentane did not prevent feeding by *Anthonomus grandis grandis*.[290]

Calendula officinalis **L.**, pot-marigold: A methanol extract of the leaves did not deter feeding by *Bombyx mori* larvae, but all of the larvae died after feeding.[82]

Callistephus chinensis **(L.) Nees,** china-aster: Extracts of the flowers tested against *Sitophilus oryzae* larvae on filter papers were only moderately active as feeding deterrents.[251,252]

Carlina acaulis **L.:** Carlina oxide, an acetylenic epoxide isolated from the tops of the plant, was only slightly phytotoxic to *Aedes astropalpus* mosquitos.[290]

Carthamus tinctorius **L.**, safflower: A methanol extract of the leaves did not deter *Bombyx mori* from feeding,[82] and safflower oil failed to prevent development of *Callosobruchus chinensis* on redgram seeds.[291]

Chamaemelum nobile **(L.) All.**, chamomile: Chamomile oil failed to repel *Balttella germanica* on filter papers.[61]

Chrysanthemum balsamita **L.**, costmary: Erivanin, a sesquiterpene lactone isolated from this plant, failed to deter feeindg by larvae of *Sitophilus granarius, Togeoderma granarium,* and *Tribolium confusum,* as well as adult *T. confusum*.[255]

Chrysanthemum coronarium **L.:** A spiro enol ether isolated from the leaves and stems inhibited feeding by 5th instar *Bombyx mori* larvae.[292]

Chrysanthemum macrophyllum **W. & K.:** A methanol extract of the flowers deterred feeding by adult *Sitoophilus granarium* and *Tribolium confusum,* and by larvae of *Trogoderma granarium*.[282]

Chrysanthemum monilifera **(L.) Norl.:** Ether extracts of the combined woody stems and stem bark deterred feedingy by adult *Acalymma vittatum* at 0.5% but not at 0.1%.[89]

Chrysanthemum parthenium **Bernh.:** Extracts of the whole plant did not repel *Polillia japonica* adults.[89]

Chrysopsis mariana **(L.) Nutt.**, golden aster:

Extracts of this plant did not repel *Popillia japonica* adults.[126]

***Chrysopsis villosa* (Pursh.) Nutt.:** Aqueous extracts of the whole plant did not prevent feeding by larvae of *Melanoplus differentialis* but did not repel feeding by *M. sanguinipes* larvae.[275]

***Chrysothamnus nauseosus* (Pall.) Britt.,** rubber rabbitbush: An aqueous extract of the whole plant strongly deterred feeding by adult *Leptinotarsa decemlineata*. A mixture of diterpene acids contains the active components(s).[281]

***Cichorum intybus* L.,** chicory, blue sailors: Extracts of this plant did not repel *Popillia japonica* adults.[126] In two-choice and no-choice tests. the sesquiterpene lactones 8-deoxylactucin, lactupicrin, and cichoriin isolated from the roots and leaves significantly reduced feeding by adult *Schistocerca gregaria*.[293]

***Cirsium discolor* (Muhl.) Spreng**

***Cirsium flodmani* (Rydb.):** The bracts of these two species of thistle produce a sticky exudate that physically traps some insects and may deter other physically or chemically.[294]

***Cirsium undulatum* Spreng:** Aqueous extracts of the whole plant did not deter feeding by larvae of *Melanoplus differentialis* or *M. sanguinipes*.[275]

***Cleistanthus collinus* Benth. & Hook.,** karada: Extracts of the bark prepared with petroleum ether, ethanol, and water gave no protection to timbers from termites.[120]

***Cnicus benedictus* L.,** blessed thistle: Extracts of the whole plant did not repel *Popillia japonica* adults.[126]

***Conyza lyrata* H.B.K.:** The fresh plant is tied above the head of the bed in El Salvador at night as a mosquito repellent.[295]

***Coreopsis grandiflora* Hogg.,** big coreopsis: Extracts of the entire plant were repellent to *Popillia japonica* adults.[126] An aqueous extract of the stems, leaves, and flowers was toxic to *Blattella germanica, Periplaneta americana,* and *Oncopeltus fasciatus*.[25]

***Crassocephalum chepidioites* S. Moore:** A methanol extract of the leaves did not deter feeding by *Bombyx mori* larvae.[82]

***Dolichos lablab* L.,** hyacinth bean, field bean: An acetone extract of the whole plant deterred feeding by *Oncopeltus fasciatus*.[25] The pod exudate acts as a repellent for some species of insects (unidentified).[296]

***Echinacea pallida* (Nutt.) Britton,** hedgehog coneflower: An extract of the whole plant did not repel *Popillia japonica* adults.[126]

***Eclipta alba* (L.) Hassk.,** eclipta: An aqueous extract of the whole plant deterred feeding by *Attagenus megatoma* larva. An extract of the fresh roots and shoots was impregnated on filter papers upon which some eggs of *Sitotroga cerealella* were place for 8 d; 99% of the eggs failed to hatch.[297] An aqueous extract of the whole plant was highly toxic to *Periplaneta americana* when injected into the bloodstream, but *Blattella germanica* and *Oncopeltus fasciatus* were unaffected after immersion in the extract. Petroleu, ether and chloroform extracts were toxic to *Attagenus megatoma* larvae but not to *Blattella germanica, Oncopeltus fasciatus, Tribolium confusum,* and larvae of *Tineola bisselliella* and *Aedes* mosquito.[25]

***Encelia farinosa* Gray,** brittlebush: A chloroform extract of the leaves deterred larvae of *Heliothis zea* from feeding in choice tests,but farinosin, a eudesmanolide sesquiterpene lactone from the leaves did not.[298] The chromene, encecalin, showed both antifeedant and insecticidal properties toward larvae of *Peridroma saucia* and *Plusia gamma*.[299]

***Erechtites hieracifolia* (L.) Raf. ex DC,** burnweed, fireweed: Extracts of the whole plant were not repellent to *Polillia japonica* adults.[126]

***Erlangea cordifolia* S. Moore:** The germacranolide, eriancorin, obtained from the leaves of this plant is an antifeedant for larvae of *Spodoptera exempta* and *Epiolachna varivestis*.[300]

***Erigeron annuus* (L.) Pers.,** daisy, fleabane

***Erigeron canadensis* L.,** horseweed

***Erigeron pulchellus* Michx.,** poor-robins-plantain: Extracts of *E. annuus* and *E. pulchellus* did not repel *Popillia japonica* adults, but extracts of the fresh leaves and heads of horseweed were repellent.[126]

Erigeron philadebphicus L.: A methanol extract of the leaves was moderately effective as a deterrent for *Bombyx mori* larvae.[82]

Eupatorium adenophorum Spreng.: A cadinene compound isolated from this plant is deterrent to feeding by *Philosamia ricini*.[301]

Eupatorium aromaticum L., white sknaeroot: An aqueous extract of the whole plant deterred feeding by *Attagenus megatoma* larvae.[25]

Eupatorium cannabinum L.: The sesquiterpene lactones, eupatoriopicrin and eupatomolide, deterred feeding by *Stiophilus granarius* and *Trogoderma granarium* larvae.[302]

Eupatorium capillifolium (Lam.) Small, cypress weed, dogfennel: This plant was used to repel insect by strewing it on the floors of cellars and dairies.[303] The powdered whole plant was not toxic to *Spodoptera eridania, Diaphania hyalinata,* and *Spoladea recurvalis*,[128] and to *Ostrinia nubilalis* larvae. An acetone extract was somewhat toxic to *Cydia pomonella* but the combined petroleum ether, ethyl ether, chloroform, and ethanol extracts were not effective against this insect and *Musca domestica*, although it did show some toxicity to *Blattella germanica*.[2]

Eupatorium coelestinum L.

Eupatorium hyssopifolium L., thoroughwort

Eupatorium japonicum Thunb.

Eupatorium maculatum L.: Extracts of the leaves and flowers of *E. hyssopifolium* only were repellent to *Popillia japonica* adults.[126] Coumarin and euponin (a lactone), two inhibitors of development of *Drosphila melanogaster* fruit flies, were isolated from the leaves of *E. japonica*.[304]

Eupatorium odoratum L.: This plant is used by farmers in the Congo to protect beans from attack by *Acanthoscelides obtectus* and *Callosobruchus maculatus,* and to protect groundnuts from *Caryedon serratus*.[305] However, the powdered green tops had no deleterious effects on adult *C. serratus*.

Farfugium japonicum (L.) Kitamura: A methanol extract of the leaves did not deter feeding by *Bombyx mori* larvae.[82]

Galinsoga parviflora Cav., galinsoga

Gnaphalium obtusifolium L., sweet everlasting

Grindelia camphorum Greene, grindelia: Extract of these three plants were not repellent to *Popillia japonica* adults.[126]

Helenium amarum (Raf.) H. Koch, bitterweed, bitter sneezeweed: Helenalin obtained from this plant repels insects.[306] This compound also proved to be an excellent feeding deterrent for adult *Sitophilus granarius* and *Tribolium confusum* and for the larvae of *Trogoderma granarium* and *T. confusum*.[169] Linifolin A, another sesquiterpene lactone isolated from this species, is a good repellent for these stored grain pests.[307] Tenulin, another sesquiterpene lactone, proved to be a potent feeding deterrent for *Ostrinia nubilalis:* it also reduced growth and delayed larval development of this insect and of *Peridroma sancta*.[308] The lethal dose by injection into *Melanoplus sanguinipes* was 0.88 mmol/insect, which could be antagonized by coadministration of cysteine.

Helenium quadridentatum Labille, sneezeweed: An acetone extract of the whole plant deterred feeding by *Attagenus megatome*.[25]

Hellianthus annuus L., sunflower: An acetone extract of the whole plant was not repellent to *Popillia japonica* adults.[126] An acetone extract of the stems at 50 mg/ml deterred feeding by larvae of *Mythimna unipuncta* but not by *Leptinotarsa decemlineata*.[92] An infusion or decoction of the flower heads is used as a fly killer.[116] Sunflower oil was used with stored redgram seed as a protectant against infestation by *Callosobruchus chinensis*.[291] Evaluation of 29 species of cultivated sunflower in the laboratory showed that these species significantly retarded the development of *Empoasca abrupta*.[309] Seven perennial species of wild *Helianthus* were more resistant to the aphid, *Masonaphis masoni*, than 5 wild annual species.[310] Nine species of *Helianthus* were highly resistant to *Bithynus gibbosus*.[311] Helenalin and bisabolangelone were toxic by topical application to larvae of *Tribolium confusum* and *Trogoderma granarium;* application to adult *Sitophilus oryzae* caused fewer eggs to be laid and reduced the lifespan.[159] Resistance of the sunflower to *Homoesoma electellum* is due to the content of trachyloban-19-oic acid and (-)-kaur-16-en-19-oic acid in the florets. The larval weights were dramatically reduced by incorporation of these compounds into the diet.[23,312-315] The terpenoid chemistry of *Helianthus* spp. has been treated by Gershenzon et al.[316] and Mitscher et al.[317]

***Helianthus occidentalis* (T. & G.) Heiser:** The leaves of this plant completely deterred feeding by larvae of *Zygogramma exclamationis* and killed the larvae. Two diterpenoic acids, (-)-cis-and (-)-trans-ozoic acids, were isolated from the leaves and florets and myay be responsible for these effects.[318]

***Helianthus petiolaris*Nutt.:** An aqueous extract of the leaves deterred feeding by *Melanoplus sanguinipes* larvae but not by *M. differentialis* larvae.[275] *M. femurrubrum* grasshoppers failed to reach adulthood when reared on a diet containing extracts of this plant.[319]

***Heliopsis helianthoides*(L.),** sweet sunflower: Extracts of the whole plant did not repel *Popillia japonica* adults.[126]

***Heliopsis helianthoides* var. scabra (Pers.) Fern.,** cuague, peritre del pays, oxeye daisy: Acetone extracts of the roots and of the combines stems and leaves, tested at 0.1 and 0.5%, failed to deter feeding by *Diabrotica undecimpunctata howardi* and *Acalymma vittatum*.[52]

***Heliopsis parvifolia* Gray:** Acetone extracts of the roots did not deter feeding by *Diabrotica undecimpunctata howardi* and *Acalymma vittatum* adults.[52]

***Hemizonia fitchii* A. Gray,** tarweed, Fitch's spikeweed: The aerial portions of this plant yielded 1,8-cineole (eucalyptol) as the component responsible for the repellency to feeding and oviposition by adult *Aedes aegypti*. Several chromenes previously isolated from this plant have been shown to be responsible for mosquito larvicial activity.[321]

***Hieracium pratense* Tausch.,** hawkweed:

Extracts of the whole plant were repellent to *Popillia japonica* adults.[126]

***Homogyne alpina* (L.) Cass.:** Four sesquiterpene lactones and their adducts deterred feeding by adult *Sitophilus granarius* and *Tribolium confusum* as well as larvae of *T. confusum* and *Trogoderma granarium*.[302] Bakkenolide A was an excellent deterrent for feeding by *T. confusum*.[259]

***Inula graveolens* (L.) Desf.,** stinkwort: An aqueous extract of the entire plant did not repel *Lucilia cuprina*.[322] The plant is mixed with cereals by Greek farmers to protect them from attack by *Sitophilus* spp.[28]

***Inula helenium* L.,** eleocampane: This species was reported to protect clothing from attack by *Tineola bisselliella*. Extracts of the flowers collected in England did not kill larvae of the birch mocha moth.[323] Alantolactone significantly reduced the feeding and survival of *Tribolium confusum*.[3] Alantolactone, isoalantolactone, and *ent*-isoalantolactone were equally deterrent to adult *Sitophilus granarius* and larvae of *Tribolium confusum* but the *ent*- isomer was less effective for *Trogoderma granarium*.[325] Extracts of the whole plant were not repellent to *Popillia japonica* adults.[126] A strong concentration of an extract of the roots killed 100% of the mosquito larvae tested, and a 0.2% concentration killed 34 to 40% of *Aphis fabae*.[164]

***Inula magnifica* Lipsky:** The lactone fraction of an extract of this plant was moderately deterrent to feeding by adult *Sitophilus granarius* and strongly deterred adult *Tribolium confusum* and *Trogoderma granarium* larvae.[282]

***Isocoma wrightii* (Gray) Rbd.,** rayless goldenrod: The volatile oil inhibited feeding by *Spodoptera frugiperda* larvae at 20,000 and 50,000 ppm.[326]

***Ixeris dentata* Nakai:** Ethanol extracts of the leaves were accepted for feeding by *Bombyx mori* larvae.[82]

***Lactuca canadensis* L.,** wild lettuce: Extracts of the whole plant did not repel *Popillia japonica* adults.[126]

Lactuca sativa L., garden lettuce: Lettuce was acceptable to *Manduca sexta* larvae after 2 h of contact, but development was slow, with high mortality.[327]

Lasianthaea fruticosa DC: Lasidiol angelate, a sesquiterpenoid isolated from the leaves, repels leafcutter ants.[328]

Liatris spicata (L.) Willd., spike gayfeather, blazing star: An extract of the whole plant did not repel *Popillia japonica* adults.[126]

Libocedrus yateensis Guillaumin: Yatein, a sesquiterpene lactone isolated from this plant, was strongly deterrent to feeding by adult *Sitophilus granarius* and *Trogoderma granarium*, and by adults and larvae of *Tribolium confusum*.[302]

Ligularia tussilacinea Makino: Leaves of this plant tested in a Petri dish were not consumed by *Spodoptera littoralis* larvae.[91]

Matricaria chamomilla L., wild chamomile, false chamomile: Extracts of the whole plant were not repellent to *Bombyx mori* larvae and *Popillia japonica* adults.[126] The flowers were ineffective against bedbugs and roaches.[157]

Matricaria inodora L., scentless mayweed: An acetone extract of the stems strongly deterred feeding by *Mythimna unipuncta* and *Leptinotarsa decemlineata* larvae.[92]

Parthenium hysterophorus L., carrot grass, wild quinine: This plant is used in India to protect paddy, sorghum, and pulses from insect attack.[28] An acetone extract of the whole plant deterred feeding by *Melanoplus sanguinipes*.[25] Extracts of the leaves deterred feeding by *Spodoptera litura* larvae.[329] Leaf extracts applied topically to *Tribolium castaneum* caused complete knockdown.[330] Topical application of acetone extracts to *Dysdercus cingulatus* nymphs resulted in mortality and malformed adults. Malformed males were unable to mate, and female ovaries were reduced in size. Malformed females produced using 300 and 1000 mg of extract were completely sterile.[331] A petroleum ether extract of the whole plant, as well as a methanol etract of the defatted plant, did not synergize pyrethrum in tests against *Tribolium castaneum*.[200]

Perezia multiflora Less.: A petroleum ether extract of the aerial portion effectively deterred feeding by *Sitophilus granarius* adults, adult *Tribolium confusum*, and *Trogoderma granarium* larvae.[282]

Petasites albus (L.) J. Gaertn.: The sesquiterpenoid, a-hydroxyeremophilanolide, did not deter feeding by adults and larvae of *Sitophilus granarius* and *Tribolium confusum* or by larvae of *Trogoderma granarium*.[259]

Petasites fragrans Presl.: An acetone extract of the stems failed to deter feeding by *Leptinotarsa decemlineata* and *Mythimna unipuncta*.[92]

Petasites hybridus (L.) Gaertn.: The sesquiterpenoid, 2 b-angeloyleremophilanolide, was a potent antifeedant for adults of *Sitophilus granarius*, adults and larvae of *Tribolium confusum*, and larvae of *Trogoderma granarium*.[259]

Petasites japonicus (Sieb. & Zucc.) Fr. Schmidt, butterbus: A methanol extract of the leaves did not deter feeding by *Bombyx mori* larvae.[82]

Petasites kablikianus Tausch. ex Bercht.: The sesquiterpenoid, isopetasin, deterred feeding by larvae of *Tribolium confusum* but not by larvae of *Trogoderma granarium* or adults of *Sitophilus granarius* and *T. confusum*.[259]

Piptocoma antillana Urb.: An ether extract of the combined leaves and stems at 0.1 and 0.5% did not deter feeding by adult *Acalymma vittatum*.[91]

Ratibida columnifera Wooten & Standley: An acetone extract of the whole plant, incorporated into the basic diet, was readily accepted by *Melanoplus sanguinipes* grasshoppers.[275]

Rudbeckia hirta L., black-eyed susan: Extracts of the whole plant repelled adult *Popillia japonica*.[126]

Santolina virens Mill.: The juice of the fresh leaves and young shoorts deterred feeding by *Melanotus communis* larvae in choice and no-choice laboratory tests.[228]

Saussurea lappa (Decaisne) C. B. Clarke, costus: This plant has been used to repel

Tineola bisselliella and the leaves have been used a a wrapping for shawls.[332] The roots have also been used as an insect repellent to protect fabrics.[93] Wheat containing 2% of powdered costus roots was damaged by *Trogoderma granarium*.[333] An acetone extract of the stems failed to deter feeding by *Leptinotarsa decemlineata* and *Mythimna unipuncta*.[92] Any repellent action that costus root may have could be due to its content of the lactone, costunolide.[334]

Schkuhria pinnata (Lam.) O. Kuntze: The germacranolides, schkuhrin-I and schkuhrin-II, isolated from the whole plant, exhibit antifeedant activity against *Spodoptera exigua* and *Epilachna varivestis,* antimicrobial activity against some Gram-positive bacteria, and cytotoxicity (carcinostats).[330,335]

Senecio aureus L., groundsel

Senecio vulgaris L., groundsel: Extracts of the entire plants were not repellent to *Popillia japonica* adults.[126]

Senecio cinerarea DC: Larvae of *Plutella xylostella* would not feed on this plant.[195]

Senecio ehrenbergianus Klatt.: The roots have no repellent or toxic action against adult ants, *Pheidole dentata*.[26]

Seriocarpus asteroides (L.) B.S.P., whitetop aster

Silphium laciniatum L., compass plant: Extracts of the whole plants were not repellent to *Popillia japonica* adults.[126]

Solidago juncea Ait., early goldenrod: An extract of the entire plant did not repel *Popillia japonica* adults.[126]

Solidago missouriensis Nutt.: Leaves of this plant were readily consumed in the field by *Melanoplus sanguinipes* and *M. differentialis*.[19]

Solidago virga-aurea L.: An acetone extract of the stems failed to deter feeding by *Leptinotarsa decemlineata* and *Mythimna unipuncta*.[92]

SONCHUS OLERAEUS L., sowthistle: An extract of the whole plant was not repellent to *Popllia japonica* adults.[126]

Spilanthes acmella var. Oleraceae Clarke.: An acetone extract of the leaves inhibited the growth of *Bombyx mori* larvae. The active compound was isolated and identified as spilanthol, to which the structure N-isobutyl-E,E-4,6-decadienamide was initially assigned.[336] This was subsequently corrected to N-isobutyl-E,Z,E-decatrienamide.[337] Fourth instar *Bombyx mori* fed an artificial diet containing 200 ppm of spilanthol showed 100% mortality in 6 d.[83]

Stevia rebaudiana (Bertoni) Bertoni: Several derivatives of the sweet compounds stevioside and rebaudioside A present in this herb exhibited strong feeding deterrent effects against *Schizaphis graminum*.[333]

Tagetes lucida Cav., sweet-scented marigold: A petroleum ether extract of the aerial portion of this plant deterred feeding by adult *Sitophilus granarius* and *Tribolium confusum,* as well as larvae of *Trogoderma gtanarium*.[282]

Tagetes minuta L., wild marigold, Mexican stinking roger marigold: The oil distilled from the leaves, flowers, and seeds of the plant is repellent to *Lucilia cuprina*.[339] The leaves and flowers are repellent and toxic to *Anopheles* and *Culex* mosquitos,[340] and the leaves contain a mosquito larvicidal component designated 5E-ocimenone.[341] The pungent odor of the oil protects stored maize from insect attack,[28] but it is not repellent to *Aedes* mosquitos.[342] *Aedes* larvae and pupae coming contact with the oil do not appear to be affected adversely.[343] Topical application of 0.08 ml of the distilled oil to *Dysdercus koenigii* nymphs causes mortality, but application to *Musca domestica* pupae causes no deleterious effect.[343] Extracts of the leaves are not repellent to *Popillia japonica* adults.[126]

Tanacetum vulgare L., common tansy: Broccoli leaf disks painted with an aqueous extract of fresh tansy leaves deterred feeding by larvae of *Pieris rapae crucivora* and *Plutella xylostella;* larvae of *Trichoplusia ni* were not affected. Painting the extract on the insect's body surface had no deleterious effect.[344] *Leptinotarsa decemlineata* were repelled by tansy oil and by headspace volatiles from intact tansy plants. Thirteen components of the oil were identified: a-terpinene, thujone, carvone, and dihydrocarvone produced definite avoidance behavior, but g-terpinene had a lesser effect.[345] In laboratory tests tansy oil was strongly repellent to colonies of *Lasius alienus,*[346] but it was of no value as a repellent for *Cochliomyia hominivorax.*[155] Although aqueous extracts of the plant were not toxic to *Musca domestica,* extracts prepared with buffer solution (pH 4 or 9) were toxic to this insect.[347]

Taraxacum officinale **Wiggers,** common dandelion: Extracts of the whole plant were not repellent to *Popillia japonica* adults.[126]

Tarchonanthus camphoratus **L.:** This plant is reported to be repellent to insects.[116]

Tithonia diversifolia **(Hemsl.) Gray,** Mexican sunflower: Tagitenin A and C and Hispidulin, isolated from the aerial portion of the plant, were potent feeding deterrents fro 4th instar larvae of *Philosamia ricini;* tagitenin F was not deterrent.[348] Reacting tagitenin C with 10-deoxycyclotagitenin gave 10-deoxytagitenin C (removal of tertiary hydroxy group, designated with an asterisk). Both deoxy compounds were 10 times less active than the parent compounds as feedng deterrents.[349]

Tussilago farfara **L.,** coltsfoot: Sixth instar larvae of *Choristoneura fumiferana* exposed to an ethanol extract of the roots in the laboratory were deterred from feeding. The responsible component is the alkaloid senkirkine.[350]

Venidium hirsutum **Harv.:** Extracts of the aerial portion of the plant were satisfactory deterrents of feeding by adult *Sitophilus granarius* and *Tribolium confusum* and by the larvae of *Trogoderma granarium.*[282]

Vernonia amygdalina **Del.,** ironweed: Ether extracts of the combined woody stems and stem bark inhibited feeding by adult *Acalymma vittatum* at both 0.1 and 0.5% per only 6 h.[89] Elemanolide lactones (three) isolated from the bitter-tasting leaves, especially 11,13-dihydrovernodalin, inhibited feeding by larvae of *Spodoptera exempta* using a leaf-disk assay.[351]

Vernonia spp.: The sesquiterpene lactone,

glaucolide-A, deters feeding by the larvae of various lepidopterous insects.[352] Incorporation of this compound into the diet significantly reduced the rate of growth and the feeding level by larvae of *Spodoptera eridania, S. frugiperda,* and *S. ornithogalli.* The number of days to pupation was increase. *Spilosoma virginica* and *Trichoplusia ni* were not affected by digestion of the diet.[353]

Xanthium canadense **Mill.,** burweed: An acetone extract of the whole plant deterred feeding by larvae of *Attagenus megatoma.*[25] Tridec-1-ene-3,5,7,9,11-pentayne from the roots was ovicidal to *Drosophila melanogaster* and *Musca domestica.*[354]

Xanthium strumarium **L.,** common cocklebur: Leaves of this plant are normally accepted as food by larvae of *Spodoptera littoralis* in the laboratory.[91]

Xanthocephalum microcephalum **(DC) Shinners:** Ether extracts of the combined roots and stems did not deter feeding by adult *Acalymma vittatum.*[89]

Xeranthemum cylindraceum **Sibth. & Sm.:** A chloroform extract of the flowers strongly deterred feeding by adult *Sitophilus granarius* and *Tribolium confusum* and by larvae of *Trogoderma granarium.* The lactonic fraction was responsible for the activity.[282]

FAMILY ATHYRIACEAE

Onoclea orientalis **Hook:** An extract of the entire plant deterred feeding by larvae of *Eurema hecabe mandarina.*[90]

FAMILY AVICENNIACEAE

Avicennia germinans **(L.) L.,** black mangrove: An ethanol extract of the foliage incorporated into the laboratory diet did not deter feeding by larvae of *Lymantria dispar.*[84]

FAMILY BALSAMINACEAE

Impatiens balsamina **L.,** garden balsam

Impatiens biflora **Walt.,** spotted snapweed: The

seed oil of *I. balsamina* did not deter feeding by *Anthonomus grandis grandis* adults.[356] Extracts of these two plants did not repel *Popillia japonica* adults.[126]

Impatiens flamingii **Hook. f.:** The seed oil did not prevent feeding by *Anthonomus grandis grandis* adults.355

Impatiens sultani **L. Hook,** *Plutella xylostella* larvae would not feed on this plant.[195]

FAMILY BERBERIDACEAE

Berberis aquifolium **Pursh.,** Oregon holly grape: Extracts of the whole plant did not repel *Popillia japonica* adults.[126]

Caulophyllum thalictroides **(L.) Michx.,** blue cohosh

Jeffersonia diphylla **(L.) Pers.,** twinleaf: Extracts of these two plants did not repel *Popillia japonica* adults.[126]

Nandina domestica **Thunb.:** An ethanol extract of the foliage incorporated in the laboratory diet did not deter feeding by *Lymantria dispar* larvae.[84]

Podophyllum peltatum **L.,** mayapple, wild jalap, mandrake: An acetone extract of the whole plant deterred feeding by larvae *Attagenus megatoma*.[25] Extracts of the entire fresh plant were slightly repellent to *Popillia japonica* adults.[126] Ether extracts of each part of the plant repelled *Acalymma vittatum* adults.[89] Podophyllotoxin, a known mitotic poison, added at the level of 0.122% (by weight) to CSMA rearing medium completely inhibited the larval development of *Musca domestica*. There was no ovicidal action.[356] Newly emerged female flies were fed until sexual maturity on skimed milk duluted with water (1:1) containing 0.8 mg/ml of podophyllotoxin. The flies were permitted to mate with untreated males and were then fed again on the treated milk. On the 6th day egg-laying was induced by providing suitable oviposition sites. There was no apparent effect of the podophyllotoxin when compared with untreated controls, and the ovaries of treated flied showed no obvious abnormalities.[357] Mayapple has been known for many years for its use in the treatment of cancerous conditions.[358,359]

FAMILY BETULACEAE

Alnus incana **(L.) Moench.,** white adler: The odor of alder leaves was neither attractive nor repellent to adult *Leptinotarsa decemlineata*.[360]

Alnus rubra **Bong,** red alder: A methanol extract of the wood did not deter feeding by *Incisitermes minor* termites.[361]

Alnus rugosa **(DuBoi) Spreng,** speckled alder: An ethanol extract of the leaves deterred feeding by *Lymantria dispar* larvae.[84]

Betula davurica **Pall. Fl. Ross:** The leaves are highly resistant to damage by *Fenusa pusilla*.[362]

Betula lenta **L.,** sweet birch, black birch: An ethanol extract of the foliage did not deter feeding by *Lymantria dispar* larvae,[84] but birch oil strongly repelled *Blattella germanica* adults.[61]

Betula papyryfera **Marsh,** paper birch: *Spodoptera eridania* larvae fed on the leaves of this tree.[363]

Betula pendula **Roth,** European white birch: *Anacanthotermes ahngerianus* termites fed readily on the wood of this tree.[364]

Carpinus caroliniana **Walt.,** American hornbeam, blue beech: An ethanol extract of the foliage did not deter feeding by *Lymantria dispar* larvae.[84]

Ostrya virginiana **(Mill.) Koch,** hophornbeam: An acetone extract of the whole plant did not repel *Popillia japonica* adults.[126]

FAMILY BIGNONIACEAE

Campsis radicans **(L.) Seem. ex Bur.,** trumpet vine: An acetone extract of the whole plant did not repel *Popillia japonica* adults.[126]

Catalpa bignonioides **Walt.,** southern catalpa: An acetone extract of the plant did not repel *Popillia japonica* adults,[126] and the seed oil did not prevent feeding by these insect.[355]

Catalpa speciosa **(Warder ex Barney):** An ethanol extract of the foliage deterred feeding by *Lymnatria dispar* larvae.[84] The major active component in the leaves, designated "speciocide", was isolated and identified as an iridoid glycoside. A minor active component was named "catalposide".[365] An antifeedant for larvae of the eastern spruce budworm was isolated from the leaves and identified as specioin, an iridoid.[366] Larvae of *Ceratomia catalpae* fed nectar from the flowers regurgitated and showed reduced locomotion.[367]

Jacaranda sp.: An ethanol extract of the foliage of an unidentified species of *Jacaranda* deterred feeding by *Lymantria dispar* larvae.[84]

Paratecoma peroba (**Record.**) **Kuhlm.**: Filter paper impregnated with an ethanol extract of the wood and exposed to *Reticulitermes flavipes* termites killed 100% of the insect. Lapachanone and lapachol repelled R. lucifugus and are responsible for the resistance of the wood to termite attack.[368]

Stereospermum suaveolens **DC.**: Filter paper impregnated with an ethanol extract of the wood and exposed to *Reticulitermes flavipes* termites killed 40% of the insects and damges 30%. Lapachol from the wood was repellent to *R. lucifugus*.[368]

Tabebuia capitata (**Bur. & K. Schum.**) **Sandw.**: The wood is quite resistant to termites.[118]

Tabebuia flavescens **Benth. & Hook f.**: Lapachanone and lapachol from the wood are repellent to *Reticulitermes lucifugus* termites.[368]

Tabebuia guayacan **Hemsl.**: Extracts of the wood prepared with acetone, pentane, and a mixture (54:44:2) of acetone-hexane-water and exposed to *Coptotermes formosanus* termites gave complete kill.[369]

Tabebuia rosea **DC**: Extracts of the flowers tested on filter papers did not deter feeding by *Sitophilus oryzae* larvae.[251,252]

Tabebuia serratifolia (**Vahl.**) **Nicholson,** pau d'arco: The wood is resistant to attack by *Cryptotermes brevis* termites.[86]

Tecoma indica **DC**: An extract of the flowers tested on filter papers killed most of the *Sitophilus oryzae* larvae.[251,252]

FAMILY BIXACEAE

Bixa orellana **L.**, annatto, lipsticktree: The seed pulp was used by the American Indians to paint their bodies and prevent mosquito bits.[125] The fresh ground fruit pulp is applied to the skin as a paint in El Salvador to act as a repellent.[295] An ethanol extract of the leaves was somewhat repellent to *Bombyx mori* leaves.[82]

FAMILY BOMBACACEAE

Montezuma opeciorissima (**Moc. & Sesse.**), maga: The wood is highly repellent and resistant to attack by *Cryptotermes brevis* termites.[86]

Ochroma pyramidalis (**Cav.**) **Urban,** balsa, corkwood: The wood is susceptible to attack by drywood termites.[86,187,370]

FAMILY BORAGINACEAE

Anchusa angustifolia **DC,** Cantaloupe leaf disks were dipped in an acetone solution of petroleum ether extract of the whole plant and exposed to adult *Acalymma vittatum* and *Diabrotica undecimpunctata howardi* in Petri dishes at 0.1 and 0.5%. Feeding was not deterred by either concentration.[52]

Borago officinalis **L.,** common horage: A methanol extract of the whole plant failed to repel *Aedes aegypti* larvae.[245] An aqueous extract of the whole plant did not repel *Popillia japonica* adults.[126]

Cordia alliodora **Cham.,** Spanish elm: Extracts of the wood prepared with pentane, acetone, and acetone-hexane-water (54:44:2) deterred feeding by *Coptotermes formosanus* termites and gave complete kill.[369] The heartwood was very resistant to termites.[371]

Cynoglossum officinale **L.,** common houndstongue: Extracts of the whole plant did not repel *Popillia japonica* adults.[126]

Ehretia canarensis **Miq.**: An acetone extract of the aerial portion deterred feedingy by *Spilosoma obliqua* larvae.[194]

Lithospermum arvense **L.,** corn gromwell: A

methanol extract of the whole plant failed to repel *Aedes aegypti* larvae but it was highly toxic to these insects.[245]

***Pulmonaria officinalis* L.**, common lungwort: Extracts of the whole plant were not repellent to *Popillia japonica* adults.[126]

***Symphytum officinale* L.**, common comfrey: An extract of the dry roots was somewhat repellent to *Popillia japonica* adults.[126] A methanol extract of the leaves failed to deter feeding by *Bombyx mori* larvae.[82]

FAMILY BRASSICACEAE

***Arabidopsis thaliana* (L.) Heynh.**, mouse-ear cress: An extract of the whole plant was not repellent to *Popillia japonica* adults.[126]

***Armoracia rusticana* Gaertn. Mey. & Scherb.**, horseradish: Horseradish oil was repellent to *Blattella orientalis*[162] but no to *Popillia japonica*.[12]

***Barbarea vulgaris* R. Br.**, wintercress, yellow rocket: An extract of the whole plant did not repel *Popillia japonica* adults.[126]

***Brassica juncea* (L.) Gaertn. & Goss.**, Indian mustard, mustard greens: Treating a host leaf with mustard oil did not deter feeding by larvae of *Amsacta moorei*.[372] Field tests with different varieties of *Brassica* showed that *B. juncea* is more resistant to *Lipaphis erysimi* than is *B. rapa*.[373] Sprays of the seed oil were not effective in killing or repelling mosquitos.[374] Sinigrin, a mustard oil glucoside highly phagostimulatory for *Lipaphis erysimi*, deterred feeding by *Aphis fabae*[375] and larvae of *Papilio polyxenes asterius*.[376] Mustard oil glucosides stimulated feeding by adult *Phyllotreta cruciferae*[377] and by *Locusta migratoria*.[238] Mustard oil was an effective surface protectant of redgram seeds from *Callosobruchus chinensis*.[291] *Amrasca devastans* was not deterred from oviposition on cotton leaves exposed to mustard volatiles.[378]

***Brassica napus* L.**, rape, colza: An ethanol extract of the seeds prepared following defatting with ethyl ether repelled adult *Tribolium castaneum* and *T. confusum*.[379]

***Brassica nigra* (L.) Koch.**, black mustard: Extracts of the whole plant did not repel *Pop-*

illia japonica adults.[126] Aqueous and methanolic extracts of the seed killed the larvae of *Aedes aegypti*.[164,245]

***Brassica oleracea capitata* L.**, cabbage: Extracts of the whole plant did not repel *Popillia japonica* adults.[126]

***Brassica rapa* L.**, turnip: 2-Phenylethyl isothiocyanate isolated from this plant was a fair deterrent to feeding by *Tribolium confusum* larvae.[380]

***Capsella bursa-pastoris* (L.) Medik.**, shepherd's-purse: Extracts of the entire plant repelled *Popillia japonica* adults.[126] A methanol extract of this plant did not kill or delay development of *Aedes aegypti* larvae.[245]

***Cheiranthus allioni* Hort.**: Cardenolides present in this plant discourage feeding by *Phyllotreta memorum, P. terrastigma*, and *Phaedon cochleariae*.[381]

***Erysimum perofskianum* Fisch. & Mey.**, Afghan bittercress: Extracts of the whole plant repelled *Popillia japonica* adults.[126] Teh cabbage butterfly, *Pieris rapae*, avoids laying eggs on the leaves of this plant.[382]

***Lepidium campestre* (L.) R. Br.**, pepperweed: a methanol extract of the whole plant did not repel *Aedes aegypti* larvae.[245]

***Lunaria annua* (L.)**, honesty: Extracts of the leaves prepared with several oganic solvents deterred feeding by adult *Phyllotreta striolata*.[383]

***Raphanus sativus* L.**, radish: Glucosinolates present in the leaves repel adult *Delaia radicum* seeking to oviposit.[384]

***Sisymbrium altissimum* L.**, tumble mustard: A methanol extract of the entire plant did not repel *Aedes aegypti* larvae.[245]

***Thlaspi arvense* L.**, field pennycress: Extracts of the leaves prepared with several organic solvents deterred feeding by adult *Phyllotreta striolata*.[382] A methanol extract of the plant significantly reduced adult emergence of *Aedes aegypti* mosquitos.[245]

FAMILY BURSERACEAE

***Boswellia carteri* Birdw.**, frankincense: Frankincense is burned in houses to repel mosquitos.[385] Gun from the bark is believed to provide protection from termites, and to repel mosquitos, flies, and pests of clothing.[121]

***Bursera simaruba* (L.) Sarg.**, gumbo-limbo: An ethanol extract of the leaves deterred feeding by *Lymantria dispar* larvae.[84,386]

***Canarium commune* L.**, The wood is reported to be resistant to *Anacanthotermes ochraceus*, *Psammotermes fuscofemoralis*, and *P. assuarensis termites*.[387]

***Canarium schweinfurthii* Engler**, hediwonua: The wood is not resistant to attack by *Coptotermes formosanus* termites.[388]

***Commiphora abyssinica* Engl.**: This plant yields myrrh, which is well known in tropical countries as a mosquito repellent; it is usually burned as an incense stick.[116]

***Commiphora africana* Engl.**: The gum resin is used in West Africa as a termite repellent.[116]

***Commiphora boiviniana* Engl.**: An ether extract of the twigs was chromatographed on a silica gel column, eluting with hexane and increasing concentrations of ether in hexane. The hexane eluate completely repelled *Musca domestica*, but the other eluates did not.[209]

***Protium heptaphyllum* (Aubl.) March:** The wood is resistant to termites.[124]

FAMILY BUXACEAE

***Buxus sempervirens* L.**, common boxwood: Leaf extracts were effective repellents for *Popillia japonica* adults.[126] An aqueous extract of the branchlets and leaves was nontoxic to *Periplaneta americana*.[25] Larvae of *Plutella xylostella* would not feed on this plant.[25,195]

FAMILY CACTACEAE

***Opuntia humifusa* Raf.**, pricklypear: Extracts of the whole plant did not repel *Popillia japonica* adults.[126]

FAMILY CALYCANTHACEAE

***Meratia praecos* Rehder & Wilson:** A methanol extract of the leaves deterred feeding by *Bombyx mori* larvae.[82]

FAMILY CAMPANULACEAE

***Lobelia nicotianafolia* Heyne,** wild tobacco: The dried chopped leaves are mixed at 2% with rice and wheat to repel stored product insects.[389]

***Lobelia cardinalis* L.**, cardinal flower

***Lobelia inflata* L.**, Indian tobacco: Acetone extracts of these plants were not repellent to *Popillia japonica* adults.[126]

FAMILY CANELLACEAE

***Canella winterana* (L.) Gaertn.**, canella, wild cinnamon: An ethanol extract of the trunk bark was not repellent to *Popillia japonica* adults,[126] but it was highly deterrent to feeding by *Spodoptera littoralis* larvae.[390] The active compound, canellal, was isolated, identified, and synthesized;[391] it also proved to have antimicrobial properties.[390]

***Warburgia salutaris* (Bartol. f.) Chiev.:** An extract of the bark prepared with 60% ethanol deterred feeding by larvae of *Spodoptera littoralis* and *S. exempta* but not by *S. eridania, Manduca sexta,* or *Schistocerca vaga*.[300,392] The active compounds have been isolated and identified as warburganal, polygodial, and ugandensidial.[393] A comparison of the antifeedant activity of a number of unsaturated natural and synthetic 1,4-dialdehydes showed that the activity against *Leptinotarsa decemlineata* and *Spodoptera littoralis* is dependent upon the distance between the two aldehyde groups.[394] The mode of action of the dialdehyde antifeedants from *W. salutaris* and *W. stuhlmanni* against the armyworm and the larvae of *Heliothis virescens* and *H. ammigera* has been studied intensively.[395-398] An ether extract of the ground roots and bark of *W. salutarisa* at 0.1 and 0.5% was strongly deterrent to *Acalymma vittatum* adults.[51,89] Muzigadial and warburganal were found to have potent antifungal activity,[399] and polygodial and mukaadial, another sesquiterpene from *Warburgia* spp., are toxic to molluscs.[400,401] The reactivity of warburganal with sulfhydryl groups and of polygodial with primary amines have been studied.[402,403] Numerous procedures have been reported for synthesizing *dl*-warburganal,[404,411] (-)-warburganal,[412-417] and polygodial.[418-420]

FAMILY CANNACEAE

Canna indica L., Extracts of the flowers failed to deter feeding by *Sitophilus oryzae* larvae.[251,252]

FAMILY CAPPARACEAE

Cleome serrulata **Pursh.:** Aqueous extracts of the foliage maintained the growth and development of *Melanoplus femurrubrum* and *M. sanguinipes*,[244] but not of *M. differentialis*.[319] Extracts of the roots, capsules, and seeds, as well as the fruits were nontoxic to *Blattella germanica, Periplaneta americana,* and *Oncopeluts fasciatus*.[25]

FAMILY CAPPARIDACEAE

Capparis flexuosa L., mosto: The fruits of this Venezuelan shrub are used as an insect repellent.[421]

Capparis schinkei **Macbride:** An ether extract of the roots tested at 0.1 and 0.5% was somewhat repellent to *Acalymma vittatum* adults.[89]

Isomeris arborea **Nutt.:** An ether extract of the combined twigs, leaves, and flowers failed to deter feeding by adult *Acalymma vittatum* at concentrations of 0.1 and 0.5%.[89]

FAMILY CARPRIFOLIACEAE

Lonicera japonica **Thunb.**, Japanese honeysuckle: The leaves are refused as food by *Lymantria dispar* larvae[84,422] but not by larvae of *Spodoptera frugiperda*[204] and adult *Acalymma vittatum*[52] and *Poiliia japonica*.[126]

Sambucus canadensis L., American elder, elderberry: An ethanol extract of the leaves did not deter feeding by *Lymantria dispar* larvae.[84]

Sambucus nigra L., European elder, dwarf elder: Elder leaves are repellent to *Musca domestica* and *Trichoplusia ni*, and the branches repel turnip flies, *Phyllotreta* sp.[13]

Sambucus steboldiana **Blume ex Miq.:** A benzene extract of the whole plant failed to deter feeding by *Spodoptera litura* larvae.[258]

Sambucus simpsoni **Rehder:** An ethanol extract of the foliage failed to deter feeding by *Lymantria dispar* larvae.[84]

Viburnum acerifolium L.

Viburnum obovatum **Walt.**

Virurnum suspendum **Lindl.**

Viburnum tinus L.: An ethanol extract of the foliage of each of these species deterred feeding by *Lymantria dispar* larvae.[84]

Viburnum dentatum L., arrowwood: Extracts of the entire plant did not repel *Popillia japonica* adults.[126]

Viburnum japonicum **Spreng.:** Chavicol isolated from the leaves inhibited the growth of larval *Drosphilia melanogaster*.[423]

Viburnum nervosum **D. Don:** An acetone extract of the foliage deterred feeding by *Spilosoma obliqua* larvae only slightly.[194]

FAMILY CARICACEAE

Carica papaya L., papaya: An ointment prepared from the latex is used in El Salvador against chiggers, pubic lice, and toe-burrowing jiggers.[295]

FAMILY CARYOCARACEAE

Caryocar brasiliense **Camb.**, butternut tree, pequi: The wood is resistant to *Cryptotermes brevis* termites.[86]

Caryocar coctaricense **Donn Smith:** The wood is resistant to *Coptotermes formosanus* termites.[369]

Caryocar villosum **Pers.**, Brazilian butternut: The wood is resistant to termites.[124]

FAMILY CARYOPHYLLACEAE

Cypsophila paniculata L., Filter papers impregnated with 0.05 or 0.5% solutions of the saponin isolated from the wood inhibited feedig by *Reticulitermes flavipes* termites but failed to kill them; 3% solutions were completely without effect.[229] Air-dried leaves, roots, and stems were highly toxic to *Aedes aegypti* larvae.[189]

Saponaria officinalis L., bouncing-bet, soapwort: A methanol extract of the leaves did not repel *Bombyx mori* larvae or *Popillia japonica* adults.[126]

Silene alba (Mill.) Krause, white campion, whitecockle: A methanol extract of the whole plant had no effect on *Aedes aegypti* larvae.[245]

Silene antirrhina L., sleepy catchfly: Extracts of the whole plant were not repellent to *Popillia japonica* adults.[126] A methanol extract of the whole plant had no effect on *Aedes aegypti* adults.[245]

Stellaria media (L.) Vill., common chickweed: Extracts of the whole plant were not repellent to *Popillia japonica* adults.[126]

FAMILY CELASTRACEAE

Cassine transvaalensis (Burt Davy) Codd.: Ether extracts of the woody stems with bark did not deter feeding by adult *Acalymma vittatum*.[89]

Celastrus angulatus Maxin., bitter tree, anglestem bittersweet: This plant is used by Chinese farmers fo rcontrolling vegetable insects. Pulverized root bark and aqueous suspensions thereof were effective as a repellent and stomach poison against *Malacosoma pluviale*. Crystalline and noncrystalline fractions of a petroleum ether extract of the root bark were repellent and toxic to the larvae.[424] Aqueous extracts of the roots and root bark were toxic to *Spolades recurvalis, Ostrinia nubilalis* larvae, and *Blattella germanica* adults, but no to *Spodoptera eridania, Diaphania hyalinata, Herpetogramma bipunctalis, Urbanus proteus, Periplaneta americana, Ocopeltus fasciatus,* and *Cydia pomonella* larvae, and *Evergestis rimosalis* grasshoppers.[25,26,128]

Euonymus alata (Thunb.) Sieb., winged eounymus: The insecticidal alkaloid, wilfordine, and another alkaloid, alatamine, were isolated from the fruits of this plant.[425]

Goupia glabra Aubl.: The wood is resistant to termites.[124]

Lophopetalum wightianum Arn., banati: The wood is very resistant to *Microcerotermes beesoni* termites.[426] Ether extracts of the roots and of the combined twigs and leaves did not deter feeding by *Acalymma vittatum* adults.[89]

Maytenus rigida Mart.: The alkaloid, wilforine, was isolated from the roots of this plant and found to be deterrent to feeding by nymphs of *Melanoplus sanguinipes* and to larvae of *Pieris rapae*.[427]

Maytenus undatus (Thunb.) Blakelock: Ether extracts of the combined leaves and twigs did not deter feeding by *Acalymma vittatum* adults at 0.1 or 0.5%, but an extract of the stem bark deterred feeding at 0.5%.[89]

Putterlickia verrucosa Sim.: An ether extract of the combined twigs and leaves did not deter feeding by *Acalymma vittatum* adults at 0.1% but it was deterrent at 0.5%.[89]

Tripterygium forrestii Loesener: Larvaee of *Bombyx mori, Epilachna varivestis, Tricholplusia ni, Archips cerasi, Aphis fabae,* and *Paleacrita vernata* did not feed when placed on food plants treated with the powdered roots of this plant. The roots showed considerable toxicity to *Cydia pomonella* larvae.[428] Dusting of the powdered root bark on food plants strongly repelled melon leaf beetles and cabbage flea beetles.[429] Powdered root bark repelled the bean plataspid, *Spodoptera litura,* cabbage leaf beetles, and *Acalymma vittatum* adults.[424]

Tripterygium wilfordii Hook f., thundergod vine: An ether extract of the fruits was strongly deterrent to feeding by adult *Acalymma vittatum* at 0.5% but not at 0.1%.[89] The root bark is repellent to *Plutella xylostella* and toxic to numerous species of vegetable insects.[26,27,430] Highly stoxic alkaloidal mixtures were isolated from the root bark,[431] and these were separated into the alkaloids wilforine, wilfordine, wilforgine, wilfortrine, and wilforzine, all of which were identified.[432-435]

FAMILY CHENOPODIACEAE

Anabasis aphylla L., tree tobacco: Smearing 5

to 10% solutions of anabasine sulfate on the body did not repel malaria mosquitos but did protect against their bites. Solutions of 1 to 3% were ineffective.[435]

Atriplex lentiformis **S. Wats.**: A petroleum ether extract of the leaves used as a surface protectant on cowpea seed completely deterred feeding by adult *Callosobruchus chinensis*. A concentration of 1 or 2.5% was slightly less effective, 0.5% was highly effective, and 0.1% was moderately effective.[436]

Chenopodium album **L.**, lambsquarters: Juice pressed from the fresh leaves did not deter feeding by larval *Melanotus communis*.[190]

Chenopodium ambrosioides **L.**, Mexican tea wormseed: An aqueous extract of the whole plant deterred feeding by *Attagenus megatoma*.[25] Juice pressed from the fresh seeds failed to deter feeding by *Heliothis zea* larvae.[190] The powdered seed was effective as a repellent for *Cochliomyia hominivorax* for only 2 days,[155] and extracts of the whole plant were not repellent to *Popillia japonica* adults.[126] A solution of the methanol extract of the seeds in water significantly increased larval development and inhibited pupal development of *Aedes aegypti*.[245] Wormseed oil at 25 ppm killed 90 to 100% of mosquito larvae.[164]

Chenopodium anthelminticum **L.**: A methanol extract of the leaves added to the diet retarded the growth and development of *Bombyx mori* larvae.[82]

Chenopodium botrys **L.**, Jerusalem-oak goosefoot: Women in France preserve their clothes and linen with this herb because it repels *Tineola bisselliella* larvae and gives the clothing a pleasant odor.[437]

FAMILY CHRYSOBALANACEAE

Licania densiflora **Kleinh.**: The wood is resistant to termites.[124]

Licania heteromorpha **Benth. var. glabra (Mart. ex Hook.) Prance**: An ether extract of the leaves failed to deter feeding by adult *Acalymma vittatum*.[89]

FAMILY CISTACEAE

Cistus villosus **L.**, The phenolic dilactone, ellagic acid, was isolated from hot methanol extracts of the plant. It inhibits the growth of *Heliothis virescens* larvae.[438]

Helianthemum canadense **(L.) Michx.**, sunrose: Extracts of this plant were not repellent to *Popillia japonica* adults.[126]

FAMILY CLETHRACEAE

Clethra barbinervis **Sieb. & Zucc.**: A methanol extract of the leaves did not deter feeding by *Bombyx mori* larvae.[82]

FAMILY CLUSIACEAE

Calophyllum brasiliense **Camb.**, ucaria, jacaruba: The wood is resistant to termites.[370] An ether extract of the stems with bark deterred feeding by *Acalymma vittatum* adults at 0.5% but not at 0.1%.[89]

Calophyllum inophyllum **L.**, Indian laurel, laurelwood, undi: This wood is moderately resistant to termites.[188] Rice plants sprayed with the seed oil (5% in water) inhibited feeding by larvae of *Lerodea eutela*.[439] Greengram seed coats treated with a 0.5% solution of the oil in petroleum ether were protected from feeding by *Callosobruchus chinensis*.[440]

Mammea africana **Sabine**, African mammy-apple: An ether extract of the stems with bark deterred feeding by adult *Acalymma vittatum* at 0.5% but not at 0.1%.[189] The wood is resistant to attack by *Coptotermes formosanus*[388] and *Reticulitermes lucifugus* termites.[441]

Mammea americana **L.**, mamey, mammy apple: The wood is susceptible to attack by termites.[124] When used against fleas and ticks on dogs, an infusion of 1 lb of the half-ripe fruits/gallon of water was as effective as a 1% suspension of DDT.[442] The leaves have been used for many years in Puerto Rico as a wrapping around the trunks of newly set garden plants to prevent attack by garden insects. The powdered seeds are very toxic to numerous species of insects.[26,27] The active principle is mameyin (mammein), a coumarin-like compound averaging 0.19% of the seed weight.[443] It was identified as 4-*n*-propyl-5,7-dihydroxy-6-isopentenyl-8-isoavalerylcoumarin.[444-447] A related compound, 4-phenyl-5,7-dihydroxy-8-isopentenyl-6-isovalerylcoumarin, was isolated from the fruit pulp and shown to be toxic to insects.[448,449]

Messua ferrea L., The wood is resistant to attack by the subterranean termites *Anacanthotermes ochraceus, Psammotermes fuscofemoralis*, and *P. assuarensis*.[387]

Platonia esculenta (Arr. Cam.) Rickett & Stafl., bacury: The wood is resistant to *Cryptotermes brevis* termites.[86]

Psorospermum febrifugum Spach.: An ether extract of the roots deterred feeding by adult *Acalymma vittatum* at 0.5% but not at 0.1%.[89]

FAMILY COMBRETACEAE

Anogeissus leiocarpus (Guill. & Perr.) Benth., kane: The wood completely deterred feeding by *Coptotermes formosanus* termites.[388]

Buchenavia capitata (Vahl) Eichl., granadillo: The wood is resistant to *Cryptotermes brevis* termites.[86]

Bucida buceras L., An ethanol extract of the foliage deterred feeding by *Lymantria dispar* larvae.[84]

Combretodendron africanum (Welw. ex Benth. & Hook f.) Exell., esia: The wood is reported to be resistant to termites in Nigeria.[370,388]

Combretum apiculatum Sond.: The wood is not attacked by termites.[116]

Combretum cafrum (Eckl. & Zeyh.) Kuntze: An ether extract of the combined twigs, leaves, and fruits did not deter feeding by adult *Acalymma vittatum*.[89]

Combretum ghasalense Engl. & Diels.: An ether extract of the stem bark did not deter feeding by adult *Acalymma vittatum*.[89]

Conocarpus erectus L., button mangrove: An ethanol extract of the foliage deterred feeding by *Lymantria dispar* larvae.[84] An ace-

tone extract of the wood repelled *Coptotermes formosanus* termites.[369]

Labuncularia racemosa (L.) Gaertn. f., white mangrove: The wood is resistant to termites.[124] An acetone extract of the wood repelled *Coptotermes formosanus* termites.[369]

Terminalia alata Heyne, Indian laurel
Terminalia amazonia (Grnel.) Excell., amarillo, guayabon
Terminalia arjuna Wight & Arn, arjuna: The wood of these three species is resistant to *Cryptotermes brevis* termites.[86,115]

Terminalia bialata Steud., white chuglam, Indian silver-gray wood: The wood is susceptible to termite attack.[370]

Terminalia catappa L., tropical almond, Indian almond: The wood is very susceptible to termite attack.[85,124] An aqueous extract of the entire plant is an effective repellent for *popillia japonica* adults.[126] An ether extract of the woody stems with bark did not repel *Acalymma vittatum* adults.[99] An acetone extract of the meal was nontoxic to mosquito larvae,[167] and an aqueous extract of the leaves was toxic to *Periplaneta americana* but not to *Blattella germanica* or *Oncopeltus fasciatus*.[25]

Terminalia chebula Retz.: An acetone extract of the aerial portion deterred feeding by *Spilosoma obliqua* larvae.[194]

Terminalia coriacea (Roxb.) Wight & Arn.
Terminalia crenulata Heyne: The heartwood of these two species is reported to be durable in India, remaining unattacked by subterranean termites for 4 to 5 years in India and for 5 years in Malaya.[370]

Terminalia ivorensis A. Chev., idigbo, emeri: The wood is moderately deterrent to *Coptotermes formosanus* termites[388] and susceptible to attack by *Cryptotermes brevis*.[86]

Terminalia superba Engl. & Diels., korina, limba: The wood is susceptible to attack by *Reticulitermes lucifugus* termites.[441]

FAMILY COMMELINACEAE

Commelina communis L., dayflower: The leaves are susceptible to feeding by larvae of *Spodoptera littoralis*.[91]

FAMILY CONNARACEAE

Spiropetalum heterophyllum (**Bak.**) **Gilg.:**
An extract of the stem bark did not deter
feeding by Acalymma vittatum adults.[89]

FAMILY CONVOLVULACEAE

Argyreia nellycherya **Choisy:** An acetone
extract of the aerial portion was slightly
deterrent to *Spilosoma obliqua* larvae.[194]

Convolvulus soldanella **L.,** An ether extract
of the combined stems, leaves, and fruits did
not deter feeding by adult *Acalymma vittatum.*[89]

Ipomoea coccinea **L.,** An ether extract of the
seed did not deter feeding by adult *Acalymma vittatum.*[89]

Ipomoea cornea **L. O. Williams:** The powdered leaves (8 parts) mixed with gram seed
(1 part) did not repel *Callosobruchus chinensis.*[73]

Ipomoea hederaceae (**L.**) **Jacq.,** ivyleaf
morning-glory: The plant is rejected by larvae of *Manduca sexta.*[327]

Ipomoea jalapa (**L.**) **Pursh.,** jalap: An aqueous extract of the whole plant was not repellent to *Popillia japonica* adults.[126]

Ipomoea muricata **Jacq.:** The plant is used in
India as a repellent for many household insects.[450]

Ipomoea palmata **Forsk.:** A petroleum ether
extract of the leaves tested as a surface protectant for stored cowpea seeds afforded
complete protection from *Callosobruchus
chinensis.*[196]

Ipomoea pandurata (**L.**) **G. F. W. Mey.,**
bigroot morning-glory: An acetone extract
of the whole plant did not repel *Popillia
japonica* adults.[126]

Ipomoea purpurea (**L.**) **Roth,** morning glory:
The plant is rejected by *Manduca sexta* larvae.[327]

FAMILY CORNACEAE

Cornus florida **L.,** flowering dogwood: An
ether extract of the twigs completely deterred feeding by *Spodoptera frugiperda*

larvae, but a methanol extract was only
slightly deterrent.[204] The ehter extract did
not deter feeding by adult *Acalymma vittatum*
or *Diabrotica undecimpunctata howardi,* but
the water-soluble fraction of a methanol
extract of the twigs completely deterred
feeding by these beetles.[52] Extracts of the
leaves did not deter feeding by *Melanotus
communis*[247] or larvae of *Lymantria dispar.*[84]
Extracts of the whole plant were not repellent to *Popillia japonica* adults.[126]

Cornus nuttalli **Audubon:** An ethanol extract
of the leaves failed to deter feeding by *Lymantria dispar* larvae.[84]

Cornus officinalis **Sieb. & Zucc.:** A methanol extract of the leaves strongly deterred
feeding by *Bombyx mori* larvae.[82]

Cornus Stolonifera **Michx.:** An ethanol extract of the leaves failed to deter feeding by
Lymantria dispar larvae.[84]

Mastixia tetrandra **Clarke:** An ether extract
of the whole plant deterred feeding by adult
Acalymma vittatum at 0.5% but not at 0.1%.[89]

FAMILY CRASSULACEAE

Kalanchoe daigremontiana **Ham.:** An acetone extract of the leaves completely inhibited feeding by *Schistocerca greparia.*[99]

Kalanchoe somaliensis **Baker:** The juice of
the foliage is used as an insect repellent in
East Africa.[116]

Kalanchoe tubiflora (**Harv.**): An acetone
extract of the leaves at 0.5% completely
inhibited feeding by *Schistocerca gregaria.*[99]

Sedum kantschaticum **Fisch. & Mey.:** A
methanol extract of the leaves did not deter
feeding by *Bombyx mori* larvae.[82]

Sempervivum tectorum **L.,** hen-and-chick-
ens, house leek: The classicial Greeks advocated soaking vegetable seeds in house leek
juice prior to sowing in order to repel insects,
and adding the dried leaves to granaries to
protect stored wheat and barley.[13]

FAMILY CUCURBITACEAE

Cayoponia ficifolia (**Lam.**) **Cogn.:** An ethanol extract of the whole plant did not deter

feeding by *Acalymma vittatum* adults or *Diabrotica undecimpunctata howardi* adults.[52]

***Citrullus colocynthis* (L.) Schrad.**, colocynth: Classicial methods involved the use of the fruit juice to protect germinating seeds from insect attack.[13] Decoctions of the plant were recommended in France against leafeating caterpillars on fruit trees.[451] Cucurbitacins, a large series of which has been isolated from cucurbits, are known to be the bitter substances active as beetle attractants and feeding deterrents. *Diabrotica undecimpunctata howardi* responds to 1 ng of cucurbitacin B with compulsive feeding, while other cucurbitacins repel these insects.[34]

***Citrullus lanatus* (Thunb.) Matsum. & Nakai,** watermelon: Four varieties of watermelon yielded cucurbitacins E and I, which elicited feeding responses by species of *Diabrotica* beetles.[452] A bitter-fruited mutant strain of watermelon was high in cucurbitacin content, which caused the bitter taste. These compounds were extremely attractive to cucumber beetles, but *Apis mellifera* and *Dolichovespula arenaria* wasps were repelled.[453]

***Cucumis anguria* L.,** West Indian gherkin

***Cucumis melo* L.,** muskmelon, canteloupe

***Cucumis sativus* L.,** cucumber: Although *C. sativus* contains cucurbitacin C, the other two species contain only cucurbitacin B.[452] Larvae of *Plutella xylostella* would not feed on *C. sativus*. The juice of cucumbers is said to banish fish moths and wood lice and to kill cockroaches; the green perl strewn on the floor at night for 3 or 4 nights is also effective. The cockroaches collect around the pieces and devour them, with fatal results.[116] Fresh cucumber juice was fairly deterrent to larvae of *Melanotus communis*.[247] Of nine species of *Cucumis* tested, six were fairly resistant to injury by *Tetranychus urticae*.[454]

***Cucurbita maxima* Duch.,** pumpkin, squash

***Cucurbita moschata* Duch. ex Poir,** pumpkin, squash

***Cucurbita pepo* L.,** pumpkin, squash: Four varieties of *C. maxima* and three varieties of *C. moschata* contained cucurbitacin B, and

23 varieties of *C. pepo* contained cucurbitacins B and D. No feeding by *Diabrotica* beetles was observed on two varieties of *C. maxima*, two varieties of *C. moschata*, and ten varieties of *C. pepo*.[452] When leaves of *C. moschata* are experimentally damaged, substances, are mobilized to the damaged region within 40 min which stimulate feedng by *Acalymma vittatum* adults and inhibit feeding by *Epilachna tredecimnotata*.[135,455]

***Ecballium elaterium* (L.) A. Rich.**, squirting cucumber: Elaterin, derived from this plant, was ineffective for mothproofing.[456]

***Lagenaria ciceraria* (Mol.) Standl.**, white-flowered gourd, calabash gourd: This plant is avoided by larvae of *Diaphania nitidalis*.[457]

***Luffa acutangula* (L.) Roxb.**, angled luffa, singkwa towel gourd: Extracts of the flowers were not fed upon by *Sitophilus oryzae* larvae; the larvae were killed upon contact with filter papers dipped in solutions of the extact.[251,252]

***Luffa aegyptiaca* Mill.**, luffa, smooth loofah: The plant is used in Nigeria to protect stored products from insects.[458]

***Momordica cochinchinensis* (Lour.) Spreng:** The seed oil deterred feeding by adult *Anthonomus grandis grandis*.[355]

***Momordica foetida* Schum.:** In Tanganyika the fruit pulp is regarded as poisonous to weevils, moths, and ants and is used as a repellent.[116]

***Trichosanthes anguina* L.,** snake gourd: The seed oil failed to deter feeding by adult *Anthonomus grandis grandis*.[355]

FAMILY CUPRESSACEAE

***Calocedrus decurrens* (Torr.) Florinm**, incense cedar: The wood is resistant to *Anacanthotermes ochraceus*, *Psammotermes fuscofemoralis*, and *P. assuarensis* termites.[387]

***Chamaecyparis formosensis* Matsum.:** The wood is initially resistant to *Cryptotermes brevis* termites but becomes susceptible to feeding after 1 year of exposure.[86] The active compound has been isolated, identified as chamaecynone, and synthesized.[459,460]

Chamaecyparis lawsoniana (**A. Murr.**) **Parl.,** Port Orford cedar, Lawson cypress: The wood is completely resistant to *Anacanthotermes ochraceus, Psammotermes fuscofemoralis,* and *P. assuarensis* termites for 4 months,[387] as well as to *Reticulitermes flavipes* and *R. virginicus.*[461,462] Topical application of ether extracts of the wood to larvae of *Malacosoma pluviale* produced typical morphological abnormalities.[463]

Chamaecyparis nootkatensis (**D. Don.**) **Spach.,** yellow cedar, Alaska cedar: The wood is fairly resistant to *Cryptotermes brevis* termites.[86,387]

Chamaecyparis fisifera **D. Don.,** Sawara wood: The strong termiticidal activity of the wood is due to its high content of chamaecynone and isochamaecynone.[464]

Chamaecyparis formosensis **Matsum.:** The wood is highly resistant to feeding by *Reticulitermes flavipes* and *R. lucifugus* termites, owing to its content of *l*-citronellic acid.[465]

Chamaecyparis thyoides (**L.**) **B. S. F.,** Atlantic white cedar, southern white cedar, yellow cedar: The wood is susceptible to termites,[124] but an ethanol extract of the leaves deterred feeding by *Lymantria dispar* larvae.[84] An acetone extract of the wood did not repel *Popillia japonica* adults.[126]

Cupressus arizonica **Greene,** Arizona cypress: An ether extract of the seeds deterred feeding by adult *Acalymma vittatum* at 0.5% but not at 0.1%.[89] No pests have been reported from plantations of this species in Southern Rhodesia.[119]

Cupressus lusitanica **Mill.,** Mexican cypress: The wood is resistant to *Oemida gahani.*[119] An ether extract of the seeds deterred feeding by adult *Acalymma vittatum* at 0.5% but not at 0.1%.[89]

Cupressus macrocarpa **Hartw.,** Monterey cypress: The wood is resistant to termites in Nigeria.[119]

Cupressus sempervirens **L.,** cypress: Wooden stakes buried in Egyptian soil withstood attack by *Anacanthotermes ochraceus, Psammotermes fuscofemoralis,* and *P. assuarensis* for 4 months.[387] The smoke from burning cypress cones and twigs was reported to repel mosquitoes.[437]

Juniperus communis **L.,** common juniper: Juniper oil is strongly repellent to *Blattella germanica*[162] but an acetone extract of the plant did not repel *Popillia japonica* adults.[126]

Juniperus gracilior **Pilger,** sabina: The wood is moderately repellent to *Cryptotermes brevis* termites.[86]

Juniperus hispanica **Mill.:** The wood is resistant to *Anacanthotermes ochraceus, Psammotermes fuscofemoralis,* and *P. assuarensis* termites.[387]

Juniperus oxycedrus **L.,** cade: Cade oil was considered to be one of the best repellents for *Cochliomyia hominivorax.*[155] The oil has been patented for use as a mothproofing agent.[166]

Juniperus procera **Hochst. ex Endl.,** East African juniper, African pencil cedar: The heartwood is reported to be immune from termite attack.[207] The wood is resistant to *Nasutitermes exitiosus, Coptotermes lacteus,* and *C. acinacitermis* termites.[467]

Juniperus recurva **Buch.:** The powdered heartwood is an excellent fumigant due to its content of thujopsene and 8-cedren-13-ol.[468]

Juniperus rigida **Sieb. & Zucc.:** A methanol extract of the leaves failed to deter feeding by *Bombyx mori* larvae.[82]

Juniperus virginiana **L.,** Eastern red cedar: Many reports attest to the value of red cedar chests as repellents to *Tineola bisellella* larvae.[154,157,409,470] Oil of the cedar obtained from the leaves sprinkled between the bedsheets gave some degree of protection from fleas.[471] Termite-susceptible wood impregnated with 1% cedar oil was repellent to termites, but the effect was only temporary. The oil was more repellent than the heartwood from which it was obtained.[118] A product for painting on closet walls to repel moths contains a powder derived from cedar oil plus a plastic binder.[472] Exposure to the vapor of red cedar oil at 0.6 mg/l for 1 week killed 91% of half-grown *Tineola bisselliella* larvae; concentrations of 1 to 2 mg/l acted more rapidly (8 to 24 h).[473] The vapor of the powdered wood and leaf oil was not an effective repellent or fumigant against larvae or adults of *T. bisselliella, Attagenus* *piceus*, and *Anthrenus vorax*.[474] The wood is recorded from the West Indies as being susceptible to *Cryptotermes brevis* attack[86] but resistant to *Anacanthotermes ochraceus, Psammotermes fuscofemoralis*, and *P. assuarensis* termites. *Reticulitermes flavipes* termites refused to feed on the wood in both choice and force-fed tests.[462,463] *Incisitermes minor* termites refused to feed on the wood and its extracts.[361] Cosecutive extraction of the dried leaves and twigs with ether and methanol was followed by exposure of *Spodoptera frugiperda* larvae to the extracts; little feeding resulted.[204] Very minor feeding resulted when ether and methanol extracts of the leaves were exposed to *Acalymma vittatum* and *Diabrotica undecimpunctata howardi* adults at 0.5% more moderate feeding resulted at 0.1% concentration.[52] The leaf essential oil was moderately repellent to *Apis florea* bees.[475]

Libocedrus bidwilli Hoof. f., New Zealand mountain cedar: Feeding tests with the powdered dried leaves and with methanol extracts thereof showed them to be toxic to larvae of *Musca domestica* and *Cydia pomonella* but not to *Austrotortrix postvittana*. The active compound is β-peltatin A methyl ether.[476]

Thuja occidentalis L., northern white cedar, arborvitae: Extracts of the whole plant did not repel *Popillia japonica* adults.[126]

Thuja plicata Donn. ex Don, western red cedar: The wood is susceptible to attack by *Reticulitermes flavipes* termites,[461,462] but fairly resistant to attack by *Incisitermes minor*.[361] The heartwood was accepted by *R. virginicus* termites under both choice and force-feeding conditions.[87] The leaf oil is deterrent to feeding by *Pissodes strobi* and *Altica ambiens* due to its content of (+)-3-thujone and (-)-3-isothujone; it also deterred oviposition by *Delia antiqua*.[477]

FAMILY CYCADACEAE

Zamia debilis L., Ether extracts of the roots bark failed to deter feeding by adult *Acalymma vittatum*.[89]

FAMILY CYPERACEAE

Cyperus rotundus L., nutsedge, nutgrass, purple nutsedge: The bulbous ends of the roots gave a camphoraceous odor, which probably accounts for their use as an insect repellent. The chief constituent of the volatile oil is α-cyperone.[116] Gram seed or mung seed mixed with the powdered root of this plant was somewhat toxic to adult *Callosobruchus chinensis*.[150] A 1% aqueous solution of an acetone extract of the stems completely destroyed *Bagrada cruciferarum* in both the laboratory and field.[478]

FAMILY DATISCACEAE

Datisca cannabina L., alkalbir: An ethanol extract of the whole plant did not deter feeding by adult *Acalymma vittatum* and *Diabrotica undecimpunctata howardii*.[52]

FAMILY DILLENIACEAE

Doliocarpus dentatus (Aubl.) Standl.: An ether extract of the woody stems with bark deterred feeding by *Acalymma vittatum* adults at 0.5% but not at 0.1%.[89]

FAMILY DIOSCOREACEAE

Dioscorea oppositifolia L., Chinese yam, cinnamon tree: Leaves of this plant were not accepted for feeding by the larvae of *Spodoptera littoralis* and *Trimeresia miranda*.[91] The plant was recommended as a repellent for fleas on man.[479]

FAMILY DIPSACEAE

Scabiosa atropurpurea L., sweet scabiosa: An extract of the whole plant was not repellent to adult *Popillia japonica*.[126]

FAMILY DIPTEROCARPACEAE

Balanocarpus heimii King, chengal: The wood is not attacked by termites and has been described as "one of the most durable woods of the world."[480]

Cotylobium melanoxylon Pierre: The heartwood was resistant to 20 species of termites.[481]

Hopea helferi (Dyer) Brandis, giam
Hopea nutans Ridley
Hopea semicuneata Sym.: The wood of these species is sufficiently durable to be safe for use against termites.[480]

Parashorea plicata Brandis, white seraya: The wood is susceptible to termite attack.[188,480]

Shorea robusta Gaertn. f., sal: Stored greengram seeds were protected from attack by *Callosobruchus maculatus* for 5 to 6 months by mixing them with sal oil at 5%.[482] Aqueous and alcoholic extracts of deoiled sal meal were not toxic to *Drosophila melanogaster* and *Spodoptera litura* larvae; the aqueous extract did not kill *Tribolium castaneum* but the alcoholic extract was toxic to this insect at 1%.[483]

Shorea stenoptera Burck.: The heartwood was resistant to 20 species of termites,[481] but the sapwood was highly susceptible to *Reticulitermes lucifugens* and *R. flavipes* termites.[484]

Upuna borneensis Sym., penyau: The seasoned wood is highly resistant to termite attack.[485]

Vatica cineria King
Vatica cuspidata (Ridley) Desch.
Vatica odorata (Griff.) Sym.: These three species are resistant to termite attack.[480]

FAMILY DROSERACEAE

Drosera rotundifolia L., roundleaf sundew: An acetone extract of the whole plant did not repel adult *Popillia japonica*.[126]

FAMILY EBENACEAE

Diospyros crassiflora Hiern., African ebony: The wood of this tree in Nigeria is very resistant to termite attack.[370]

Diospyros ebenum Koenig ex Retz., Ceylon ebony: The wood is immune to attack by *Anacanthotermes ochraceus, Psammotermes*

fuscofemoralis, and *P. assuarensis* termites.[370,387]

Diospyros melanoxylon Roxb., Indian ebony: The wood is fairly resistant to termites in India.[370]

Diospyros virginiana L., common persimmon: Filter paper impregnated with an ethanol extract of the wood did not repel *Reticulitermes flavipes* termites.[388] However, termites survived poorly when exposed to the wood. The toxic components were isolated and identified as 7-methyljuglone (5-hydroxy-7-methyl-1,4-naphthoquinone). Components that were lacking toxicity that were also isolated were 4,8-dihydroxy-6-methyl-1-tetralone and scopoletin (7-hydroxy-6-methylcoumarin).[486] Cotton plants imbibing alcoholic extracts of the wood failed to exhibit repellency or toxicity to larvae or adults of *Anthonomus grandis grandis*.[487]

FAMILY ELAEOCARPACEAE

Elaeocarpus canitrus Roxb.: An ether extract of the fruit did not repel adult *Acalymma vittatum*.[89]

Elaeocarpus dolichostylus Schltr.: An ether extract of the stem bark was somewhat deterrent to feeding by adult *Acalymma vittatum* at 0.5% but not at 0.1%.[89]

FAMILY ERICACEAE

Arctostaphylos glauca Lindl.: An ethanol extract of the leaves did not deter feeding by *Lymantria dispar* larvae.[84]

Arctostaphylos uva-ursi (L.) Spreng., bearberry: An alcohol extract of the shrub neither repelled nor was toxic to larvae or adults of *Anthonomus grandis grandis*.[487]

Arctostaphylos viscida Parry: An ether extract of the leaves deterred feeding by *Lymantria dispar* larvae.[84]

Azalea nudiflora L., pinxterbloom
Chimaphila umbellata (L.) Nutt., pipsissewa: Acetone extracts of these plants did not repel *Popillia japonica* adults.[126]

Enkianthus perulatus C. R. Schneider: A benzene extract of the whole plant did not deter feeding by the larvae of *Spodoptera litura*.[258]

Epigaea repens **L.**, trailing arbutus, mayflower: An acetone extract of the whole plant did not repel *Popillia japonica* adults.[126] An alcohol extract of the plant exhibited neither repellency nor toxicity to the larvae or adults of *Anthonomus grandis grandis*.[487]

Gaultheria procumbens **L.**, wintergreen: An acetone extract of the whole plant did not repel *Popillia japonica* adults.[126] Diluted oil of wintergreen was attractive to *Blatta orientalis*.[162]

Kalmia latifolia **L.**, mountain laurel: Sequential extracts of the combine leaves and twigs were prepared with ether and methanol and these were tested at 15,000 ppm with *Spodoptera frugiperda* larvae; neither deterrency nor toxicity was shown.[204] An ethanol extract of the leaves deterred feeding by *Lymantria dispar* larvae.[84] An acetone extract was not repellent to *Popillia japonica* adults.[126] Ten grayanoid diterpenes were isolated from an ethanol extract of the leaves and identified. The major antifeedants for *Lymantria dispar* larvae were two kalmitoxins and grayanotoxin-III.[488] A number of phenolic flavonoids were also identified.[489]

Leiophyllum buxifolium **(Berg) Ell.**, sandmyrtle: An alcoholic extract of the whole plant was neither repellent nor toxic to larvae or adults of *Anthonomus grandis grandis*.[487]

Leucothoe catesbaei **A. Gray:** A methanol extract of the leaves failed to deter feeding by *Bombyx mori* larvae and was nontoxic to this insect.[82] Grayanotoxins I and III, which were toxic to *Chilo suppressalis* and *Tetranychus teralius*, were isolated from the extract.[490]

Lyonia mariana **(L.) D. Don**, stagger-bush: An alcohol extract of the whole plant exhibited neither repellency nor toxicity to *Anthonomus grandis grandis* adults and larvae.[487]

Oxydendrum arboreum **(L.) DC**, sourwood: An acetone extract of the whole plant did not repel *Popillia japonica* adults.[126]

Pieris japonica **(Thunb.) D. Don ex G. Don**, Japanese andromeda, asebo, asemi: An ethanol extract of the leaves deterred feeding by *Lymantria dispar* larvae.[84]

Rhododendron canadense **(L.) Torr.**, rhodora, Canadian rhododendron: An alcoholic extract of this shrub was neither repellent nor toxic to larvae of *Anthonomus grandis grandis*.[487]

Rhododendron maximum **L.**, rosebay rhododendron: An ethanol extract of the leaves deterred feeding by *Lymantria dispar* larvae.[84]

Rhododendron metternichii **Sieb. & Zucc., var** *hondoense:* A methanol extract of the leaves did not deter feeding by *Bombyx mori* larvae.[82]

Rhododendron spp. The juice expressed from the fresh leaves of an unidentified species of this genus strongly deterred feeding by *Melanotus communis* larvae.[247] Tests conducted with the foliage of 104 varieties of *Rhododendron* showed that several were strongly resistant to feeding by *Sciopithes obscurus*, a pest of small fruit and ornamental crops; 16 varieties showed moderate resistance and 73 varieties showed low resistance; 8 varieties were highly susceptible.[491] Investigation of the basis of resistance in several varieties led to the isolation of germacrone, the sesquiterpene responsible for the repellency in *R. edgeworthii* leaves.[493] It is highly likely that this compound is also responsible for the deterrency to this insect by other species of the genus.

Vaccinium arboreum **Marsh:** An ethanol extract of the leaves failed to deter feeding by *Lymantria dispar* larvae.[84]

Vaccinium myrsinites **Lam.:** An ethanol extract of this shrub failed to exhibit repellency or toxicity to larvae or adults of *Anthonomus grandis grandis*.[487]

FAMILY ERYTHROXYLACEAE

Erythroxylum areolatum **L.**, false cocaine: The wood is very resistant to termites.[124]

FAMILY EUCOMMIACEAE

Eucommia ulmoides **Oliv.:** A methanol extract of the leaves deterred feeding by *Bombyx mori* larvae.[82]

FAMILY EUPHORBIACEAE

Acalypha virginica **L.**, Virginia copperleaf: An alcohol extract of the whole plant did not repel larvae or adults of *Anthonomus grandis grandis.*[487]

Actinostemon concolor **(Spreng.) Arg.**, larenjeira: The wood is highly susceptible to attack by *Cryptotermes brevis* termites.[86]

Alchornea latifolia **Sw.**: The wood is very susceptible to attack by *Cryptotermes brevis* termites.[86,124]

Alchornea triplinervia **(Spreng.) Muell. Arg.**: An ether extract of the leaves was somewhat deterrent to feeding by adult *Acalymma vittatum* at 0.5% but not at 0.1%.[89]

Aleurites fordii **Hemsl.**, tung tree: A 1% pentane extracts of the nuts considerably deterred feeding by *Anthonomus grandis grandis* adults, a 0.1% ether extract of the defatted nuts reduced feeding by *Acalymma vittatum* and *Conotrachelus nenuphar* adults, and a concentration of 0.2% reduced feeding by *Acalymma vittatum*. These extracts, as well as ethanolic and aqueous extracts, deterred feeding by the larvae of *Cydia pomonella* and *Argyrotaenia velutinana.*[49] Crude tung meal is itself repellent to *Anthonomus grandis grandis* adults.[494] Removal of the antennal clubs of the weevils did not eliminate the deterrent effect, which may thus be detected by the labial and maxillary palps. α-Eleostearic acid and *erythro*-9,10-dihydroxy-1-octadecanol acetate have been identified as the components responsible for the feeding deterrency of tung oil to adult *A. grandis grandis,* and methods have been developed for isolating large quantities of the acid and for synthesizing its methyl ester.[495] Although α-eleostearic acid is too unstable for practical as a feeding deterrent under field conditions, its methyl ester is much more stable and equally effective as a deterrent.[495,496] This ester has been synthesized in three steps using readily available starting materials.[497] Pentane and ether extracts of tung nuts failed to deter feeding by adult *Acalymma vittatum* or *Diabrotica undecimpunctata howardi* at 0.1 or 0.5%.

Ethanolic and aqueous extracts were likewise ineffective.[89]

Aleurites molucanna **(L.) Willd.**, candlenut, candleberry: Ether extracts of the stems failed to deter feeding by adult *Acalymma vittatum* and *Diabrotica undecimpunctata howardi* at 0.1 or 0.5%.[89]

Antidesma platyphyllum **H. Mann.**, hame, haa: The wood is resistant to attack by *Cryptotermes brevis* termites.[89]

Bridelia micrantha **(Hochst.) Baill.**: An ether extract of the stem bark did not deter feeding by adult *Acalymma vittatum* and *Diabrotica undecimpunctata howardi* at 0.1 and 0.5%, but an extract of the woody stems with bark was effective at both concentrations.[89]

Cleistanthus collinus **(Roxb.) Benth. & Hook**: White ants are repelled by the bark of this tree.[498]

Colliguaia brasiliensis **(Willd. ex A. Juss.) Muell. Arg**: An ether extract of the woody stems with bark deterred feeding by adult *Acalymma vittatum* and *Diabrotica undecimpunctata howardi* at 0.1 and 0.5%.[89]

Croton eleuteria **(L.) Swartz.**, cascarilla: An aqueous extract of the leaves sprayed on rice seedlings deterred feeding by *Nephotettix virescens.*[137] Cascarilla bark oil was slightly repellent to *Blattella germanica.*[59]

Croton flavens **Michx.**: The plant is reported to be repellent to insects.[500]

Croton glandulosus **L.**: An alcohol extract of the whole plant did not repel the larvae or adults of *Anthonomus grandis grandis.*[488]

Croton texensis **Muell. & Arg.**, skunkweed: An acetone extract of the whole plant deterred feeding by *Attagenus megatoma.*[25]

Endospermum formicarium **Becc.**, basswood *Endospermum medullosum* **L. S. Smith**: The wood of these two species is not resistant to attack by termites and fungal decay.[188]

Euphorbia antiquorum **L.**, Stems of this plant stored in the water inlet of rice paddy fields repelled *Spodoptera mauritia.*[499]

Euphorbia ipecacuanna **L.**, spurge: An aqueous extract of the whole plant was not repellent to *Popillia japonica* adults.[126]

Euphorbia lathyris **L.**, caper spurge, mole plant

Euphorbia poinsettiana **Buist.:** *Plutella xylostella* larvae refused to feed on these two species.[175]

Euphorbia pulcherrima **Willd. ex. Kl.**, poinsettia: An extract of the flowers repelled *Sitophilus oryzae* larvae on filter papers.[251,252]

Euphorbia royleana **Boisd.:** An ether extract of the leaves deterred feeding by larvae of *Athalia proxima.*[213]

Euphorbia serpene **H.B.K.:** An alcoholic extract of the leaves failed to deter feeding by the larvae or adults of *Anthonomus grandis grandis.*[487]

Euphorbia splendens **Boj.:** Larvae of *Plutella xylostella* refused to feed on this plant.[195]

Euphorbia tirucalli **L.**, milkbush: The tree is regarded as a mosquito repellent in Tanganyika.[116]

Hevea brasiliensis **(Willd. ex A. Juss.) Muell.-Arg.**, rubber tree: The wood is resistant to the subterranean termites *Anacanthotermes ochraceus, Psammotermes fuscofemoralis,* and *P. assuarensis.*[387] Rubber seed oil at 1% deterred feeding by *Rhyzopertha dominica* and *Sitotroga cerealella.*[233] An ether extract of the twigs did not deter feeding by *Acalymma vittatum* and *Diabrotica undecimpunctata howardi* adults at 0.1 or 0.5%.[89]

Hippomane mancinella **L.**, manchineel: The wood is very susceptible to termite attack.[86,124]

Hura crepitans **L.**, monkey pistol, sandbox: The wood is susceptible to termite attack.[124]

Hyeronima caribea **Urban.**, tapana

Hyeronima clusioides **(Tul.) Muell.-Arg.**, cedro macho

Hyeronima laxiflora **Muell.-Arg.**, suradan: The wood of these three species is susceptible to termite attack.[124]

Mallotus philippinensis **(Lam.) Muell.-Arg.**, monkeyface tree: An extract of the whole plant did not repel *Popillia japonica* adults.[126]

Nealchornia yapurensis **Huber.:** An ether extract of the twigs deterred feeding by larvae or adults of *Acalymma vittatum* and *Diabrotica undecimpunctata howardi* at 0.1 and 0.5%, but an extract of the woody stems with bark was not effective.[89]

Phyllanthus acidus **(L.) Skeels,** Indian gooseberry: An ether extract of the combined stems, leaves, and fruits did not deter feeding by *Acalymma vittatum* or *Diabrotica undecimpunctata howardi* adults.[89]

Ricinus communis **L.**, castorbean plant: A review lists 52 published papers on the insecticidal uses of this plant.[500] The seed oil has been used to protect stored grains from insect attack,[150] but it did not deter feeding by *Callosobruchus chinensis* when mixed at 1% with redgram seeds.[271] Aqueous and alcoholic extracts of the deoiled seed cake were not toxic to *Drosphila melanogaster* or *Tribolium castaneum.*[483] Castor bean is not a preferred host for *Amsacta moorei.*[372] An ether extract of the stems deterred feeding by *Acalymma vittatum* and *Diabrotica undecimpunctata howardi* at 0.1 and 0.5%.[89] A methanol extract of the leaves strongly deterred feeding by *Bombyx mori* larvae.[82] Castor oil mixed with green gram seed at 0.3% inhibited multiplication of *Callosobruchus chinensis.*[237] A methanol extract of the whole plant was highly toxic to *Aedes aegypti* larvae.[245] Oviposition by mated females of *Earias fabia* was sharply reduced when okra fruits were coated with castor bean leaf juice fed to the insects.[501] Neither ricin nor ricinine, two alkaloids obtained from castor bean seeds, was toxic to *Musca domestica,*[502] but ricinine was highly toxic to *Cydia pomonella* larvae.[503] Extracts of the leaves and stalks were effective against *Ctenocephalides canis, C. felis, Echidnophaga gallinacea, Cuclogaster heterographus,*[126] and mites.[504]

Sapium laurocerasum **Desf.:** The wood is susceptible to attack by *Cryptotermes brevis* termites.[86]

Spirostachys africanus **Sond.:** Containers made of the wood of this tree have been packed with clothing to repel insects.[116] An alcohol extract of the whole plant did not repel larvae or adults of *Anthonomus grandis grandis.*[487]

Stillingia sylvatica **L.**, stillingia: An extract of the whole plant did not repel *Popillia japonica* adults.[126]

Tragia incana **Klotsch. ex. Baill.:** A petroleum ether extract of the whole plant did not deter feeding by *Acalymma vittatum* adults.[52] *Trewia nudiflora* **Wight.**, false white teak: An ethanol extract of the seeds was an effective antifeedant for larvae of *Ostrinia nubilalis*[46] and *Acalymma vittatum* but not for *Diabrotica undecimpunctata howardi* . The extract was toxic to *Argyrotaenia velutinana* and *Conotrachelus nenuphar* and gave 100% control of *Menacanthus stramineus*. Fractionation of the extract resulted in the isolation of at least six maytansinoid compounds.[505] Trewiasine, isolated from the seeds, was both deterrent to feeding and toxic to adult *Diabrotica undecimpunctata howardi* as well as *Cydia pomonella* larvae at dosages as low as 0.32 ppm incorporated in the diet; topical application to 4th instar larvae of *C. pomonella* caused mortality or delayed maturity.[506]

Addendum — Jacobson et al.[507] prepared ether and ethanol extracts of 44 species of the family Euphorbiaceae for testing as feedng deterrents for *Spodoptera frugiperda* larvae and adults of *Anthonomus grandis grandis* and *Acalymma vittatum*. Meyer et al.[508] and Farriani[509] have described novel bioassay methods for Euphorbiaceae using brine shrimp and potato disks.[509,510]

FAMILY FABACEAE

Acacia acuminata **Benth.**, raspberry-jam tree
Acacia benthamii **Willd.**
Acacia caffra **Willd.**: The wood of these three species is very resistant to termite attack.[116,188,307]
Acacia catechu (**L. f.**) **Willd.**, black, cutch, catechu: A patented insect repellent contains tannic acid from the wood and pods of this plant to be used in a paint or coating for application to underwater structures to prevent decay and ravages by marine life, insects, vermin, and rodents.[511] An aqueous extract of the plant was not repellent to *Popillia japonica* adults.[126]
Acacia concinna **DC**, soapnut, shikai, soap pod: An acetone extract of the seeds deterred feeding by *Rhyzopertha dominica*,[29] *Sitophilus oryzae*,[512] and *ribolium castaneum*.[512]

A 2% admixture of the ground seeds with stored rice did not deter feeding by *Sitotroga cerealella* and *Rhyzopertha dominica*.[513]
Acacia farmesiana (**L.**) **Willd.**, sweet acacia, huisache, cassie: The wood is resistant to *Cryptotermes brevis* termites.[86] An ethanol extract of the leaves did not prevent feeding by *Lymantria dispar* larvae.[84]
Acacia giraffe **Willd.**: The heartwood is not attacked by termites.[116]
Acacia homalophylla **A. Cunn. ex Benth.**: The wood is resistant to termite attack.[188]
Acacia melanoxylon **R. Br.**, blackwood: The wood is resistant to termites in Tanganyika.[117]
Acacia nigrescens **Oliver:**
Acacia nilotica (**L.**) **Del.**, babul, bulbul: The wood of these trees is resistant to termites.[116,117]
Adenanthera favonina **L.**, An ether extract of the seeds did not prevent feeding by *Acalymma vittatum* and *Diabrotica undecimpunctata howardi* adults.[89]
Afzelia africana **Sm.**, afzelia
Afzelia bella **Harms.**, papao
Afzelia bipindensis **Harms.**
Afzelia caudata **Hoyle**
Afzelia pachyloba **Harms.**
Afzelia quanzensis **Welw.**, mahogany-bean: The wood of these six species is resistant to termite attack.[117,370,388]
Albizzia anthelmintica **Brogn.**: Filter papers impregnated with 3% of musennin, a triterpene saponin isolated from the root bark, were toxic to *Reticulitermes flavipes* and inhibited feeding by this insect; 0.05 and 0.5% solutions were not effective.[229]
Albizzia ferruginea (**Guill. & Perr.**) **Benth.**, albizia: The wood is resistant to termites in Nigeria.[370,388]
Albizzia julibrissin **Durazz.**, silktree, mimosa
Albizzia lebbeck (**L.**) **Benth.**, lebbeck: Seeds of *A. julibrissin,* which contain the alkaloid, albizine, fed to larvae of *Spodoptera eridania* caused high mortality.[514] The wood of *A. lebbeck* is resistant to termite attack for at least 3 to 4 years.[370] The seed oil did not repel *Apis florea* bees at concentrations below 5 g/l.[149] Ethanol extracts of the leaves of both species do not prevent *Lymantria dispar* larvae from feeding.[84]

Amorpha canescens **Nutt:** An aqueous extract of the leaves incorporated into an artificial diet was fed upon *Melanoplus differentialis*. About 91% of *M. fermurrbrum* reached adulthood when reared on a diet containing the extract.[319]

Amorpha fruticosa **L.,** false indigo: Larvae of *Melanoplus sanguinipes, Leptinotarsa decemlineata,* and *Pieris brassicae* were deterred from feeding on host leaves sprayed with a 1% solution of a benzene extract of the fruits. *Messor structor* were repelled by the treated leaves. Three active compounds were isolated from the leaves (*p*-cymol, terpinen-4-ol, and α-terpineol).[515] However, these compounds did not account for the activity observed with *Megoura viciae* aphids and *Tetranychus urticae*.[515,516]

Andira surinamensis **(Bond) Splitg.,** botseed: The wood is resistant to termites.[116]

Arachis hypogaea **L.,** peanut, groundnut, goober: Several species of peanut plants have been shown to be resistant to *Spodoptera frugiperda,*[11] *Empoasca kerri, Aphis brassivora,* or several species of termites.[517] Peanut oil at 5 mg/kg completely protected stored cowpea seeds from *Callosobruchus maculatus* for up to 180 d without affecting cooking time, taste, or germination of the seed, even after 6 months' storage.[518] Stored greengram mixed with 0.5% peanut oil inhibited the multiplication of this insect.[237] On stored bambara groundnut the oil reduced oviposition of the insect at 3 ml/kg seed.[235] Ground peanut meal rapidly killed young and adult *Plodia interpunctella* and *Cadra cautella*.[520] A 10% concentration of the oil killed 90% of *Urentius echinus* adults in 72 h; 5 and 7% were almost as effective but 1 and 3% were ineffective.[521]

Baikaea plurijuga **Harms.,** Rhodesian teak, Zambesi redwood: The wood is resistant to *Cryptotermes brevis* termites.[86,116,370]

Baptisia tinctoria **(L.) R. Br.,** yellow wild indigo: Plants placed in the harness kept flies from the horses.[522] Aqueous extracts of the whole plant were not repellent to *Popillia japonica* adults.[126]

Berlinia acuminata **Soland ex Hoof. f.,** berlinia

Berlinia grandiflora **(Vahl.) Hutchinson & Dalziel,** berlinia: Tests conducted in Nigeria and the Gold Coast show that these woods are moderately resistant to termites.[370] Shavings of *B. acuminata* wood were neither repellent nor resistant to *Reticulitermes lucifugus* termites.[187]

Bowdichia braziliensis **Ducke,** sapupira: The wood is resistant to termites.[370]

Bowdichia nitira **Spruce,** sapupira: The wood is resistant to *Cryptotermes brevia* termites.[86]

Bussea mascaiensis **Harms.:** The timber is exceedingly hard and resistant to termites.[116]

Butea frondosa **Roxb.,** palas: Cantaloupe leaf disks dipped in an ethanol extract of the whole plant were accepted by adult *Acalymma vittatum* and *Diabrotica undecimpunctata howardi* at 0.1 and 0.5%.[52]

Caesalpinia pulcherrima **(L.) Sw.,** paradise flower, Pride-of-Barbados, peacock flower: Extracts of the flowers offered to *Sitophilus oryzae* larvae on filter paper did not deter feeding.[251,252]

Calpocalyx klainei **Pierre:** The wood shavings were resistant and repellent to *Reticulitermes lucifugens*.[187]

Canavalia ensiformis **(L.) DC.,** jackbean, horsebean, swordbean: This plant has long been known for its resistant and toxicity to many species of pest insects.[523] The active component, *l*-canavanine, has been isolated from the stems of this plant as well as several varieties of alfalfa *(Medicago sativa* subsp. *sativa), clover,* and other members of the same family.[524-529] An aqueous extract of the whole plant deterred feeding by *Attagenus piceus*.[25] *Spodoptera litura* larvae that fed on castor leaves sprayed with 50 to 1000 ppm of *l*-canavanine showed retarded growth at 200 ppm.[530] Leafcutting ant colonies (*Atta* spp.) supplied nightly for 3 consecutive nights with 5 to 15 kg of jackbean leaves showed complete cessation of ant activity for from 4 months to 5 years.[531] When larvae of *Musca domestica, M. autumnalis, Haematobia irritans,* and *Stomoxys calcitrans* were reared in media containing 800 ppm of *l*-canavanine, at least 70% of all of these species except *M. domestica* showed 18% mortality.

H. irritans and *M. autumnalis* showed growth deformations in the larval and pupal stages.[532] Disruption of insect growth and development was caused by feeding or injecting *l*-canavanine into *Manduca sexta*,[533-535] *Bombyx mori*,[525,533] and *Callosobruchus maculatus*,[536] and antifertility effects resulted from feeding the compound to *Periplaneta americana*[537] and *Dysdercus koenigii*.[538] Larvae of the bruchid beetle, *Caryedes brasiliensis*, are able to adapt to the presence of *l*-canavanine.[539]

***Cassia alata* L.,** ringworm bush: The plant is place around huts in Tanganyika as an ant repellent.[191]

***Cassia botrya* Fres.:** The leaf has a peculiar odor that is said to repel bees (unidentified). Natives in Tanganyika use the foliage as a bee repellent and burn it to smoke out bees.[191]

***Cassia fasciculata* Michx.,** partridge pea: An acetone extract of the whole plant did not repel *Popillia japonica* adults.[126]

***Cassia fistula* L.,** Indian laburnum, golden shower, purging cassia: An ethanol extract of the wood killed 100% of exposed *Reticulitermes flavipes* termites.[229] An ethanol extract of the leaves did not deter feeding by *Lymantria dispar* larvae.[84] Extracts of the wood and part prepared with methanol-chloroform followed by acetone-hexane, applied topically to the abdominal tergites of *Dysdercus koenigii*, induced juvenilization.[540]

***Cassia grandis* L. f.:** A diet containing 5% of the seed powder was accepted normally by *Spodoptera eridania* larvae.[514]

***Cassia hebecampa* Fern.,** wild senna: An acetone extract of the whole plant did not repel *Popillia japonica* adults.[126]

***Cassia nigricans* Vahl.,** An ethanol extract of the seeds controlled *Acanthoscelides obtectus* in stored cowpeas.[541]

***Cassia siamea* Lam.,** Siamese senna: The wood is very resistant to termites in Nigeria,[117] but it is susceptible to attack by *Cryptotermes brevis* termites.[86]

***Castanospermum australe* A. Cunn. & Fraser,** black bean: The wood is reported to be termite-resistant in Australia.[370]

***Cercis canadensis* L.,** eastern redbud: An ethanol extract of the leaves added to the diet increased feeding by *Lymantria dispar* larvae.[84]

***Clitoria ternata* L.,** butterfly pea: An ethanol extract of the whole plant was somewhat deterrent to feeding by *Acalymma vittatum* adults at 0.5% but not at 0.1%.[52]

***Copaifera lansdorfii* Desf.,** copaiba: Copaiba oil exhibited good repellent action on *Cochliomyia hominivorax* for 1 or 2 d only.[155] Caryophyllene, present in the wood, offers little protection against termite attack at low concentrations, but 5% caryophyllene prevented termites from eating treated wood for almost 4 weeks.[118]

***Coronilla varia* L.,** crownvetch: The compound, β-nitropropionic acid, present in the leaves may act as a feeding deterrent for *Trichoplusia ni* larvae.[542] An acetone extract of the whole plant deterred feeding by *Attagenus megatoma*.[25]

***Crotalaria spectabilis* Roth,** showy crotalaria: The alkaloid, monocrotaline, isolated from this plant exhibited modest feeding deterrency to *Acyrthosiphon pisum*.[543]

***Cylicodiscus cabunensis* (Taub.) Harms,** okan: The wood is resistant to termites in Nigeria.[370]

***Cynometra inaezualifolia* A. Gray,** kekatong: The wood is resistant to termites.[480]

***Dalbergia latifolia* Roxb. ex DC.,** Indian rosewood: The wood may remain in the ground for four years without being attacked by termites.[370]

***Dalbergia retusa* Hemsl.,** cocobolo: *Coptotermes formosanus* termites could not survive on a diet containing extracts of the wood prepared with pentane or acetone.[369] An orange pigment designated "obtusaquinone" has been isolated from the wood; it is an effective larvicide for the marine borer, *Lyrodus pedicellatus*.[544] This led to the discovery that several related compounds (derivative of alkylphenols and 1,3-benzodioxoles) are promising beetle repellents, mosquito growth inhibitors, and fly sterilants.[545]

***Dalbergia stevensonii* Standl.,** Honduras rosewood: The wood is reported to be resistant to termites in Honduras.[370] Due to the content of flavanoids designated "dalbergiones", woods of the genus *Dalbergia* are toxic to *Reticulitermes flavipes*.[546]

Daniellia oliveri (Rolle) Hutch. & Dalz., copaiba balsam: The wood resin of the tree protects it from termites.[116,121]

Delonix regia (Boj. ex Hook.) Raf., flame tree, flamboyant tree, royal poinciana, Pride-of-Barbados: Extracts of the flowers tested against *Sitophilus oryzae* on filter papers were not effective as feeding deterrents.[251,252] An acetone extract of the leaves disrupted growth and development of *Tribolium castaneum* larvae.[547]

Desmodium racemosum DC.: Leaves of this plant were accepted only slightly for feeding by *Spodoptera littoralis* larvae.[91]

Detarium senegalense J. F. Gmel.: The gum resin is used as a fumigant for garments and houses in Africa.[121]

Dialium patens Baker
Dialium platysepalum Baker
Dialium wallichii Prain, keranji: The wood of these trees is resistant to termites.[480]

Dichrostachys cinerea Miq.: The wood is termite-proof.[116]

Dicorynia guianensis Amsh., basralocus, angelique: The wood is susceptible to feeding by *Cryptotermes brevis* termites,[86] but it is extremely resistant to *Teredo navalis* borers due to its content of the alkaloid, tryptamine.[546,548]

Diplotropis purpurea (Rich.) Amsh., tatabu: The wood is resistant to *Cryptotermes brevis* termites.[86,124]

Dipteryx odorata (Aubl.) Willd., tonka bean: Coumarin obtained from this plant and added to the standard rearing diet at 0.25% completely inhibited larval development of *Musca domestica* without ovicidial action.[356]

Distemonanthus benthamianus Baill., ayan: The wood is moderately resistant to *Cryptotermes brevis*[86,370] and *Reticulitermes lucifugus* termites,[441] and completely resistant to *Coptotermes formosanus*.[388]

Entandrophragma ancolense (Welw.) C. DC., adinam

Entandrophragma candollei Harms, candollei

Entandrophragma cylindricum (Sprague) Sprague, penkua

Entandrophragma utile (Dawe & Sprague), Sprague, utile: The wood of these four species suffered only slight damage from feeding by *Coptotermes formosanus* termites.[388]

Enterolobium cyclocarpum (Jacq.) Griseb., guanacaste, earpod tree: The wood is resistant to *Cryptotermes brevis* termites.[86,115] An ethanol extract of the leaves deterred feeding by *Lymantria dispar* larvae.[84]

Erythrina senegalensis DC.: An alcohol extract of the whole plant did not repel *Anthonomus grandis grandis* larvae or adults.[487]

Erythrophleum africanum Harms, utile: The timber is termite-proof.[116]

Erythrophleum guineense G. Don.: The leaves are placed among stored corn to repel insects.[116] The compound, erythrophleine, is responsible for this activity.[121] Shavings of the wood are resistant and repellent to *Reticulitermes lucifugus* termites.[187]

Erythrophleum ivorense A. Chev., potrodon: The wood is resistant to *Coptotermes formosanus* termites.[388]

Erythrophleum laboucherii f. Muell. ex Benth., ironwood: The wood is resistant to termites in Australia.[188]

Galea officinalis L., goat's-rue, galesa
Cenista tinctoria L., dyer's greenwood: Extracts of these plants did not repel *Popillia japonica* adults.[126]

Gleditsia triacanthos L., honey locust: An ethanol extract of the leaves did not deter feeding by *Lymantria dispar* larvae.[84]

Glycine max (L.) Merr., soybean: A standard diet containing 0.4 mg/g of an alcoholic or aqueous extract of soy meal deterred feeding by *Bombyx mori* larvae; the few larvae that fed showed retarded growth.[549] Stored wheat with 5 ml of soybean oil/kg repelled adult *Sitophilus granarius* and reduced the number of insect progeny; 10 ml/kg gave complete control for at least 60 d, but reduced seed germination.[19] Pinitol, the compound responsible for the resistance of the leaves of several soybean cultivars to *Heliothis zea* larvae,[550] was isolated; it is a monomethyl ether of *chiro*-inositol.[14,551] The pinitol content increases with the age of the plant.[4] In choice tests, potato slices treated with the fresh juice of soybean leaves and seedling strongly repelled *Melanotus communis* larvae.[247] Soybean saponin and its calcium salt were highly toxic to *Sitophilus oryzae*, giving a high degree of protection to wheat dusted with 300 ppm of the saponin; 600 ppm gave 100% mortality in 1 week.[552] No connection was found between the sterol content of soybean leaves and the plant's resistance to insects.[553]

***Glycyrrhiza glabra* L.,** licorice: Filter papers impregnated with 3% solutions of glycyrrhizin, isolated from the roots of this plant, were toxic to *Reticulitermes flavipes* termites and inhibited their feeding; 0.05 and 0.5% solutions were not effective.[229]

***Glycyrrhiza lepidota* Pursh,** wild licorice: Only 53% of *Melanoplus femurrubrum* reached adulthood when reared on a diet containing an extract of the leaves,[319] but 85% of *M. sanguinipes* reached adulthood.[275]

***Gossweilerodendron balsamifera* (Vermoesen) Harms,** agba, white tola: The wood is resistant to *Cryptotermes brevis* termites in Nigeria.[86,370]

***Guiboutia demensei* J. Leonard:** The wood is resistant to *Reticulitermes lucifugus* termites.[441]

***Haematoxylum campechianum* L.,** logwood: Submersing a termite-susceptible wood for 10 min in a 2% solution of hematoxylon, a dye obtained from the heartwood of this tree, will protect it from termite attack for nearly 3 weeks.[118]

***Hardwickia binata* Roxb.,** anjan: The wood is strongly resistant to *Microcerotermes beesoni* termites.[427]

***Hardwickia mannii* Oliver:** Caryophyllene, present in the wood, offers little protection against termite attack at low concentrations, but 5% caryophyllene did prevent termites from eating treated wood for almost 4 weeks.[124]

***Hymenaea courbaril* L.,** West Indian locust tree: The heartwood is very resistant to *Cryptotermes brevis* termites,[86] but the sapwood is very susceptible to *Reticulitermes*

lucifugus and *R. flavipes*.[484] The leaf resin deters feeding by *Spodoptera exigua* larvae.[554] The presence of caryophyllene epoxide in the leaves prevents feeding by *Atta cephalotes*.[555]

***Inga laurine* (Sev.) Willd.,** guama, jina

***Inga vera* Willd.,** guaba, inga: The wood of these two species of trees is very susceptible to attack by *Cryptotermes brevis* termites.[86]

***Inocarpus edulis* Forst.:** An ether extract of the leaves did not deter feeding by *Acalymma vittatum*.[89]

***Intsia bijuga* (Colbr.) O. Kuntze,** kwila, ipil: The wood is very resistant to *Cryptotermes brevis* termites.[86,188] An ether extract of the stems with bark did not deter feeding by *Acalymma vittatum*.[89]

***Intsia palembauica* Miq.:** The wood is resistant to termites.[480]

***Koompassia excelsa* (Becc.) Taub.,** tapang

***Koompassia malaccensis* Maing.,** kempas: The wood of these two species is readily destroyed by termites,[480,485] although Becker[481] reported that the heartwood of *K. malaccensis* was resistant to 20 species of termites.

***Lonchocarpus sericeus* H. B. K.:** The compound, 2,5-dihydroxymethyl-3,4-dihydroxypyrrolidine, isolated from the seed (as well as from the leaves of *Derris elliptica*[556]) inhibited feeding by larvae of *Spodoptera littoralis, S. exempta,* and *Heliothis virescens* and nymphs of *Schistocerca gregaria* and *Melanoplus sanguinipes*.[557]

***Lotus corniculatus* L.,** birdsfoot trefoil: *Spodoptera eridania* larvae were not deterred from feeding on the leaves of this plant, despite their content of cyanogenic glycosides.[558]

***Lotus pedunculatus* Cav.:** Addition of an extract of the roots to the diet strongly inhibited feeding by larvae of *Costelytra zealandica*.[559] The active component proved to be 3R-(-)-vestitol,[560] a phytoalexin which is also effective as an antifeedant for larvae of *Heteronychus arator*.[561]

***Lupinus angustifolius* L.,** European blue lupine: Extracts of the roots deterred feeding by the larvae of *Costelytra zealandica*[562] and

Heteronychus arator.[563] A number of isoflavones were isolated and shown to be responsible for the activity.[562]

Lupinus perennis L., sun-dial lupine: An acetone extract of the whole plant was not repellent to *Popillia japonica* adults.[126]

Lupinus polyphyllus (Lindl.): The leaves, which are feeding deterrents for *Choristoneura fumiferana* larvae, yielded 8 alkaloids, of which 13-*trans*-cinnamoyloxylupanine and 13-hydroxylupanine were active at 25 µg/feeding disk.[564]

Lysiloma latisiliqua (L.) Benth., tabernau, caracoli: The wood is resistant to *Cryptotermes brevis* termites.[86,115]

Markhama lanata K. Schum.: Filter papers impregnated with an ethanolic extract of the wood did not affect *Reticulitermes flavipes* termites.[368]

Markhamia stipulata Seem.: Filter papers impregnated with an ethanolic extract of the wood and exposed to *Reticulitermes flavipes* termites killed 100% of the insects.[368]

Medicago ciliaris (L.) Krock.: The plant is susceptible to attack by *Hypera brunneipennis*.[565]

Medicago sativa L., alfalfa: *Plutella maculipennis* larvae fed readily on this plant,[195] which also attacked by *Sitona cylindricollis* and *Therioaphis riehmi*.[566] Compounds occuring in alfalfa that are repellent to *Bruchophagus roddi* are butyric acid, succinic acid, xanthophyll, shikimic acid, malic acid, betaine, and coumarin.[567] The plant is susceptible to attack by *Hypera brunneipennis* and *H. postica*.[565] However, certain varieties are resistant to *Bruchophagus roddi*.[568] The leaves are attacked by *Melanoplus sanguinipes* grasshoppers.[319] Several species of *Medicago* were found to be toxic to *Hypera postica* when the insects were placed thereon; the larvae did not survive.[569] The exudate from secretory trichomes of *M. disciformia* DC was also toxic to the larvae.[1] Immature *Megachile rotundifolia* were killed by exposure to alfalfa leaves and their extracts, as well as by injection of the extract into cells of the hive, probably due to the content of water-soluble saponins in the leaves.[570]

Melilotus alba Medik., white sweetclover: An acetone extract of the whole plant is not repellent to *Popillia japonica* adults.[126] *Plutella xylostella* larvae fed readily on this plant,[571] as did *Sitona cylindricollis* and *Therioaphis riehmi*.[566]

Melilotus infesta Guss., sweetclover: The plant is resistant to *Sitona cylindricollis* but not to *Therioaphis riehmi*.[566] The active component was isolated and identified as ammonium nitrate.[572]

Melilotus officinalis Lam., yellow sweetclover: Larvae of *Plutello xulostella* fed readily on this plant.[195] The plant is also susceptible to attack by *Sitona cylindricollis* and *Therioaphis riehmi*, although several varieties of the plant are resistant.[566] Natives of Bessarabia kept their houses free of moths by keeping bunches of this plant in all the rooms.[573] An extract of the leaves strongly deterred feeding by *Melanotus communis* when it was mixed with corn seeds and seedlings.[190]

Melilotus speciosa Dur.

Melilotus suaveolens Ladeb.

Melilotus sulcata Desf.

Melilotus taurica (Sieb.) Ser.

Melilotus wolgica Poir.: These five species are susceptible to attack by *Sitoma cylindricollis* and *Therioaphis riehmi*.[566]

Mora excelsa Benth., mora

Mora gongrijpii (Kleinh.) Sandw., morabukea: The wood of *M. excelsa* is susceptible to attack by *Cryptotermes brevis* termites, but the wood of *M. gongrijpii* is resistant to these insects.[86,124]

Mundulea sericea (Willd.) A. Cheval, mundulea: The powdered bark gave complete protection from brunchids when it was scattered thinly over grain in bins.[574] The fresh or dried leaves of this plant were also used for this purpose.[127]

Myroxylon balsamum (L.) Harms., Tolu balsam: The seed oil was moderate repellent for *Blattella germanica* adults.[61]

Onobrychis viciifolia Scop., mainfoin: Extracts of the roots deter feeding by *Costelytra zealandica*.[559] The active components, identified as two 2-arylbenzofurans, have been isolated.[575,576]

Ononis adenotricha **Boiss.:** The plant is resistant to attack by *Sitona cylindricollis* and *Therioaphis riehmi.*[566]

Pachyrhizus erosus **(L.) Urb.,** jicama, yam bean: Reports of feeding deterrency and, mainly, toxicity of the seeds for numerous species of insects have been published.[24,26,27,29] The responsible components have been identified as rotenone[57] and several rotenoids.[578]

Paramachaerium gruberi **Ducke:** An acetone extract of the wood repelled *Coptotermes formosanus* termites.[369]

Paratecoma peroba **(Record) Kuhlm.:** Filter papers impregnated with an ethanol extract of the wood and exposed to *Reticulitermes flavipes* killed 100% of the termites. Lapachanone and lapachol from the wood repelled *R. lucifugus* termites and are responsible for the resistance of the wood.[368]

Parkia bicolor **A. Chev.,** acoma: The wood is resistant to *Coptotermes formosanus* termites.[388]

Parochetus communis **Buch.-Ham.:** The wood is resistant to *Sitona cylindricollis.*[566]

Peltogyne lecontei **Ducke,** pau rozo

Peltogyne porphyrocardia **Griseb.,** purpleheart

Peltogyne pubescens **Benth.,** purpleheart: The wood of all three species is resistant to *Cryptotermes brevis* termites.[86]

Pentaclethra macroloba **(Willd.):** Ether extracts of the stems, bark, and twigs did not deter feeding by *Acalymma vittatum* adults.[89]

Pericopsis elata **Harms.:** The wood is resistant to *Coptotermes formosanus* termites.[288]

Pericopsis mooniana **Thw.:** Ether extracts of the twigs and bark did not deter feeding by *Acalymma vittatum.*[89]

Petalostemon villosus **Nutt.:** *Melanoplus sanguinipes* larvae did not feed upon this plant.[319]

Phaseolus vulgaris **L.,** greenbean, kidney bean: The leaves of this plant were resistant to feeding by *Empoasca krameri* in the greenhouse and in the field.[579-581] Larvae of *Callosobruchus chinensis,*[582] *Leptinotarsa decemlineata, Epilachna vigintioctomaculata,* and *E. vigintioctopunctata* could not survive on this plant.[583] Larvae of *Costelytra zealan-*

dica and *Haemonchus arator* would not feed on the roots of this plant. Phytoalexin flavonoids are responsible for this effect.[561] Arcelin, a seed protein in the beans, is toxic to *Zabrotes subfasciatus* and *Callosobruchus maculatus.*[534] Phytohemaglutinin extracted from this plant reduced the fecundity of these insects on cowpeas.[585]

Physostigma venenosum **Balf.,** calabar bean: An aqueous extract of the whole plant did not repel *Popillia japonica* adults.[126]

Piptadenia africana **Hook. f.,** dahoma: The wood is resistant to *Reticulitermes lucifugus* termites.[441]

Piscidia communis **I. M. Johnston:** An ethanol extract of the foliage deterred feeding by *Lymantria dispar* larvae.[84]

Piscidia erythrina **L.,** Jamaica dogwood: An aqueous extract of the whole plant did not repel *Popillia japonica* adults.[126]

Piscidia grandifolia **(Donn. Smith) I. M. Johnston:** The crushed bark and fresh leaves are used in El Salvador against bedbugs.[295]

Pisum sativum **L.,** pea, green pea: *Plutella xylostella* larvae fed readily on this plant.[195] (+)-Pisatin incorporated into the diet at 200 μg/ml deterred feeding by *Costelytra zealandica* larvae but not by *Heteronychus arator* larvae.[561] Productivity of *Sitophilus oryzae* was markedly reduced on a diet of yellow split peas and wheat.[586] An extract of the seedlings deterred feeding by adult *Oulema melanopus.*[587]

Pithecellobium jupunba **(Willd.)** Urban, puni

Pithecellobium racemosum **Mez.,** Surinam snakewood: The wood of the former is susceptible to feeding by *Cryptotermes brevis* termites but the wood of the latter is not.[86]

Pithecellobium mangense **Mohlenbr.:** The acetone extract of the wood deterred feeding by *Coptotermes formosanus* termites.[369]

Platymenia reticulata **Benth.,** vinhatico: The wood is totally repellent to *Cryptotermes brevis* termites.[86]

Platymiscium pinnatum **(Jacq.) Dugand:** Pentane and acetone extracts of the wood deterred feeding by *Coptotermes formosanus* termites.[369]

Platymiscium trinitatis **Benth.,** roble

Platymiscium ulei **Harms.,** Yama rosewood,

letterwood: The wood of these trees is susceptible to attack by *Cryptotermes brevis* termites.[86]

Pongamia pinnata (L.) **Pierre,** (synonym *P. glabra* Vent.), pongram, karanja, Indian beech: The seed oil tested as a surface protectant for redgram seeds was quite effective in preventing the reproduction of *Callosobruchus chinensis*.[291] Extracts of the flowers tested on filter papers did not repel *Sitophilus oryzae* larvae.[251,252] Ether extracts of the twigs, stem bark, and leaves deterred feeding by *Acalymma vittatum* adults at 0.1 and 0.5%.[89] The seed oil was very toxic to *Periplaneta americana* but not to *Blattella germanica* and *Musca domestica*.[588] The active component is karanjin.[588] Toxicity was also shown to *Monomorphus villiger, Opatroides frater,* and *Seleron latipes*.[589] Karanjin is also a potent nitrification inhibitor.[590] The seed oil deterred feeding by *Amsacta moorei*.[372] Juvenile hormone activity was shown by the oil against larvae of *Tribolium castaneum*.[593] The oil is also an effective synergist for chlorinated insecticides[591] and pyrethrins.[592]

Psoralea corylifolia **L.,** Bakuchiol, a meroterpenoid isolated from the seeds of this plant, is a strong juvenile hormone for *Dysdercus koenigii* nymphs.[594]

Psoralea pedunculata (Mill.) **Vail.,** Sampson snakeroot: Extracts of the whole plant did not repel *Popillia japonica* adults.[126]

Pterocarpus erinaceus **Poir. ex DC.,** African kino, barwood: The wood repels *Coptotermes formosanus* termites.[388]

Pterocarpus marsupium **Roxb.:** An acetone extract of the stems applied topically to the abdominal termites of *Dysdercus cingulatus* nymphs induced juvenilization.[595]

Pueraria lobata (**Willd.**) **Ohwi,** kudzu: A benzene extract of the whole plant was dissolved in acetone and applied to filter papers, then presented to *Spodoptera litura* larvae. No untoward effect was observed at concentrations of 1, 2.5, and 5%.[258]

Robinia pseudoacacia **L.,** black locust: Extracts of this plant were not repellent to *Popillia japonica* adults.[126] An ethanol extract of the leaves did not deter *Lymantria*

dispar larvae from feeding.[84] The wood did not deter feeding by *Reticulitermes flavipes* termites[462] although Wolcott[86,124] and Scheffer and Duncan[371] reported that it is resistant to termites (not identified).[310] Shavings of the wood were both resistant and repellent to *Reticulitermes lucifugus* termites.[187] This tree was not infested during an outbreak of *Malacosoma disstria* in southern Louisiana.[88]

Sesbania bispinosa (**Jacq.**) **W. F. Wight:** The natives in West Africa claimed that animals washed in water in which the leaves of this shrub has been pounded could safely traverse a tsetse fly belt.[596]

Sesbania punctata **DC,** sabral: The natives in Africa used a decoction of the leaves for washing animals to prevent bites of the tsetse fly.[597]

Sophora microphylla **Ait.,** The isoflavonoid, (-)-naackiain, isolated from the wood strongly deterred feeding by *Costelytra zealandica* and *Heteronychus arator* larvae at 10 µg/ml.[561]

Spartium juncium **L.,** Spanish broom: Three alkaloids isolated from this plant (sparteine, nicotine, and quinine) deterred feeding by *Entomosclis americana*.[598,599]

Stahlia monosperma (**Tul.**) Utban, cobana, caobanillo: The wood is resistant to *Cryptotermes brevis* termites.[86]

Stylosanthes biflora (L.) **B.S.P.,** pencil flower: Extracts of this plant did not repel *Popillia japonica* adults.[126]

Swartzia leiocalycina **Benth.,** wamara: The wood is resistant to *Cryptotermes brevis* termites.[86,370]

Swartzia madagascariensis **Desvaux:** The powdered fruit is used to line storage bins for millet and as a termite repellent.[116] Triterpenoid saponins isolated from the fruit possessed high molluscidal activity against schistosomiasis-transmitting snails, *Biomphalaria glabrata* and *Bulinus globosus*.[600] The powdered fruit is used to stupefy fish and repel termites.[601]

Sweetia panamensis **Benth.,** guayacan: The heartwood is very resistant to termites.[371] The wood, as well as pentane and acetone extracts thereof, were highly deterrent to *Coptotermes formosanus* termites.[369]

Tabebuia guayacan **Hemsl.**: This plant yielded derivatives of lapachol toxic to termites.[602]

Tamarindis indica **L.**, tamarind: The wood is susceptible to feeding by *Cryptotermes brevis*[86] but resistant to *Anacanthotermes ochraceus, Psammotermes fuscofemoralis,* and *P. assuarensis* termites.[387]

Tephrosia elata Deflers: Tephrosin, isolated from the seed pods, strongly deterred feeding by *Spodoptera exempta* larvae. Isopongaflavone deterred feeding by *Maruca testulalis* and *Eldana saccarina* larvae. Rotenone from this plant was active against all these species.[603]

Tephrosia hildebrandtii **Vatke:** Hildecarpin, a 6a-hydroxylated pterocarpan from the roots, deterred feeding by *Maruca testulalis* and showed antifungal activity against *Cladosporium cucumerinum.*[604]

Trifolium aureum **Poll.,** hop clover

Trifolium arvense **L.,** rabbitfoot clover: Extracts of these plants did not repel *Popillia japonica* adults.[126]

Trifolium pratense **L.,** red clover: Larvae of *Plutella xylostella* fed readily on this plant,[195] but the isoflavonoid, formonoetin, isolated from the leaves deterred feeding at 200 μg/ml by *Costelytra zealandica* and *Heteronychus arator* larvae.[561]

Trifolium repens **L.:** A methanol extract of the roots greatly reduced feeding by *Heteronychus arator* larvae. Medicarpin and vestitol were isolated from the extract.[605]

Trigonella arucata **Meyer:** The plant is resistant to *Therioaphis riehmi* but not to *Sitona cylindricollis.*[506]

Trigonella foenum-graecum **L.,** fenugreek: An extract of the whole plant did not repel *Popillia japonica* adults.[126] The dried plant is mixed with grain in India as an insect repellent.[116] A petroleum ether extract of the fenugreek leaves repelled adult *Tribolium castaneum, Sitophilus granarius,* and *Rhyzopertha dominica* when mixed with cereal grain.[606]

Trigonella gladiata **Stein**

Trigonella monantha **C. A. Mey.**

Trigonella monspeliaca **L.:** *T. gladiata* is resistant to *Sitona cylindricollis* but not to

Therioaphis riehmi. T. monantha is resistant to *T. riehmi* but not to *S. cylindricollis,* and *T. monospecliaca* is resistant to both insect species.[566]

Trigonella noweana **Boiss.**

Trigonella polycerata **L.**

Trigonella radiata **(L.) Boiss.**

Trigonella spicata **Sibth. & Sm.:** *T. Noweana* and *T. Polycerata* are resistant to *Therioaphis riehmi* but not to *Sitona cylindricollis. T. spicata* is resistant to *S. cylindricollis* but not to *T. riehmi. T. radiata* is resistant to both insect species.[566]

Vicia faba **L.,** broadbean, horsebean: *Plutella xylostella*[195] and *Callosobruchus chinensis* larvae fed on this plant.[582]

Vicia villosa **Roth.,** winter vetch, hairy vetch: An alcohol extract of the whole plant did not repel *Anthonomus grandis grandis* adults,[487] but the juice expressed from the fresh leaves was fairly effective as a feeding deterrent for *Melanotus communis.*[247] Extracts of the whole plant did not repel *Popillia japonica* adults.[126]

Zollernia guianensis **Aubl.,** letterwood: The wood is very resistant to attack by *Cryptotermes brevis* termites.[86]

FAMILY FAGACEAE

Castanea dentata **(Marsh.) Borkh.,** American chestnut: An extract of the entire plant repelled *Popillia japonica* adults.[126] An alcoholic extract was not repellent to *Anthonomus grandis grandis* adults.[487]

Castanea sativa **Mill.,** European chestnut: The wood is fairly resistant to *Reticulitermes lucifugus* termites.[441]

Fagus grandifolia **Ehrh.,** American beech: Extracts of the whole plant were not repellent to *Popillia japonica* adults.[126] Consecutive extracts of the twigs prepared with ethyl ether and methanol failed to deter feeding by *Spodoptera frugiperda* larvae.[204] Tests conducted under choice and forced-feeding showed that the wood was susceptible to feeding by *Coptotermes formosanus* and *Reticulitermes virginicus* termites.[87] Ethereal and methanolic extracts of the twigs failed to deter feeding by adult *Diabrotica undecimpunctata howardi* and *Acalymma vittatum*.[52,607] Wood exposed to *Ahacanthotermes ahngerianus* termites lost 85% of its weight in 2 years.[362]

***Nothofagus cunninghamii* Oerst.**, myrtle beech, Tasmanian beech: The wood is moderately durable to termites and fungal decay.[188]

***Nothofagus menziesii* Oerst.**, New Zealand silver beech: The wood is not resistant to termites or fungal decay.[188]

***Quercus agrifolia* Nees.**, California live oak

***Quercus alba* L.**, white oak

***Quercus douglasii* Hook. & Arn.:** Ethanol extracts of the leaves of these species stimulated feeding by *Lymantria dispar* larvae.[84] When tannin (catechin) obtained from *Q. alba* was applied to potato leaves, larvae of *Leptinotarsa decemlineata* refused to feed on the leaves.[608] This was also true of alcoholic extracts prepared from the leaves and bark of the plant. However, the extracts failed to deter feeding by adult *Epilachna varivestis*.[609] Tests conducted under choice and forced-feeding with each of the species of *Quercus* showed the wood to be deterrent to *Reticulitermes virginicus* termites.[87]

***Quercus cambelii* Nutt.:** A hot methanol extract of the aerial portion caused growth inhibition when fed in an artificial diet to *Heliothis virescens* larvae. The active component was identified as ellagic acid.[438]

***Quercus falcata* Michx.**, southern red oak, Spanish oak: An ethanolic extract of the leaves deterred feeding by *Lymantria dispar* larvae.[84]

***Quercus glandulifera* Blume**, Japanese oak

***Quercus incana* Bartr.**, bluejack oak

***Quercus laevis* Walt.**, turkey oak

***Quercus macrocarpa* Michx.**, bur oak: Only 60% of *Melanoplus femurrubrum* grasshoppers reared on a diet containing extracts of these plants reached adulthood.[319] An ethanolic extract of the leaves of *Q. laevis* deterred feeding by *Lymantria dispar* larvae.[84]

***Quercus marilandica* Muenchh.**, blackjack oak: An ethanolic extract of the leaves deterred feeding by *Lymantria dispar* larvae.[84]

***Quercus nigra* L.**, water oak, possum oak: The wood did not deter feeding by *Coptotermes formosanus* or *Reticulitermes virginicus* termites.[85,87]

***Quercus prinus* L.**, chestnut oak

***Quercus robur* L.**, English oak: The wood of these trees was more strongly attacked by *Reticulitermes lucifugus* than by *R. flavipes*.[610]

***Quercus rubra* L.**, red oak: The wood is susceptible to feeding by *Cryptotermes brevis* and *Reticulitermes flavipes* termites,[86,462] but it is moderately resistant to feeding by *Incisitermes minor* termites.[361]

***Quercus velutina* Lam.**, black oak: Extracts of the fresh leaves repelled *Popillia japonica* adults.[126]

***Quercus virginiana* Mill.**, live oak: An ethanol extract of the leaves deterred feeding by *Lymantria dispar* larvae.[84]

FAMILY FLACOURTIACEAE

***Homalium foetidum* Benth.**, malas: The wood is resistant to termites and fungal decay.[188]

***Homalium racemosum* Jacq.**, caracolillo, tostado: The wood is resistant to *Cryptotermes brevis* termites.[86,124]

***Hydnocarpus anthelmintica* Pierre ex Laness.:** A methanol extract of the leaves strongly deterred feeding by *Bombyx mori* larvae.[82]

***Hydnocarpus laurifolia* (Dennst.) Sleum:** The oil cake made from the seeds is used as an ant deterrent and for controlling *Xyloryctes jamaicensis* in coconut trees.[611]

***Hydnocarpus venenata* Gaertn.:** An ether extract of the leaves at 0.5% was somewhat deterrent to feeding by adult *Acalymma vittatum*.[89]

***Lacistemma agregatum* (Berg.) Rusby:** An ether extract of the roots did not deter feeding by adult *Acalymma vittatum* at 0.1 or 0.5%.[89]

***Lindackeria maynensis* R. & E.:** An ether extract of the roots was highly deterrent to feeding by adult *Acalymma vittatum*, but an extract of the leaves or twigs was not effective.[89]

***Ryania angustifolia* (Turcz.) Monachina**

Ryania speciosa **Vahl.**, ryania: Aqueous extracts of the whole plant deterred feeding by *Attagenus piceus, Tineola bisselliella,* and *Blattella germanica.*[25] The powdered roots and stems of *R. speciosa* stored with sorghum in bins protected the grain from insects for 9 months and gave excellent protection of bulk maize and groundnuts from *Sitophilus oryzae.*[613,614] The use of 400 g/ton protected shelled maize and wheat for 2 years.[615] Many reports attest to the toxicity of *R. speciosa* to numerous species of insects,[26] but especially to cockroaches.[616] The active alkaloid, ryanodine, was obtained from the roots and stems.[617] A 0.02% solution was toxic immediately, and repellent for 4 months, to *Cryptotermes brevis.*[618] a 0.01% solution impregnated into susceptible woods protected them from attack for 204 d; at 0.05%, the wood was not eaten for at least 18 months.[619] Aqueous and acetone solutions of the extractive were repellent to *Attagenus megatoma* larvae.[620] The ryanodine precursor, (+)-ryanodol, has been synthesized.[621]

Scottellia coriacea **A. Chev.**, odoko: The wood is susceptible to termite attack in Nigeria.[370]

Trichadenia zeylanica **Thn.**: An ether extract of the twigs did not deter feeding by adult *Acalymma vittatum.*[89]

FAMILY FUMARIACEAE

Dicentra canadensis **(Goldie) Walp.**, squirrelcorn

Fumaria officinalis **L.**, fumitory: Aqueous extracts of the whole plant were not repellent to adult *Popillia japonica.*[126]

FAMILY GENTIANACEAE

Gentiana lutea **L.**, yellow gentian

Menyanthes trifoliata **L.**, bogbean

Sabbatia angularis **(L.) Pursh.**, rose gentian: Extracts of the whole plant did not repel *Popillia japonica* adults.[126]

Swertia carolineensis **(Walt.)** Ktze, columbo: An ethanol extract of the whole plant did not repel larvae or adults of *Anthonomus grandis grandis.*[487]

Swertia chirayita **(Roxb.) Lyons,** chiretta: Extracts of the whole plant did not repl adult *Popillia japonica.*[126]

FAMILY GERANIACEAE

Erodium cicutaricum **(L.) L'Her. ex Ait.**, redstem filaree: Geraniin, the bound form of ellagic acid occurring in this plant, is a growth inhibitor for *Heliothis virescens* larvae.[438]

Geranium carolinianum **L.**, Carolina geranium, cranebill geranium: Extracts of the whole plant did not repel adult *Popillia japonica.*[126]

Geranium maculatum **L.**,wild geranium, spotted geranium: Geranium oil is moderately repellent to *Blattella germanica* adults.[61]

Pelargonium graveolens **L'Her. ex Ait.**, rose geranium: The essential oil was one of the best repellents for *Cochliomyia hominivorax,*[155] *Musca domestica,*[622,623] and *Culex fatigans* mosquitoes.[622,623] Juice expressed from the fresh leaves deterred feeding by *Melanotus communis.*[247]

Tropaeolum majus **L.**, garden nasturtium, Indian-cress: Extracts of the whole plant did not repel adult *Popillia japonica.*[126]

FAMILY GESNERIACEAE

Negelia hyacinthi **Carr.**: Larvae of *Plutella xylostella* refuse to feed on this plant.[195]

FAMILY GINKGOACEAE

Ginkgo biloba **L,** ginkgo tree, maidenhair tree: The leaves are used in China and Japan as bookmarks to repel *Lepisma saccharina* and the larvae of other insects. The wood is used to make insect-proof cabinets. *Popillia japonica* adults will not eat the leaves.[624] A methanol extract of the leaves deterred feeding by *Pieris rapae crucivora* on cabbage leaf disks treated with the extract. The active components are 6-penta-decenylsalicylic acid, 6-heptadecenylsalicylic acid, and the sesquiterpene, bilobalide.[625] An ethanol extract of the leaves stimulated feeding by *Lymantria dispar* larvae.[84]

FAMILY HERNANDIACEAE

Hernandia peltata **Meissn.:** An ether extract of the stem with bark was strongly deterrent to feeding by *Acalymma vittatum* adults.[89]
Hernandia sonora **L.,** jack-in-the-box: The wood is susceptible to attack by *Cryptotermes brevis* termites.[86]

FAMILY HIPPOCASTANACEAE

Aesculus californica **(Spech.) Nutt.,** California buckeye
Aesculus glabra **Willd.,** Ohio buckeye: An ethanol extract of the leaves of *A. californica* deterred feeding by *Lymantria dispar* larvae, but an extract of the leaves of *A. glabra* did not.[84]
Aesculus hippocastanum **L.,** horsechestnut: Filter papers impregnated with a 3% solution of aescin, a triterpene saponin from the fruits of this plant, inhibited feeding by, and were toxic to, *Reticulitermes flavipes* termites; 0.05 and 0.5% were ineffective.[229]
Aesculus octandra **Marsh.,** yellow buckeye
Aesculus pavia **L.,** red buckeye
Aesculus sylvatica **Bartr.,** painted buckeye: Alcoholic extracts of these three plants did not repel larvae of adults of *Anthonomus grandis grandis.*[487]

FAMILY HUMIRIACEAE

Humiria balsamifera **Jaume,** umiri: The wood is resistant to *Cryptotermes brevis* termites.[86]

FAMILY HYPERICACEAE

Caraipa fasciculata **Cambess:** An ether extract of the stem with bark and of the twigs failed to deter feeding by *Acalymma vittatum* adults.[89]
Hypericum perforatum **L.,** St. Johnswort: Larvae of the ticks *Ixodes redikorzevi, Haemaphysalis punctata, Rhypicephalum rossicus,* and *Dermacentor marginatus* exposed to the powdered leaf died in 100, 120, 120, and 120 min, respectively.[153]

FAMILY ICACINACEAE

Apodytes dimidiata **E. Mey. ex Arn.:** An ether extract of the woody stems with bark deterred feeding by adult *Acalymma vittatum* at 0.5% but not at 0.1%.[89]
Cantleya corniculata **(Becc..) Howard,** dedaeu: The wood is resistant to termites.[480]

FAMILY IRIDACEAE

Crocus sativus **L.,** saffron crocus: An extract of the whole plant did not repel *Popillia japonica* adults.[126]
Iris germanica **var. *Florentina* (L.) Dykes,** orris: Dusting orris roots over garments did not protect them from *Tineola bisselliella.*[62] The root oil did not repel *Blattella germanica* adults.[61]
Iris versicolor **L.,** blueflag iris
Sisyrinchium **sp.,** blue-eyed grass: Extracts of these plants did not repel *Popillia japonica* adults.[126]

FAMILY JUGLANDACEAE

Carya glabrata **(Mill.),** sweet-pignut hickory: An infusion of the leaves rubbed on a horse repels flies.[522] Extracts of the whole plant did not repel adult *Popillia japonica* .[126] The wood is very susceptible to attack by *Cryptotermes brevis.* termites.[86,124,370]
Carya illinoensis **(Wangenh.) K. Koch,** pecan: Certain genotypes of the pecan tree are resistant to insects.[627] An ethanol extract of the leaves stimulated feeding by *Lymantria dispar* larvae.[84]
Carya ovata **(Mill.) K. Koch.,** shagbark hickory: An ethanol extract of the leaves stimulated feeding by *Lymantria dispar* larvae.[84] The wood deterred feeding by *Reticulitermes virginicus* and *Coptotermes formosanus* termites.[87]
Carya tomentosa **(Poir.) Nutt.,** mockernut hickory: An ethanol extract of the leaves did not deter feeding by *Lymantria dispar* larvae.[84]
Juglans californica **S. Wats.,** California walnut: The wood is not resistant to *Incisitermes minor* termites.[86]

Juglans cinerea L., butternut: Extracts of the whole plant did not repel adult *Popillia japonica*.[126] The wood is susceptible to termite attack.[86,124]

Juglans nigra L., black wanut: The wood is susceptible to attack by *Cryptotermes brevis* termites.[86] A decoction of the leaves applied to the skin of horses repels flies.[303] An ethanol extract of the leaves deterred feeding by *Lymantria dispar* larvae.[84] Cotton caterpillars avoided cotton leaves sprayed with an alcoholic extract of the plant.[628]

Juglans regia L, English walnut, Persian walnut: An ethanol extract of the leaves did not deter feeding by *Lymantria dispar* larvae.[8]

FAMILY LABIATAE

Ajuga SPP.: On trips to African and Amazonian jungle areas, Kubo[629] became aware of the medicinal and insecticidal properties of several species of *Ajuga*. Of these, the species *remota* was well known locally for its properties of deterring feeding by *Schistocerca gregaria*. In his laboratory, Kubo tested extracts of this plant on maize leaves against larvae of *Spodoptera exempta* and obtained strongly positive results. Thus began an investigation that was to result in the isolation, identification, and eventual synthesis of numerous insect deterrent and insecticidal compounds.[300]

Ajuga chamaepitys Guss.: Two neo-clerodane compounds, ajugapitin and its dihydro derivative, were isolated from this species and their structures were elucidated.[630]

Ajuga iva Schreber: Four new clerodane diterpenoids were isolated and identified from this species.[631] These investigators subsequently isolated additional clerodane compounds from several *Ajuga* species; nine of the compounds tested by the leaf disk method deterred feeding by *Spodoptera littoralis* larvae at concentrations as low as 0.01 ppm.[632]

Ajuga nipponensis Makino: The neo-clerodane, ajugamarin, and several related compounds were isolated from the leaves of this plant.[633-635]

Ajuga pseudoiva (L.) Schreber: Two neoclerodane diterpenoids, 2-acetylivain and its C-2 epimer (14,15-dihydroajugapitin isolated from *A. chamaepitys*), were obtained from the plant.[636]

Ajuga remota Benth.: From the leaves of this species, Kubo et al.[636] isolated ajugarins I, II, and III, which deter feeding by *Spodoptera exempta* larvae.[300] These investigators also isolated from the leaves and roots the phytoecdysones, β-ecdysone and cyasterone, which strongly inhibit ecdysis of *Bombyx mori*, *Pectinophora gossypiella*, and *Spodoptera frugiperda*.[637]

Ajuga reptans L.: Two new clerodane diterpenoids, ajugareptansones A and B, were isolated from extracts of the whole plant,[638] which caused molting inhibition in *Periplaneta americana*.[639] Expressed leaf juices and methanol extracts of the roots strongly repelled female *Tetranychus urticae* and reduced their fecundity.[640] These materials also deterred feeding by *Epilachna varivestis* larvae and disrupted their development in the form of deformed wings.[641] The deterrency is probably due to ajugarins and the disruption is due to ecdysterone. This plant has been micropropagated from the shoot tips of greenhouse-grown plants.[642] The structures of the ajugarins, five of which have been isolated thus far, are described by Kubo et al.[643-645] Synthetic methods for the ajugarins and other significant feeding deterrents from *Ajuga* species have been published.[646-651]

Collinsonia canadensis L., citronella horsebalm

Cunila origanoides (L.) Britton, stonemint

Glechoma hederacea L., ground ivy: Extracts of these plants did not repel *Popillia japonica*.[126]

Hedeoma pulegioides (L.) Pers., American pennyroyal: Oil of pennyroyal has been widely used to repel fleas. It is applied to shoetops, hose, and trousers, and its use on bedding and floors has been advocated. The plant itself has also been used for this purpose.[471] The oil is also a good repellent for *Cochliomyia hominivorax*[155] and *Blatta ori-*

entalis,[162] but it does not repel *Popillia japonica* adults.[126] The oil inhibits feeding by *Spodoptera frugiperda* larvae.[326] An extract of the leaves was moderately deterrent to feeding by *Melanotus communis.*[247]

***Hyptis spicigera* Lam.:** The entire plant is burned in Africa to repel mosquitoes,[121] and it is placed in a layer below bundles of millet to repel termites.[596] Nigerian farmers protect cowpeas from insects with the pungent-smelling inflorescence.[458]

***Hyssopus officinalis* L.,** hyssop

***Lamium amplexicaule* L.,** henbit: Extracts of these two plants were not repellent to adult *Popillia japonica.*[126]

***Lavandula angustifolia* Mill.,** lavender: Oil of lavender is strongly repellent to *Blatta orientalis,*[162] *Blattella germanica,*[61] and *Apis florea* bees.[147,475] The oil also protected flannel from *Tineola bisselliella* infestation.[157] An extract of the flowers prepared with propylene glycol did not inhibit feeding by *Pieris brassicae* larvae.[652]

***Lavandula spica* L.,** spike lavender: The oil did not repel *Aedes* mosquitoes,[342] but it killed 100% in 24 h on contact.[182] However, it was not toxic to *Anopheles claviger.*[182]

***Leonurus cardiaca* L.,** motherwort: An extract of the whole plant did not repel adult *Popillia japonica.*[126] A bitter-tasting clerodane derivative related to those isolated from *Ajuga* species has been isolated from this plant.[653]

***Leucas aspera* Spreng:** Oleanolic acid and ursolic acid were isolated from the entire plant, which possesses insecticidal properties.[654] A petroleum ether extract applied topically to the abdominal termites of *Dysdercus cingulatus* nymphs caused developmental disruption.[109]

***Leucas martinicensis* R. Br.:** This strongly-scented plant is burned in African rooms to repel mosquitoes.[287,596]

***Leucas prostrata* Hook f. & Gamble:** An acetone extract of the whole plant impregnated on castor leaves did not deter feeding by *Spilosoma obliqua* larvae.[194]

***Leucas zeylanica* (L.) Benth.:** The powdered plant parts, tested separately on mung bean

seeds, did not prevent oviposition by *Calosobruchus chinensis.*[150]

***Lycopus virginicus* L.,** bugleweed: An extract of the whole plant did not repel adult *Popillia japonica* .[126]

***Marrubium vulgare* L.,** horehound: Extracts of the whole plant were not repellent to adult *Popillia japonica,*[126] but an aqueous extract of the leaves was strongly deterrent to *Pieris brassicae* larvae.[652]

***Melissa officinalis* L.,** lemon balm: An aqueous extract of the leaves at 0.5 to 2.0% was strongly repellent to *Pieris brassicae* larvae.[652] A solvent-free homogenate of the fresh leaves did not deter feeding by *Melanotus communis.*[247]

***Mentha arvensis* L.,** field mint, Japanese mint, corn mint: Extracts of the whole plant did not repel adult *Popillia japonica.*[126] The essential oil strongly repelled *Blattella germanica* adults.[61]

***Mentha piperita* L.,** peppermint: Oil of peppermint did not repel *Cochliomyia hominivorax*[135] but did repel *Blatta orientalis* adults,[162] *Blattella germanica* adults,[61] *Apis mellifera,*[475] and *A. florea.*[149] An aqueous extract of the leaves strongly repelled *Pieris brassicae* larvae.[652]

***Mentha rotundifolia* (L.) Huds.,** applemint: An extract of the leaves did not repel *Melanotus communis* larvae.[247]

***Mentha spicata* L.,** spearmint: The oil was strongly repellent to adult *Blattella germanica*[61] and to *Cochliomyia hominivorax* larvae.[155] but extracts of the whole plant were not repellent to adult *Popillia japonica.*[126] The oil was also repellent to *Blattella germanica* nymphs.[655] A mixture of wheat seeds and powdered spearmint leaves at 0.5, 1, and 2 parts/100 parts was very effective in controlling *Sitophilus oryzae* adults.[656] (-)-Carvone, obtained from spearmint oil, incorporated in an artificial diet deterred feeding by *Earias insulana* larvae and killed large numbers of them.[657] Carvone and (-)-pulegone were highly repellent to *Blattella germanica* and *Periplaneta fuliginosa* adults.[658] (-)-Methol, obtained from the oil, was toxic to *Reticulitermes speratus, Micro-*

cerotermes crassus, and *Nasutitermes nigriceps* termites, but (+)-menthol was not toxic.[659]

***Monarda punctata* L.**, spotted beebalm, horsemint: An extract of the whole plant did not repel *Popillia japonica* adults.[126] An extract of the leaves did not repel *Melanotus communis* larvae.[247]

***Nepeta cataria* L.**, catnip: Extracts of the whole plant did not repel adult *Popillia japonica.*[126] Nepetalactone, the constituent of catnip responsible for its well-known excitant effect on cats, acts as a repellent to numerous species of insects. Of 27 species of insects tested, 20 were strongly repelled and seven showed no response when exposed to the compound.[660]

***Ocimum basilicum* L.**, basil, sweet basil: The strongly scented herb is burned in African rooms to repel mosquitoes[121] and is placed under beds for the same purpose.[116] The oil has larvicidal action on *Culex fatigans* mosquitoes.[661] It was moderately repellent to adult *Blattella germanica* for 3 h but not for 24 h.[61] Fresh juice from the leaves was moderately repellent to *Melanotus communis* larvae[247] and *Apis florea* bees.[475] Two juvocimenes, potent juvenilizing compounds for *Oncopeltus fasciatus,* have been isolated from a distillate of the plant and identified.[662] They have also been synthesized.[663]

***Ocimum canum* Simms**, hoary basil: The leaves are used to protect stored products from insect attack.[458]

***Ocimum gratissimum* L.**, The plant has been used as a mosquito repellent but the action has not been confirmed experimentally.[116,148]

***Ocimum sanctum* L.**, holy basil: The leaves are used as a mosquito repellent in the Philippines.[94,664] An acetone extract of the plant applied to the abdominal termites of *Spodoptera litura* induced juvenilization.[108]

***Ocimum suave* Willd.:** The smoke from this burning plant is used as a mosquito repellent.[116]

***Origanum hirtum* Link:** The oil was of no value as a repellent for *Cochliomyia hominivorax.*[155]

***Origanum vulgare* L.**, wild marjoram, oreg-

ano: The Spanish oil was repellent to *Blattella germanica* adults for up to 170 h.[61]

***Otostegia integrifolia* Aubl.**, An ether extract of the combined stems, leaves, and flowers deterred feeding by adult *Acalymma vittatum.*[89]

***Perilla frutescens* (L.) Britt.**, perilla: An ether extract of the combined stems and leaves did not deter feeding by adult *Acalymma vittatum.*[89] The oil repelled *Blattella germanica* adults for 24 h but not for 170 h.[61]

***Plectranthus barbatus* Andr.**, A methanol extract of the whole plant deterred feeding by larvae of *Pectinophora gossypiella* and *Schizaphis graminum.*[665]

***Plectranthus excisus* var. *intermedius* (Makino) Hara:** Isodomedin, the component in the leaves responsible for the feeding deterrency toward *Spodoptera exempta* larvae, is identified as having an *ent*-kaurenoid structure. It also exhibits antibacterial and cytotoxic properties.[666]

***Pogostemon cablin* (Blanco) Benth.**, patchouly: Strewing the leaves among woolen clothes is said to keep insects away.[81,205] The leaves and tops are used in the Philippines to repel cockroaches, moths, ants, and leeches.[94] The leaves have been used for many years in China and Malaya as an insecticide, mainly against moths. Natives protect their leaf beds from attack by mixing patchouly leaves with them.[667] The oil does not repel *Blattella germanica* adults,[61] but it does repel *Apis mellifera* and *A. florea* bees.[149]

***Pycnanthemum flexuosum* (Walt.) B.S.P.:** Extracts of the whole plant did not repel adult *Popillia japonica.*[126]

***Pycnothymus rigidus* (Bartr.) Small.**, wild savory: The powdered plant was neither repellent nor toxic to *Epilachna varivestis* larvae or *Aphis fabae.*[256]

***Rosmarinus officinalis* L.**, rosemary: Branches of this shrub packed together with clothing will repel *Tineola bisselliella.*[332] Oil of rosemary neither repels nor attracts *Cochliomyia hominivorax* larvae,[155] *Blattella germanica,*[61] *Melanotus communis* larvae,[247] or *Blatta orientalis,*[162] but it is moderately repellent to *Popillia japonica* and *Lucilia cuprina* adults.[668] A methanol extract of the leaves did not deter feeding by *Acalymma vittatum* adults.[52]

Salvia officinalis L., sage: Fresh juice expressed from the leaves failed to deter feeding by *Melanotus communis* larvae on treated potato slices.[247] The development of *Aedes aegypti* larvae was not affected by exposure to an aqueous extract of the leaves.[245]

Salvia repens Burch.: The Sotha tribe in Basutoland used the smoke from the burning plant to repel insects.[116]

Salvia sclarea L., clary: Clary oil showed little or no repellency to adult *Blattella germanica*,[61] but the juice expressed from the fresh leaves was highly repellent to *Melanotus communis* larvae.[247] Corn seeds treated with the juice showed little or no deterrency to feeding by *Diabrotica undecimpunctata howardi* larvae.[228]

Salvia splendens F. Sellowe ex Roem. & Schult.*, scarlet sage

Salvia triloba L., green sage: Extracts of these two species did not repel *Popillia japonica* adults.[126]

Saturejia hortensis L., summer savory: The oil was repellent to *Blattella germanica* adults.[61] The juice expressed from the fresh leaves was moderately repellent to *Melanotus communus* larvae.[247] A methanol extract of the leaves significantly increased the larval development time of *Aedes aegypti*.[245]

Scutellaria laterifolia L., mad-dog skullcap: Extracts of the whole plant were not repellent to adult *Popillia japonica*.[126]

Stachys officinalis (L.) Franch, betony

Teucrium africanum Thunb.: The leaves of these two species are very effective in deterring feeding by 5th instar male *Locusta migratoria*. The active components were isolated and identified as two chlorinated clerodane diterpenoids.[669]

Teucrium canadense L., American germander: Extracts of the whole plant were not repellent to *Popillia japonica* adults.[126]

Thymus vulgaris L., thyme: The oil was moderately repellent to adult *Blattella germanica*[61] and highly repellent to *Melanotus communis* larvae.[247] An hour's exposure to

the oil killed larvae of *Anopheles claviger* and *Aedes cantans*.[182]

FAMILY LAURACEAE

Aniba ovalifolia Mez.: The wood is very resistant to termites.[124]

Aniba rosaeodora Ducke, Brazilian rosewood: Linalool obtained from this tree is an effective termite repellent.[118]

Cinnamomum aromaticum Nees, cassia: The leaf oil did not repel adult *Blattella germanica*.[61] However, it was moderately repellent to *Blatta orientalis*[162] and exhibited good repellency to *Cochliomyia homonivorax* for 1 to 2 d.[155]

Cinnamomum camphora (L.) J. S. Presl., camphor tree: Camphor preserved clothing and other articles against insects and worms.[191] There are at least 26 patents describing the used of camphor as an insecticide or repellent.[208] Formosan camphor oil did not repel *Lucilia cuprina*.[322] In tests of the knockdown and lethality of mosquitoes, safrole (a volatile constituent of camphor) was the most potent.[670] The wood is completely resistant to *Cryptotermes brevis*,[86] *Anacanthotermes ochraceus, Psammotermes fuscofemoralis*, and *P. assuarensis* termites.[387] This species, planted in Tanganyika, is susceptible to attack by a shoot borer.[117]

Cinnamomum tamala (Buch.-Ham.) Nees & Oberm., Indian-bark: The leaf oil is repellent to *Apis florea* bees.[149,475]

Cinnamomum verum J. S. Presl., cinnamon: The bark oil and leaf oil repelled *Blattella germanica* adults.[61] Female *Tribolium castaneum* fail to oviposit or survive for 20 d when maintained on a cinnamon oil-yeast mixture.[187] Cinnamon oil was considered to be one of the best repellents against *Cochliomyia hominivorax*. There was no emergence of these flies from meat treated with the oil. Cinnamon powder was effective for only 2 days.[155]

Cryptocarya mandioccana Meissn.: The wood is susceptible to attack by *Cryptotermes brevis* termites.[86]

Eusideroxylon lauriflora (Blanco) **J. Schulze-Motel,** Bornean ironwood: The wood is completely resistant to termites.[485]

Laurus nobilis **L.,** bay, laurel: The leaf oil is reported to be repellent to *Blattella germanica* [61] and *Periplaneta americana.*[671] The repellent components were isolated and identified as phenylhydrazine, piperidine, geraniol, and cineole.[671] Eugenol and methyleugenol, obtained from the oil, deterred feeding by *Mythimna unipuncta.*[92] An ethanol extract of the foliage failed to deter feeding by *Lymantria dispar* larvae.[84] Aqueous extracts of the whole plant did not repel *Popillia japonica* adults.[126]

Licaria canella **(Meissn.) Kosterm.,** brown silverballi: The wood is resistant to *Cryptotermes brevis* termites.[86,124]

Lindera benzoin (L.) **Bl.,** spicebush: An aqueous extract of the whole plant did not repel adult *Popillia japonica.*[126] A methanol extract of the twigs did not deter feeding by *Acalymma vittatum* adults.[52] An ether extract of the twigs deterred feeding by *Spodoptera frugiperda* larvae, but a methanol extract of the marc did not.[204] An ethanol extract of the leaves deterred feeding by *Lymantria dispar* larvae.[84]

Nectandra coriaceae **(Sev.) Griseb.,** laurel avispillo

Nectandra saligna **Neis.**

Nectandra sintanissi **Mez.,** laurel amarillo: The wood of these three species is susceptible to attack by *Cryptotermes brevis* termites.[86]

Nectandra whitei **(Woodson) C. K. Allen,** bambito Colorado: the heartwood from Panama was moderately resistant to termites.[371]

Ocotea acutangula **Mez.,** Louro Tamancao: The wood is resistant to *Cryptotermes brevis* termites.[86,124]

Ocotea arechavaletae **Mez.**

Ocotea canaliculata **(Richm.) Mez.,** white silverballi

Ocotea leucoxylon **(Sev.) Maza,** laurel geo

Ocotea moschata **(Meissn.) Mez.,** nuez moscada: The wood of these four species is highly susceptible to attack by *Cryptotermes brevis* termites.[86]

Ocotea ovalifolia **Mez.,** yellow silverballi

Ocotea rodiaei **Mez.,** greenheart

Ocotea rubra **Mez.,** louro vermelho

Ocotea usambarensis **Engl.,** East African camphorwood: The wood of these species is resistant to *Cryptotermes brevis* termites.[86,118] The wood of *O. usambarensis* is not repellent to *Tineola bisselliella* larvae.[370]

Ocotea whitei **R. E. Woodson:** An ether extract of the combined woody stems and bark completely deterred feeding by adult *Acalymma vittatum.*[89]

Parabenzoin praecox **Nakai:** A benzene extract of the leaves deterred feeding by larvae of *Spodoptera litura.* The active compounds were identified as (+)-eudesmin and (+)-epieudesmin.[672]

Parabenzoin tribolium **Nakai:** The leaves were not accepted as food by *Spodoptera littoralis* larvae. Three new sesquiterpenoids (shiromodiol, shiromodiol monoacetate, shiromodiol diacetate) were isolated and identified as the active compounds.[91] These compounds also inhibited feeding by *Trimeresia miranda.*

Persea americana **Mill.,** avocado: The wood is very susceptible to termite attack.[124] Avocado seeds are mashed, fried in oil, and applied to the head in the Dominican Republic as a shampoo to eliminate *Pediculus humanus capitis.*[673] *Bombyx mori* larvae feed on the leaves but their subsequent growth is inhibited due to the action of dimethyl sciadinonate and 1-acetoxy-2-hydroxy-4-oxoheneicosa-12,15-diene.[674,675] A methanol extract of the leaves killed *B. mori* larvae.[83] *Spodoptera littoralis* larvae that emerged from eggs laid on avocado leaves died within 48 h.[676]

Persea borbomia **Spreng:** An ethanol extract of the leaves did not deter feeding by *Lymantria dispar* larvae.[84]

Persea rigens **C. K. Allen,** pizarra: The heartwood was moderately resistant to termites.[371]

Pleurothyrium cuneifolium **Nees,** louro abacate: The wood is susceptible to attack by *Cryptotermes brevis* termites.[86]

Sassafrass albidum **(Nutt.) Nees,** sassafras: An ethanol extract of the leaves did not deter feeding by *Lymantria dispar* larvae.[84] Oil of sassafras plus petrolatum was considered to be one of the best repellents for *Cochliomyia hominivorax*. The powdered bark was effective for only 2 days.[677] A volatile oil consisting of 80% safrole, a little eugenol, pinene, phellandrene, a sesquiterpene, and (+)-camphor killed 100% of *Anopheles claviger* and *Aedes cantator* mosquito larvae in 24 h.[182]

Umbellularia californica (Hook. & Arn.) **Nutt.**, California laurel: The leaves appeared to be a good repellent for fleas.[678]

FAMILY LECYTHIDACEAE

Cariniana pyriformis Miers., albarco: The wood has little resistance to termites,[116] but only 23% of *Coptotermes formosanus* coming in contact with it survived.[369] The wood is somewhat resistant to *Cryptotermes brevis* termites.[86]

Couroupita amazonica Kunth.: An ether extract of the leaves deterred feeding by adult *Acalymma vittatum* at 0.5% but not at 0.1%.[89]

Couroupita guianensis Aubl., cannonball tree: Ether extracts of the leaves, fruits, stem bark, and flowers did not deter feeding by adult *Acalymma vittatum* at 0.1 or 0.5%.[89]

Eschweilera corrugata (Poit.) **Miers.**, wena kakeralli: The wood is susceptible to attack by *Cryptotermes brevis* termites.[86,124]

Eschweilera odora (Poeff.) **Miers.**

Eschweilera sagotiana Miers., black kakeralli: Of the two species, only the wood of *E. sagotiana* is resistant to *Cryptotermes brevis* termites.[86,124]

Lecythis paraensis (**Huber**) **Ducke**, monkeypot: The wood is resistant to *Cryptotermes brevis* termites.[86]

FAMILY LILIACEAE

Aloe barbadensis Mill., aloe: The juice pressed from the fresh leaves deterred feeding by *Athalia proxima*.[213] Aloin derived from this plant was not effective as a mothproofing agent.[456] An aqueous extract of the dry leaves was somewhat repellent to adult *Popillia japonica*.[126]

Asphodelus liburnica Scop.: An ethanol extract of the whole plant did not deter feeding by adult *Acalymma vittatum*.[52]

Colchicum autumnale L., autumn crocus, meadow saffron: An aqueous extract of the whole plant did not repel adult *Popillia japonica*.[126]

Convallaria majalis L., lily-of-the-valley: An aqueous extract of the whole plant did not repel adult *Popillia japonica*.[126] Filter papers impregnated with 0.5 or 3% of convallamaroside from the roots was toxic to *Reticulitermes flavipes* termites and prevented feeding.[229] Juice expressed from the whole fresh plant strongly deterred feeding by *Melanotus communis* larvae.[247]

Erythronium americanum Ker., trout-lily, deer's tongue: The powdered leaf was an excellent repellent for *Cochliomyia hominivorax*.[155] An aqueous extract of the whole plant was not repellent to adult *Popillia japonica*.[126]

Eucomis undulata Ait.: An alcohol extract of the whole plant did not repel *Anthonomus grandis grandis* adults.[487]

Helonias bullata L., swamp pink: An aqueous extract of the whole plant did not repel adult *Popillia japonica*.[126]

Hemerocallis dumortieri Morr.: Larvae of *Plutella xylostella* would not feed on this plant,[195] and an acetone extract of the whole plant deterred feeding by this insect.[25]

Hemerocallis fulva L., daylily: An aqueous extract of the whole plant did not repel adult *Popillia japonica*.[126] Juice expressed from the fresh leaves did not deter feeding by *Melanotus communis* larvae.[247]

Hyacinthus orientalis L., hyacinth: An aqueous extract of the whole plant did not repel adult *Popillia japonica*.[126]

Lilium gigantium Wall.

Lilium longiflorum Thunb., trumpet lily: *Plutella xylostella* larvae refused to feed on these plants.[195]

Lilium superbum L., turks-cap lily: An aqueous extract of the whole plant did not repel adult *Popillia japonica*.[126]

Maianthemum canadense Desf.:

Medeola virginiana L., cucumber root: Aqueous extracts of these plants failed to repel adult *Popillia japonica*.[126]

Ornithogalum umbellatum **L.,** star-of-Bethlehem:

Polygonatum biflorum *(Walt) Ell.,* hairy solomonseal:

Polygonatum commutalum *Dietz.,* great solomonseal: Aqueous extracts of these three species were not repellent to adult *Popillia japonica*.[126]

Schoenocaulon officinale *(Schlecht. & Cham.) A. Gray,* sabadilla: Used as a fumigant, the seed had a slight effect on *Musca domestica* and *Tineola bisselliella* and a considerable effect on mosquitoes (unidentified).[679] An aqueous extract of the seed was not repellent to adult *Popillia japonica*.[126]

Smilax aristolochiifolia Mill., sarsaparilla: Filter papers impregnated with 0.5 or 3% of parillin, a steroidal saponin from the roots, was toxic to *Reticulitermes flavipes* termites and prevented feeding; 0.05% was not effective. Sarsaparilloside from the roots was only weakly toxic at 3%.[229]

Smilax hispida *Muhl.:* An ethanol extract of the leaves did not deter feeding by *Lymantria dispar* larvae.[84]

Smilax rotundifolia **L.,** greenbrier, bullbrier: Ethereal and methanolic extracts of the twigs at 0.1 and 0.5% did not deter feeding by adult *Acalymma vittatum*.[52] Aqueous extracts of the whole plant did not repel adult *Popillia japonica*.[126]

Trillium erectum **L.,** purple trillium: An aqueous extract of the dry rhizomes and roots was somewhat repellent to adult *Popillia japonica*.[126]

Tulipa gesneriana **L.,** tulip: An aqueous extract of the whole plant did not repel adult *Popillia japonica*.[126]

Urginea maritima **(L.)** Baker, red squill, sea onion: The Arabs in Northern Africa were reported to repel mosquitoes with 1 or 2 bulbs of this plant.[680] Aqueous extracts of the bulbs did not repel adult *Popillia japonica*[126] or *Spodoptera litura* larvae.[681]

Uvularia perfoliata **L.,** wood merrybells: An aqueous extract of the whole plant did not repel adult *Popillia japonica*.[126]

Veratrum album **L.,** European white hellebore: A hot water decoction of the rhizomes, applied to the backs of cattle, was very effective in repelling warble-fly larvae[682] and mange mites.[683]

FAMILY LINACEAE

Linum ositatissimum **L.,** flax, linseed: Coating grain with linseed oil did not protect it from weevil attack.[116] Linseed oil is repellent to *Apis florea* bees.[149,475] Only 15.5% of the eggs laid on flax by *Melanoplus mexicanus mexicanus* survived.[684]

FAMILY LOASACEAE

Mentzelia decapitata *(Pursh.) Urban & Gilg.:* An ethanol extract of the foliage failed to deter feeding by *Lymantria dispar* larvae.[84]

FAMILY LOGANIACEAE

Fagraea fragrans *Roxb.,* tembusu:

Fagraea gigantea *Ridley:* The wood of these species is resistant to termites.[480]

Strychnos ignatii *Bergius,* ignatia: A mothproofing composition containing an alkaloid or alkaloids from the seeds was claimed in a British patent.[208] An aqueous extract of the dry beans did not repel adult *Popillia japonica*.[126]

Strychnos nux-vomica **L.,** strychnine tree, nux vomica: Acidulated aqueous and alcoholic extracts of the seeds did not protect timbers from termites.[120] An aqeous extract of the dry beans did not repel adult *Popillia japonica*.[126] A 1% solution of the alkaloid, brucine, from the seeds prevented termite attack on the wood for only 1 month.[118] Steam extracts of the plant had a juvenomimetic effect when applied topically to the abdominal tergites of *Dysderous similis, Spodoptera litura, Musca domestica,* and *Anopheles stephensi* larvae and a gonadotrophic effect on the females with the exception of *A. stephensi*.[685] An acetone extract of the stems at 0.5%, prepared as a dry film in Petri dishes in the laboratory or as a spray in the field was somewhat toxic to *Bagrada cruciferarum*.[478]

FAMILY LORANTHACEAE

Phoradendron trinervium Nutt.: An ether extract of the combined leaves and stems completely deterred feeding by *Acalymma vittatum* larvae at 0.5%; the effect was less severe at 0.1%.[89]

Viscum album L., mistletoe: An aqueous extract of the foliage was a moderate deterrent to feeding by *Pieris brassicae* larvae.[652]

FAMILY LYTHRACEAE

Lagerstroemia indica L.: An ethanol extract of the leaves did not deter feeding by *Lymantria dispar* larvae.[84]

FAMILY MAGNOLIACEAE

Kadsura japonica Don.: A methanol extract of the leaves prevented feeding by *Bombyx mori* larvae, all of which died from exposure to the extract.[82]

Liriodendron tulipifera L., tulip tree, yellow poplar: An ethanol extract of the foliage deterred feeding by *Lymantria dispar* larvae.[84] The sesquiterpene lactones, peroxyferolide and tulirinirol are responsible for this activity.[686,687] An ether extract of the twigs deterred feeding by *Spodoptera frugiperda* larvae, but a methanol extract of the marc did not.[204] Aqeous extracts of various parts of this tree did not repel adult *Popillia japonica*.[126] The wood is susceptible to attack by *Cryptotermes brevis* termites,[86,124] but it is resistant to *C. formosanus* and *Reticulitermes virginicus*.[87]

Magnolia grandiflora L., southern magnolia: An ethanol extract of the leaves collected in Florida deterred feeding by *Lymantria dispar* larvae, but extracts of the leaves collected in Georgia and California did not.[84]

Magnolia kobus DC.: A methanol extract of the leaves did not repel *Bombyx mori* larvae.[82]

Magnolia portoricensis Bello, mauricio, burro:

Magnolia splendens Urban: The wood of these species is susceptible to attack by *Cryptotermes brevis* termites.[86,124]

Magnolia virginiana L., sweetbay magnolia: An aqueous extract of the fresh leaves was somewhat repellent to adult *Popillia japonica*. An ethanol extract of the leaves collected in North Carolina deterred feeding by *Lymantria dispar* larvae, but an extract of the leaves collected in Florida did not.[84] An ether extract of the combined leaves and twigs deterred feeding by *Spodoptera frugiperda* larvae, but a methanol extract of the marc did not.[204]

FAMILY MALPIGHIACEAE

Byrsonima spicata (Cav.) DC: Although Wolcott[124] reported that the wood is very resistant to termites (unidentified), it has been shown to be highly susceptible to attack by *Cryptotermes brevis*.[86,370]

FAMILY MALVACEAE

Abelmoschus esculentis (L.) Moench., okra: The juice expressed from fresh okra seedlings did not deter feeding by *Melanoplus communis* larvae.[247]

Abutilon indicum (L.) Sweet: A petroleum ether extract of the combined stems and leaves had no effect on *Dysdercus cingulatus* nymphs.[109]

Abutilon pictum Walp.: *Plutella xylostella* larvae refused to feed on this plant.[195]

Gossypium hirsutum L., cotton: It has been known for many years that some varieties of cotton are more resistant to insects than other varieties. Hairless varieties are less subject to attack by *Heliothis* pests and leafhoppers.[688] The chemical resistance to several insect species was at first ascribed to the presence of gossypol in discrete glands throughout the plant, but it was subsequently found that this compound is not the main determinant *Heliothis* factor and that other factors, both chemical and physical, play important roles in resistance.[4,313,315] Condensed tannins appear to play a considerable role in determining resistance.[1,689,690] A fraction was isolated from cotton squares that repelled 100% of the adults of *Anthonomus grandis grandis* for 5 h.[691] The responsible component was identified as gossypol.[23] Cottonseed oil at 0.3% protects stored greengram from attack by *Callosobruchus maculatus* for 5 to 6 months.[482] Stored bean seeds can be protected from attack by *Zabrotes subfasciatus* with 1 ml of cottonseed oil/kg seed.[236] Acids isolated from the oil also repel adult *C. chinensis* and prevent female oviposition by mixing with the fraction (400 g/ton stored seed);[291] this procedure also prevents oviposirion by *Calandra oryzae.*[692] Little or no feeding was shown by adult *Schistocerca gregaria* and *Melanoplus sanguinipes* exposed to filter papers treated with the cottonseed oil.[238] The feeding deterrence of gossypol, as well as its toxicity to insects have been discussed in detail for *Spodoptera littoralis,*[693-695] *Earias insulana,*[696,697] *Pectinophora gossypiella,*[698,699] *Heliothis virescens,*[699,700] and many other insect species.[27] Other compounds isolated from resistant varieties of cotton are p-hemigossypolone (inhibits the growth of *Heliothis virescens*),[701] helioside H_1 (inhibits the growth of and is toxic to *H. virescens*),[702] heliocide H_2 (toxic to *H. virescens* and *H. zea*),[703] heliocide H_3 (toxic to *Heliothis* spp.),[704] and cyanidin-3β-glucoside (resistant to *H. virescens*).[705] Also, important in the composition of resistance factors is cotton condensed tannin, which is effective against *H. zea*[706,707] and *Leptinotarsa decemlineata.*[608] The tannin is also an antibiotic for *H. virescens.*[708] Other plant phenols affect the performance of pest insect larvae such as *Spodoptera eridania,*[709] as do flavonoids inhibiting larval growth in *H. zea.*[7710] Gossypol has been shown to have antifertility properties in male mondeys.[711]

***Hibiscus rosa-sinensis* L.,** Chinese hibiscus, Hawaiian hibiscus: Juice expressed from the fresh leaves failed to deter feeding by *Melanotus communis.*[247] Extracts prepared from the flowers caused 98% mortality of *Sitophilus oryzae* coming in contact with treated filter papers.[251,252] Topical application of an acetone extract of the stems to the abdominal tergites of adult female *Dysdercus cingulatus* resulted in the absence of yolk deposition in the excised ovaries.[123]

***Hibiscus syriacus* L.,** shrub althea, Rose-of-Sharon: Larvae of *Plutella xylostella* will not feed on this plant,[195] nor will *Anthonomus grandis grandis* unless the clyx is removed.[712] Fatty acids and methyl esters isolated from the calyx tissue are responsible for feeding and ovipositional deterrency to *A. grandis grandis,*[713,714] and it is possible that these are also responsible for the deterrency of *P. xylostella.* The compounds isolated and identified are pelargonic acid, myristic acid, palmitic acid, methyl palmitate, methyl linoleate, methyl stearate, and methly linolenate.[714]

***Hibiscus tiliaceus* L.,** sea hibiscus, mahoe, majagua: The wood is susceptible to attack by *Cryptotermes brevis* termites.[86]

***Hibiscus trionum* L.,** Venice mallow, flower-of-an-hour: An aqueous extract of the whole plant did not repel adult *Popillia japonica.*[126]

***Malva aegyptia* L.:** An ethanol extract of the whole plant did not deter feeding by *Acalymma vittatum* adults.[52]

***Montezuma speciosissima* DC.:** The wood is very resistant to termites.[124]

***Thespesia populmea* (L.)** *Soland ex Correa,* tulip, portia: The timber is resistant to termites.[117]

FAMILY MARANTACEAE

***Maranta arundinacea* L.,** arrowroot: Stored products treated with 0.25g of powdered arrowroot/100g of produce were resistant to feeding by *Sitophilus oryzae, Tribolium castaneum, Trogoderma granarium, Stegonium paniceum, Rhyzopertha dominica,* and *Callosobruchus chinensis.*[715]

FAMILY MELASTOMATACEAE

***Dactylocladus stenostachys* Oliv.,** jongkong: The wood is susceptible to termite attack.[485]

***Miconia eichlerii* Cogn.:**

***Miconia humilis* Cogn.:** Ether extracts of the

stems of thses plants did not deter feeding by adult *Acalymma vittatum* at 0.1% and 0.5%.[89] The quinone, primin, and its quinol, miconidin, isolated from the leaves were strongly deterrent to feeding by larvae of *Heliothis armigera, Pieris brassicae, Spodoptera exempta,* and *S. littoralis,* and to adult *Melanoplus sanguinipes* and *Schistocerca gregaria.*[716]

Tetrazygia bicolor (Mill.) Cogn.: An ethanol extract of the fresh leaves did not deter feeding by *Lymantria dispar* larvae.[84]

FAMILY MELIACEAE

Aglaia cordata Hiem.

Aglaia odoratissima Blume.: In a no-choice test, an ethanolic extract of the seeds of *A. odoratissima* deterred feeding by *Spodoptera frugiperda* larvae but an extract of *A. cordata* did not. Both extracts deterred feeding at 0.5% concentration by *Acalymma vittatum* adults.[717] Roxburghilin, a *bis*-amide of 2-aminopyrrolidine isolated from the leaves of *A. roxburghia,*[718] may be responsible for this activity. An acetone extract of the combined leaves and twigs of *A. odoratissima* prevented feeding by *Pieris rapae* and *Mythimna separata* larvae.[719]

Amoora ruhituka Wright & Arn., pithra: Extracts prepared from the dried leaves and seeds with hexane, ether, and ethanol significantly deterred feeding by *Nilaparvata lugens, Nephotettix nigropictus,* and *Dicladispa armigera* adults and larvae on rice seedlings.[131]

Aphanamixis grandifolia Bl.

Aphanamixis polystachya Wall & Parker, tiktara

Aphanamixis sinensis Bl.: An acetone extract of *A. polystachya* combined leaves, bark, and fruits prevented feeding by larvae of *Pieris rapae, Mythimna separata,* and *Spodoptera litura,* and an extract (acetone) of *A. sinensis* twigs prevented feeding by *S. litura.*[718] Aphanin, a tetracyclic triterpene hemizcetal isolated from the defatted chloroform extract of *A. polystachya* fruits, is believed to possess insect feeding inhibitory properties.[720]

Azadirachta indica A. Juss., (synonym **Melia azadirachta**), neem, nim, margosa: Leaves, fruits, and, especially, the seeds of this subtropical and tropical tree have been shown to possess repellency, feeding deterrency, and toxicity to a large variety of insects. On the basis of its pesticidal, fertilizer, antimicrobial, and numerous other uses this tree is probably the most versatile in the botanical kingdom at the present time. The oil cake of the seeds has been used in India as a fertilizer in numerous fields to control termites, but with indifferent success.[721] The wood is reported to be resistant to *Anacanthotermes ochraceus, Psammotermes fuscofemoralis,* and *P. assuarensis.*[387] The numerous reports of the tree's pesticidal properties through 1971 have been cited by McIndoo[24] and Jacobson.[26,27] Responsibility for the extensive antifeedant and repellent properties resides in the numerous compounds it contains, such as azadirachtin, meliantriol, and salannin; satisfactory reviews of the subject are those by Van Beek and De Groot,[722] Koul,[723] Das.[724] Arthropods deterred from feeding by the use of various extracts and constituents of neem are shown in Table 1. Although more than 65 different terpenoidal compounds have been isolated from various parts of the neem tree,[842] only a few of these have thus far been tested for antifeedant or toxic effects. Various parts of the tree deter oviposition of insects. Prime examples are *Lucilia cuprina,*[843] *Spodoptera litura,*[844] and *Tetranychus cinnabarinus.*[726] Neem disrupts normal growth and development in *Aedes aegypti,*[845] *Anopheles gambia,*[846] *Apis mellifera,*[847] *Athalia proxima,*[848] *Blattella germanica,*[845] *Calliphora vicina,*[849] *Callosobruchus analis,*[845] *Crocidolomia binotalis,*[848,850] *Epilachna varivestis,*[851-855] *Heliothis armigera,*[856,857] *H. zea,*[859] *Locusta migratoria migratorioides,*[860,861] *Manduca sexta,*[862,863] *Musca domestica,*[845] *Mythimna separata,*[864] *Oncopeltus fasciatus,*[43,865] *Ostrinia nubilalis,*[866] *Phormia terrae-novae,*[777] *Plutella xylostella,*[818] *Popillia japonica,*[755] *Spodoptera frugiperda,*[859] *S. littoralis,*[45] and *S. litura.*[717,867,868] Compounds of major interest isolated from neem seeds or leaves are azadirachtin,[869] 17-epiazadiradione and 17β-hydroxyazadiradione,[870] salannolide,[871] margosinolide and isomargosinolide,[872] gedunin,[873] isomeric azadirachtins,[874] numbocinone,[875] gedunin,[813] nimocinol,[876] and additional triterpenoids.[877-879] In addition, an isoprenylated flavanone has been isolated from neem leaves.[880] Recent structural studies for many neem compounds have been reported.[45,881-895] A simplified isolation procedure for azadirachtin has very recently been reported which gives this compound in crystalline form for the first time.[896]

Carapa guianensis Aubl., crabwood, cedro macho:

Carapa nicaraguensis C. DC., cedro macho:

Carapa procera DC., monkey cola, crabwood: The wood of these three species is susceptible to attack by *Cryptotermes brevis* termites.[86,115] An ethanol extract of *C. procera* seeds was susceptible to feeding by *Spodoptera frugiperda* larvae but it deterred feeding by *Acalymma vittatum* adults.[717] Ether extracts of *C. guianensis* seeds were somewhat deterrent to feeding by *Acalymma vittatum* adults at 0.1% and highly deterrent at 0.5%.[51] Extracts of the roots, twigs, stems, and leaves were highly deterrent at both concentrations. Several nortriterpenoid compounds, designated carapolides, isolated from this species[897] may be responsible for this activity. A hexane extract of *C. Procere* seeds yielded a mexicanolide limonoid deterrent to feeding by *Spodoptera frugiperda* larvae.[899]

Cedrela fissilis Vell., caroba, South American cedar:

Cedrela mexicana M. Roem., Spanish cedar: The wood of these two species is susceptible to attack by *Cryptotermes brevis* termites.[86]

Cedrela odorata L., Spanish cedar, cigar-box cedar: The wood is resistant to *Cryptotermes brevis* and *Coptotermes formosanus* termites.[86,124] An ether extract of the leaves deterred feeding by *Acalymma vittatum* adults.[51]

Cedrela toona Roxb., cedro toona: The wood is resistant to attack by *Cryptotermes brevis* termites.[86,116,188] Ether extracts of the twigs, leaves, and fruits were highly deterrent to

feeding by *Acalymma vittatum* adults in both choice and no-choice tests.[51] Cedrelone, a limonoid isolated from this plant, deters feeding by *Spodoptera litura* larvae.[900]

Chikrassia tabularis A. Juss., chittagong wood: The wood is resistant to attack by *Anacanthotermes ochraceus, Psammotermes fuscofemoralis*, and *P. assuarensis*.[387] An ethanol extract of the wood deterred feeding by *Acalymma vittatum* and *Diabrotica undecimpunctata howardi* adults,[52] and an ethanol extract of the seeds deterred feeding by *Spodoptera frugiperda* larvae and *Acalymma vittatum* adults.[717]

Chloroxylon swietenia DC, Ceylon satinwood: The wood is completely resistant to attack by *Cryptotermes brevis, Anacanthotermes ochraceus, Psammotermes fuscofemoralis*, and *P. assuarensis* termites.[387]

Dysoxylum binectariferum (Roxb.) hbok & Bedd.:

Dysoxylum malaaricum Bedd. & ex C. DC.:

Dysoxylum reticulatum King:

Dysoxylum spectabilis (Forst. F.) Hook f.: Ethanol extracts of the seeds of these four species deterred feeding by *Spodoptera frugiperda* larvae and *Acalymma vittatum* adults.[717]

Entandrophragma angolense (Welw.) C. DC., budongo mahogany:

Entandrophragma cylindricum Sprague, sapela

Entandrophragma uile (Dawe & Sprague) Sprague, utile: Of these three species, the wood of only *E. utile* is susceptible to termite attack.[370]

Guarea cedrata (A Chev.) Pellegr.: The wood is resistant to *Coptotermes formosanus* termites.[388] Two 3,4-secotirucallane derivatives and 2′-hydroxyrohitukin, isolated from the bark,[901] may be responsible for this activity.

Guarea longipetiola C. DC.: Pentane and acetone extracts of the wood deterred feeding by termites.[369]

Khaya anthotheca (Welw.) C. DC.: The wood is resistant to 21 species of termites but is susceptible to attack by *Hypsipyla grandella*.[119]

Khaya ivorensis Chev., African mahogany:

Khaya senegalensis (Desv.) A. Juss., dry-zone mahogany: The wood of these species is resistant to *Cryptotermes brevis* termites,[86,119] subterranean termites,[607] and *Hypsipyla grandella*.[902]

Lansium domesticum Corr., langsat: An ethanol extract of the seeds deterred feeding by *Spodoptera frugiperda* larvae but not by *Acalymma vittatum* adults.[717]

Lovoa trichiloides Harms., tigerwood: The wood is resistant to attack by *Coptotermes formosanus* termites.[388]

Melia azedarach L., chinaberry, Indian lilac: The wood is resistant to attack by *Coptotermes formosanus* termites.[903] Application of the leaves to the soil in wheat plots at 7 tons/acre reduced termite attack considerably.[904] However, the wood is susceptible to attack by *Cryptotermes brevis* termites.[86,124] The leaves deter feeding by *Ephestia ellutella*,[116] *Heliothis zea*,[905] and *Spodoptera frugiperda* larvae,[905] as well as by adults and nymphs of *Schistocerca canellata*,[906,907] and *S. gregaria*.[839] The active component was isolated and identified as meliantriol.[839] A methanol extract of the leaves deterred feeding by larvae of *Epilachna varivestis*,[745] *Plutella xylostella*,[817] and *Piesma quadratum*.[817] An ethanol extract of the fresh leaves deterred feeding by *Lymantria dispar* larvae.[84] The active component was isolated and identified as meliantriol.[839] A methanol extract of the seed kernels deterred feeding by *Pieris rapae*,[816] *Spodoptera litura*,[719,908] *Panonychus citri*,[719,909] *Mythimna separata*,[719] *Scirpophaga incertula*,[719] *Nilaparvata lugens*,[719,909] *Diaphorina citri*,[909] *Sogatella furcifera*,[784] *Nephotettix virescens*,[184] *Dicladispa armigera*,[132] and adult *Callosobruchus chinensis*.[132] Emulsified chinaberry seed oil was effective in the field as a feeding deterrent for *Panonychus citri* and *Aleurocanthus spiniferus*.[903] The seed oil also had a marked antifeedant effect on the larvae of *Spodoptera abyssina*.[910] Various parts of the chinaberry tree, especially the seeds, have yielded a number of triterpenoidal compounds.[911-916]

Melia dubia Cav.: Methanol extracts of the seeds deterred feeding by larvae of *Spodop-*

tera frugiperda and adult *Acalymma vittatum*.[717] The antifeedant salannin and several other tetranortriterpenoids have been isolated from the seeds, leaves, and fruits.[717,917,918]

Melia toosendan Sieb. & Zucc.: A methanol extract of the seeds deterred feeding by larvae of *Spodoptera frugiperda* and adult *Acalymma vittatum*.[717] The seed oil deterred feeding by *S. litura*,[783] *Scirpophaga incertula*,[903] *Pieris rapae*,[816] and *Mythimna separata*.[919] The components responsible for feeding deterrency are toosendanin[903] and toosendan sterols.[920] The seed kernel oil was repellent to *Panonychus citri*.[719]

Melia volkensii (Guerke): The fruits inhibit feeding by *Schistocerca gregaria* nymphs and adults,[919] and a methanol-water (8:2) extract of the kernels repels and kills *Aedes aegypti* larvae.[921] The limonoids volkensin and salannin isolated from the fruits are feeding deterrents for *Spodoptera frugiperda* larvae.[921]

Sandoricum evetjape (Burm. F.) Merr.: A methanol extract of the seeds deterred feeding by *Spodoptera litura* larvae and *Acalymma vittatum* adults.[717]

Swietenia macrophylla King

Swietenia mahagoni (L.) Jacq., West Indian mahogany: The heartwood of *S. mahagoni* is very resistant to *Cryptotermes brevis* termites, but the heartwood of *S. macrophylla* is susceptible to this species and to *Reticulitermes flavipes*.[124,462] A methanol extract of the seeds showed both species to be deterrent to feeding by *Spodoptera litura* larvae and *Acalymma vittatum* adults.[717] The tetranortriterpenoids, swietenine and swietenolide, isolated from the seeds of *S. macrophylla*, are probably responsible for the feeding deterrency of this tree.[92] The limonoid, 2-hydroxyswietenin, may also play a role here.[923] Ether extracts of the fruit, seed, and bark of *S. mahagoni* were toxic to *Drosophila melanogaster, Nasutitermes costalis* termites, and guppies *(Lebistes reticulatus)*.[924]

Teclea trichocarpa (Engl.) Engl.: Three 9-acridone alkaloids (melicopicine, tecleanthine, and 6-methoxytecleanthine) isolated from the bark of this tree deter feeding by *Spodoptera litura* larvae.[925]

***Toona sinensis** A.Juss.:* A methanol extract of the seeds deterred feeding by *Spodoptera litura* larvae and *Acalymma vittatum* adults.[717]

Toona sureni (Blume.) Merrill: The wood is resis tant to termites.[903] Toonacilin, a triterpenoid isolated from the leaves, is responsible for the antifeedant effect on *Epilachna varivestis* adults.[926]

***Trichilia hirta** L.,* broomstick: The heartwood is very resistant to attack by *Cryptotermes brevis* termites.[927]

***Trichilia pallida** Sev.,* caracolillo: The wood is repellent and very resistant to *Cryptotermes brevis* termites.[927]

Trichilia roka (Forsk.) Chiov.: This tree is resistant to insect attack.[8] A new limonoid, trichilin, was isolated from the root bark and two limonoids, sendanin and trichirokanin, were isolated from the fresh fruit.[810] Trichilins are effective against *Spodoptera eridania* and *Epilachna varivestis,* and sendanin and trichirokanin are effective against *Pectinophora gossypiella, S. frugiperda, Heliothis zea,* and *H. virescens* larvae.[717,928] An ether extract of the root bark yielded a limonoid, 7-acetyltrichilin A, deterrent to feeding by these insect species.[929]

***Xylocarpus molluscensis** Roem.:* The secoiridoid, xylomollin, isolated from the ripe fruit is an antifeedant for *Spodoptera exempta* larvae.[300,390]

FAMILY MELIANTHACEAE

***Bersama abyssinica** Fresen.:* The compound, abyssinin, isolated from the root bark is a potent antifeedant for *Heliothis zea* larvae.[931,932]

FAMILY MENISPERMACEAE

***Anamirta cocculus** (L.) Wight & Arn,* fishberry: An ether extract of the fruits did not permit feeding by *Acalymma vittatum* adults.[89] The defatted seeds were extracted with ethanol to obtain picrotoxin,[923] an equimolar mixture of picrotoxinin and picrotin.[934] Picrotoxinin is highly toxic to *Musca domestica,*[935] but its activity as an antifeedant is not known.

***Cocculus trilobus** DC.:* Due to their dontent of isoboldine, the leaves are not palatable to the larvae of *Spodoptera littoralis* and *Trimeresia miranda.*[91,936]

FAMILY MORACEAE

***Antiaris toxicaria** Lesch.:* An ether extract of the stems did not deter feeding by *Acalymma vittatum* adults.[38]

Artocarpus altilis (Parkins.) Fosb., breadfruit: The odoriferous flowers of the male trees were placed among clothes in the Dutch East Indies to repel clothes moths.[937] The wood is susceptilbe to attack by *Cryptotermes brevis* termites.[86]

***Artocarpus heterophyllus** Lam.,* jackfruit, jack: The wood is resistant to *Anacanthotermes ochraceus, Psammotermes fuscofemoralis,* and *P. assuarensis* termites.[387]

***Artocarpus incisa** L.:* An ethanol extract of the fresh leaves deterred feeding by *Lymantria dispar* larvae.[84]

***Brocimum paraense** Huber,* cardinalwood: The wood is highly resistant to *Reticulitermes flavipes* and *Cryptotermes brevis* termites.[86,548]

***Cannabis sativa** L.,* hemp, marijuana: Leaves of this plant scattered among bags of stored products had little effect on *Sitophilus cryzae* larvae[938] but did affect weevils.[214,939] An acetone extract of the whole plant repelled *Popillia japonica* adults,[126] and *Bombyx mori* larvae refused to feed on the leaves.[327] The leaves or the whole plant scattered under the bedsheet kept bugs and fleas away in India, and an alcoholic extract was markedly repellent to *Aedes aegypti* mosquitoes.[79] Larvae of the ticks *Ixodes redikorzevi, Haemaphysalis punctata, Rhipicephalus rossicus,* and *Dermacentor marginatus* exposed to the powered leaf quickly died.[153]

Chlorophora excelsa (Welw.) Bentham & Hook f., African teak, iroko: The wood is highly repellent to *Cryptotermes brevis* and *Coptotermes formosanus* termites.[86,388] Immersion of flamboyan wood in a 0.2% solution of chlorophorin from this tree resisted termite attack for 234 d; at 0.5% the protection lasted at least 3 years.[940] Chlorophorin is also effective against *Reticulitermes lucifugus* termites.[941]

Chlorophora tinctoria (L.) *Candich. ex Benth. & Hookf.*, fustic-mulberry, Dyer's mulberry: A commercial extract of the whole plant repelled *Popillia japonica* adults. The tree is practically immune from attack by *Cryptotermes brevis* termites,[86] and an ethanol extract of the wood killed *Reticulitermes flavipes* termites.[368]

Dorstenia contrajerva L.: *Bombyx mori* did not feed on the leaves of this plant.[32]

Ficus aurea Nutt.

Ficus carica L., fig: An ethanol extract of the fresh leaves of *F. carica* deterred feeding by *Lymantria dispar* larvae, but an extract of *F. aurea* did not.[84] A benzene extract of the whole fig plant completely deterred feeding by *Spodoptera litura* larvae.[258]

Ficus elastica Roxb. ex Hornem, rubber plant: The wood is very susceptible to attack by *Cryptotermes brevis* termites.[86,124] *Bombyx mori* larvae fed on this plant but did not survive.[327]

Ficus erecta Thunb.

Ficus hirta Vahl.: *Bombyx mori* larvae fed on the leaves of these plants but did not survive.[327] An acetone extract of *F. hirta* did not deter feeding by *Spilosoma obliqua* larvae.[194]

Ficus laevigata Vahl., wild fig

Ficus pumila L., creeping fig: The wood of these species is attacked by *Cryptotermes brevis* termites.[86] *Bombyx mori* larvae fed on the leaves of *F. pumila* but did not survive.[327]

Ficus retusa L., Indian laurel fig: The wood is attacked by *Cryptotermes brevis* termites.[86] *Bombyx mori* larvae fed on the leaves but did not survive.[327]

Ficus rostrata Lam.: An acetone extract of the whole plant offered some protection from *Spilosoma obliqua* larvae.[194]

Ficus sycamorus L., sycamore: Stakes of the wood buried in Egyptian soil withstood attack by *Anacanthotermes ochraceus, Psammotermes fuscofemoralis,* and *P. assuarensis* termites for 6 months.[387]

Ficus wightiana Wall.: *Bombyx mori* larvae did not feed on the leaves of this plant.[327]

Humulus japonicus Sieb & Zucc.

Humulus lupulus L., hops: *Bombyx mori* larvae refused to feed on the leaves of these plants.[327] Lupulin powder repelled *Cochliomyia hominivorax* larvae for 3 d,[155] but extracts of the hops did not repel *Popillia japonica* adults.[126] Hop oil did not repel *Blattella germanica* adults.[61]

Maclura aurantiaca Nutt.: *Bombyx mori* larvae fed on the leaves and survived.[327]

Maclura pomifera (Raf.) Schneid., osage orange: An ethanol extract of the fresh leaves deterred feeding by *Lymantria dispar* larvae.[84] The roots, wood, and bark are reported to repel insects.[677] The tree is practically immune from attack by *Cryptotermes brevis* termites.[86,462] Pomiferin, a constituent of the plant, at 0.01-0.05% protected *Poinciana regia* wood from bark beetles for almost a year; derivatives of this compound were less effective.[942] Solutions of osagin were somewhat repellent to termites, but pomiferin was much more effective.[940] A methanol extract of the plant did not repel *Culex* or *Aedes* mosquitoes.[245] Lupeol was isolated from the fresh fruit.[943]

Morus alba L., white mulberry: *Bombyx mori* larvae fed readily on this plant.[327] Methyl caffeate, ethyl β-resorcylate, and 5,7-dihydroxychromone, isolated from diseased leaves of this tree, were tested against *B. mori* larvae; only 5,7-dihydroxychromone deterred feeding.[944] Mulberry stakes buried in Egyptian soil withstood attack by *Anacanthotermes ochraceus, Psammotermes fuscofemoralis,* and *P. assuarensis* termites for 6 months.[387] An ethanol extract of the fresh leaves did not deter feeding by *Lymantria dispar* larvae. An acetone extract of the stems applied topically to allatectomized *Dysdercus cingulatus* female adults elicited juvenomimetic activity.[123]

Morus mesozygia Stapf., wonton: *Coptotermes formosanus* termites did not feed on the wood of this tree.[388]

Morus rubra L., red mulberry: An ethanol extract of the fresh leaves did not deter feeding by *Lymantria dispar* larvae.[84] The wood is susceptible to attack by *Cryptotermes brevis* termites.[86]

***Musanga smithii* R. Brown:** Wood shavings of this tree were neither repellent nor resistant to *Reticulitermes lucifugus* termites.[187]

***Myrianthus arboreus* P. Beauv.:** Wood shavings from this plant were resistant and repellent to *Reticulitermes lucifugus* termites.[187]

***Sorocea ilicifolia* Miq.:** The wood is susceptible to *Cryptotermes brevis* termites.[86]

***Trymatococcus amazonicus* R. & E.:** An ether extract of the combined twigs and leaves did not deter feeding by *Acalymma vittatum* adults, but an extract of the roots deterred feeding.[89]

FAMILY MUSACEAE

***Heliconia bihai* L.,** plantain, oja de fopocho: In laboratory tests, the leaves did not protect sugar from ants or flies, and they were only slightly repellent to cockroaches. A petroleum ether extract and the combined ether, chloroform, and alcohol extracts of the leaves were not repellent to ants, cockroaches, *Aedes* and *Anopheles* mosquitoes, and *Tribolium confusum* adults. The combined extracts were neither repellent nor toxic to *Attagenus megatoma* larvae.[26]

***Musa paradisiaca* L.,** banana: Banana leaves did not repel ants and were only slightly repellent to *Blattella germanica* adults.[26] Banana oil was attractive to *Blatta orientalis*.[162]

FAMILY MYRICACEAE

***Myrica carolinensis* Mill.,** northern bayberry: An acetone extract of the whole plant did not repel *Popillia japonica* adults.[126]

***Myrica cerifera* L.,** southern waxmyrtle: The Welsh people laid the branches of this plant upon and under the beds to repel fleas and moths.[383] An acetone extract of the whole plant did not repel *Popillia japonica* adults,[126] and an ethanol extract of the fresh leaves did not deter feeding by *Lymantria dispar* larvae.[84]

***Myrica rubra* S. & Z.:** A methanol extract of

the leaves did not affect *Bombyx mori* larvae.[82]

FAMILY MYRISTICACEAE

***Myristica fragrans* Houtt.,** nutmeg, mace: Nutmeg oil was considered to be one of the best repellents for *Cochliomyia hominivorax* larvae,[155] but it was attractive to cockroaches.[162] Thirteen phenylpropenoids isolated from nutmeg inhibited the growth of *Bombyx mori* larvae.[945]

***Pycnanthus angolensis* (Welw.) Exell.,** ilomba: The wood is reported to be susceptible to termite attack in Nigeria and the Gold Coast.[370]

***Staudtia stipitata* Warb.,** niove: A chloroform extract of the wood was only slightly repellent to *Reticulitermes lucifugus* termites, but a crystalline substance isolated from the wood was quite repellent and the ethereal oil of the wood was very effective.[941]

***Virola calophylla* (Spruce) Warb.,** virola: A methylene chloride extract of the wood from Peru gave five compounds with antifeedant activity against *Anthonomus grandis grandis* adults. These compounds were nicotine, β-sitosterol, friedelan-3-one, 5-methoxy-*N,N*-dimethyltryptamine, and 2-methyl-6-methoxytetrahydro-β-carboline.[946]

***Virola koschnyi* Warb.,** tapsava, banak: The wood is prone to attack by termites in Central America.[370]

***Virola mycetia* Pulla.,** Dutch mahogany: The wood is susceptible to attack by *Cryptotermes brevis* termites.[86]

FAMILY MYRSINACEAE

***Embelia ribes* Burm.:** The powdered leaves of this plant mixed with stored wheat inhibited infestation by *Corcyra cephalonica* larvae.[220]

***Myrsine africana* Linn.:** An ether extract of the combined roots, stems, leaves, and fruits did not deter feeding by adult *Acalymma vittatum*.[89]

***Rapanea ferruginea* Meq.:** Ether extracts of the stem bark and of the twigs did not deter feeding by adult *Acalymma vittatum*.[89]

FAMILY MYRTACEAE

Anomis grisea (Kiaersk.), limoncillo, pimienta: The wood is resistant to termites.[124]
Backhousia myrtifolia Hook.& Harv.: The essential oil was both repellent and toxic to *Aedes* and *Anopheles* mosquitoes, but it did not repel bush flies. The active constituent is probably elemicin,[342] which is present in the oil to the extent of 75-80%.[180]
Callistemon citrinus (Curt.) Stapf., crimson bottlebrush: An ethanol extract of the fresh leaves failed to deter feeding by *Lymantria dispar* larvae.[84]
Callistemon lanceolatum DC., bottlebrush: The essential oil effectively repelled *Apis florea* bees.[475] The effective compound is eugenol.[149] The oil has been shown to cause juvenilization in *Dysdercus koenigii* when applied topically to the last instar larva.[947]
Eucalyptus camaldulensis Dehnhardt: An ethanol extract of the fresh leaves did not deter feeding by *Lymantria dispar* larvae.[84] The wood is resistant to termites but susceptible to attacks by *Phoracantha semipunctata*.[119]
Eucalyptus camphora R. T. Baker: An ethanol extract of the fresh leaves deterred feeding by *Lymantria dispar* larvae.[84]
Eucalyptus citriodora Hook.: The wood is highly susceptible to termite attack.[86,124] The essential oil did not repel *Lucilia cuprina* but was attractive to it.[322]
Eucalyptus cladocalyx F. v. M., sugar gum
Eucalyptus cloeziana F. v. M., Queensland messmate
Eucalyptus consideniana Maiden, New South Wales messmate
Eucalyptus crebra F. v. M., narrow-leaved houbark
Eucalyptus deglupta Bl., murdanao gum
Eucalyptus diversicolor F. v. M., karri: The wood of these six species is resistant to termites.[188]
Eucalyptus dives Schauer: The essential oil did not repel *Aedes* mosquitoes or *Lucilia cuprina*,[342] although it is used in sheep fly preparations.[948]
Eucalyptus dumosa A. Cunn.: The essential

oil was repellent to *Aedes* mosquitoes but not to *Lucilia cuprina*.[322]
Eucalyptus fastigiata Deane & Maiden, brown barrel
Eucalyptus gigantea Hook. f., alpine ash, Tasmanian oak: The wood of only *E. fastigiata* is resistant to termites.[188]
Eucalyptus globulus Labill., blue gum: The essential oil in a cold cream base was highly effective as a repellent for adult *Musca domestica* and *Culex fatigans*.[949] An ethanol extract of the fresh leaves did not deter feeding by *Lymantria dispar* larvae.[84] The vapors of the essential oil greatly inhibited oviposition by *Amrasca devastans*.[378]
Eucalyptus gummifera (Gaertn.) Hochr., bloodwood
Eucalyptus hemiphloia F. v. M., graybox
Eucalyptus leucoxylon F. v. M., yellow gum
Eucalyptus longifolia Link., woolybutt: The wood of these four species is resistant to termites.[188]
Eucalyptus macrorhyncha F. v. M., red stringy-bark: The wood is resistant to termites.[188]
Eucalyptus maculata Hook, spotted gum: The wood is not resistant to termites.[370]
Eucalyptus maidenii F. v. M., maiden's gum: The wood is moderately resistant to termites.[188]
Eucalyptus marginata Donn ex Smith, jarrah: The leaves are resistant to feeding by *Perthida glyphopa*.[950]
Eucalyptus microcorus F. Muell., tallowwood: The wood is highly resistant to termites in Australia[188] due to its content of repellent chemical extractives.[951]
Eucalyptus obliqua L'Herit., Tasmanian oak
Eucalyptus paniculata Sm., gray houbark
Eucalyptus patens Benth., western Australian blackbutt
Eucalyptus polyanthemas Schaum., redbox
Eucalyptus punctata DC., gray gum: The wood of these five species is resistant to termite attack.[188] An ethanol extract of the fresh leaves of *E. paniculata* deterred feeding by *Lymantria dispar* larvae.[84]
Eucalyptus saligna Sm., Sydney blue gum: The steam-distilled leaf oil repels insects and kills *Calliphora* flies, *Anopheles funestus*,

Cimex lectularius, and *Periplaneta orientalis.* The major oil components are 1,8-cineole and α-pinene.[952]

Eucalyptus spp.: The essential oil of the leaves of various *Eucalyptus* species is repellent to several species of herbivorous beetles which cannot detoxify the oil.[953] Leaf extracts deter feeding by *Uraba lugens*[954] and *Pieris brassicae* larvae.[652]

Eucalyptus viminalis **Labill.:** An ethanol extract of the fresh leaves deterred feeding by *Lymantria dispar* larvae.[84] The wood is resistant to termite attack.[188]

Melaleuca leucadendron **(L.),** cajeput: An ethanol extract of the fresh leaves deterred feeding by *Lymantria dispar* larvae.[84] The essential oil did not repel *Aedes* or *Anopheles* mosquitoes.[342]

Melaleuca linariifolia **Sm.**

Melaleuca uncinata **R. Br.:** The essential oil of these two species did not repel *Aedes* or *Anopheles* mosquitoes.[342]

*Pimenta racemosa (*Mill.) **J. W. Moore,** bayrum tree

Pimenta dioica **(L.) Merr.,** allspice: Powdered allspice was ineffective against bedbugs, cockroaches, *Tineola bisselliella,* and carpet beetles;[157] when dusted on clothes it was worthless against *T. bisselliella* although it has often been recommended for this purpose. The leaves of *P. racemosa* are used to repel insects in bagged unshelled pigeon peas.[28]

Psidium acutangulum **DC.:** An ether extract of the stems did not deter feeding by adult *Acalymma vittatum.*[89]

Psidium guaiava **L.,** guava: The wood is very susceptible to termites.[86,124] An ethanol extract of the fresh leaves did not deter feeding by *Lymantria dispar* larvae.[84]

Syncarpia glomulifera **(Sm.) Niedenzu,** Australian turpentine: The wood is resistant to *Cryptotermes brevis* termites.[86,119,188]

Syncarpia killii **Bail.,** satinay: The wood is resistant to termites.[188]

Syzygium aromaticum **(L.) Merr. & Perry,** clove: The essential oil is repellent to *Apis florea* bees.[149,475] *Tribolium castaneum* females exposed to the plant fail to oviposit.[181]

Syzygium cumini **(L.) Skeels,** Java plum: An acetone extract of the whole plant did not repel *Popillia japonica* adults.[126]

Syzygium jambos **(L.) Alst.,** rose-apple, jambos: The wood is susceptible to attack by termites.[86]

Tristania conferta **R. Br.,** brush box: The wood is resistant to termites.[188]

FAMILY NYCTAGINACEAE

Bougainvillea glabra **Choisy:** An ether extract of the combined stems, leaves, and flowers did not deter feeding by adult *Acalymma vittatum.*[89] Extracts of the flowers were somewhat deterrent to feeding by *Sitophilus oryzae.*[251,252]

Mirabilis jalapa **L.,** common four-o'clock: The odor of the flowers was reported to keep mosquitoes away at night.[116,596]

FAMILY NYMPHAEACEAE

Nuphar japonicum **DC:** A methanol extract of the leaves weakly retarded the growth of *Bombyx mori* larvae.[182]

Nuphar luteum **(L.) Sibth. & Sm.,** spatterdeck: An acetone extract of the whole plant did not repel *Popillia japonica* adults.[126n]

FAMILY NYSSACEAE

Camptotheca acuminata **Decne.:** A methanol extract of the leaves was moderately retardant to the growth of *Bombyx mori* larvae.[82]

Nyssa aquatica **L.,** tupelo gum

Nyssa sylvatica **Marsh. var. biflora (Walt.) Sarg.,** swamp blackgum: *Coptotermes formosanus* termites fed poorly on the heartwood of this tree.[87] The trees are susceptible to attack by *Malacosoma disstria.*[88] An acetone extract of the whole plant did not repel *Popillia japonica* adults,[126] but an ethanol extract of the fresh leaves deterred feeding by *Lymantria dispar* larvae.[84]

FAMILY OCHNACEAE

Lophira alata **Banks ex Caertn. f. Ekki,** red ironwood: The wood is resistant to attack by

Reticulitermes lucifugus and *Coptotermes formosanus* termites.[370,388,441]

FAMILY OLACACEAE

Ongokea core (**Hum.**) **Pierre,** boleko nut: *Coptotermes formosanus* termites would not feed on this plant.[388]

Ximenia americana **L.:** An ethanol extract of the leaves did not deter feeding by *Lymantria dispar* larvae.[84]

FAMILY OLEACEAE

Chionanthus virginica **L.,** white fringetree: An acetone extract of the whole plant did not repel *Popillia japonica* adults.[126]

Forsythia suspensa (**Thunb.**) **Vahl.:** An ether extract of the fruit was fairly deterrent to feeding by *Acalymma vittatum* adults.[89] An ethanol extract of the fresh leaves did not deter feeding by *Lymantria dispar* larvae.[84]

Fraxinus americana **L.,** white ash: An ethanol extract of the fresh leaves did not deter feeding by *Lymantria dispar* larvae,[84] and an acetone extract of the plant did not repel *Popillia japonica* adults.[126] The wood is susceptible to attack by *Cryptotermes brevis* and *Coptotermes formosanus* termites.[86,87]

Fraxinus excelsior **L.,** European ash: The tree is susceptible to attack by the ash budmoth.[119]

Fraxinus velutina **Torr. var. modesto:** An ethanol extract of the fresh leaves did not deter feeding by *Lymantria dispar* larvae.[84]

Jasminum officinale **L.:** The essential oil was repellent to *Apis florea* bees.[149,475]

Ligustrum japonicum **Thunb.:** An ethanol extract of the fresh leaves did not deter feeding by *Lymantria dispar* larvae.[84]

Ligustrum lucidum **Ait. f.,** glossy privet: An ethanol extract of the fresh leaves did not deter feeding by *Lymantria dispar* larvae.[84] An alcohol extract of the shrub did not repel *Anthonomus grandis grandis* adults,[487] nor did it inhibit the growth of *Bombyx mori* larvae.[82]

Ligustrum vulgare **L.,** common privet: An ethanol extract of the fresh leaves did not deter feeding by *Lymantria dispar* larvae.[84]

Linociera domingensis (**Lam.**) **Knobl.:** The wood is susceptible to termites.[124]

Olea africana **Mill.:** The wood is resistant to termites.[116]

Olea chrysophylla **Lam.:** The wood is highly resistant to termites.[370]

Olea europaea **L.,** olive: An ethanol extract of the fresh leaves did not deter feeding by *Lymantria dispar* larvae.[84] An alcohol extract of the leaves caused moderate retardation of *Bombyx mori* larval growth.[82] Both *Schistocerca gregaria* and *Locusta migratoria* fed readily on a mixture of olive oil and an acetone extract of wheat bran.[238]

Osmanthus americanus (**L.**) **Gray,** devilwood, wild olive: An alcohol extract of the tree did not repel *Anthonomus grandis grandis* weevils.[195]

Osmanthus fragrans (**Thunb.**) **Lour.,** sweet osmantha: A methanol extract of the leaves retarded *Bombyx mori* larval growth when fed.[82]

Syringa vulgaris **L.,** lilac: An acetone extract of the whole plant did not repel *Popillia japonica* adults.[126]

FAMILY ORCHIDACEAE

Bletia hyacinthiana **R. Br.:** Larvae of *Plutella xylostella* would not feed on this plant.[195]

Bletilia striata **Rehab.:** A methanol extract of the leaves was readily accepted as food by *Bombyx mori* larvae.[82]

Vanilla planifolia **Andr.,** vanilla plant: Vanilla extract repelled *Blattella germanica* for 24 hr but not for 170 hr.[61]

FAMILY OXALIDACEAE

Oxalis deppei **Lodd.:** *Plutella xylostella* larvae would not feed on this plant.[195]

Oxalis stricta **L.,** yellow woodsorrel: An acetone extract of the whole plant did not repel *Popillia japonica* adults.[126]

FAMILY PAPAVERACEAE

Argemone mexicana **L.,** Mexican prickle-poppy: The seed oil was used by tribes in

Nigeria to repel white ants[596,597] and termites.[116,121]

Chelidonium majus L., greater celandine: An acetone extract of the whole plant did not repel *Popillia japonica* adults.[126]

Sanguinaria canadensis L., bloodroot: Ethereal and methanolic extracts of the whole plant did not deter feeding by adult *Acalymma vittatum*.[52] Aqueous and alcoholic extracts of the leaves and stems were toxic to *Aedes aegypti* larvae.[189]

FAMILY PASSIFLORACEAE

Passiflora incarnata L., passion flower, maypop: An acetone extract of the whole plant did not repel *Popillia japonica* adults.[126]

Passiflora mollissima (H.B.K.) L. H. Bailey, banana passion fruit: An acetone extract of the whole plant strongly deterred feeding by *Spilosoma obliqua* larvae.[194]

FAMILY PEDALIACEAE

Sesamum indicum L., sesame: Sesame seed oil coated over stored grain in closed receptacles retarded attack by *Sitophilus oryzae* weevils as long as the grain remained moist (a few months).[97] When used on green gram, the oil caused complete mortality of *Callosobruchus chinensis* eggs and prevented oviposition.[234,237] A methanol extract of the leaves caused minimal retardation of *Bombyx mori* larval growth following feeding.[82]

FAMILY PINACEAE

Abies concolor (Gord. & Glend.) Lindl. ex Hildebr., white fir: An acetone extract of the whole plant did not deter feeding by adult *Popillia japonica*.[126] The sapwood is deterrent to feeding by *Coptotermes formosanus* termites.[87]

Abies lasiocarpa (Hook.) Nutt., alpine fir: *Reticulitermes flavipes* termites force-fed the oven-dried wood survived for a number of weeks.[461]

Callitris calcarata A. Cunn. ex Mirb., black cypress pine: The heartwood is resistant to termites and borers.[119]

Callitris columellaris F. Muell., white cypress pine: A petroleum ether extract of the wood was both repellent and toxic to *Coptotermes acinaciformis*, *Mastotermes darwiniensis*, and *Nasutitermes exitiosus* termites.[955] Components isolated from the extract were *l*-citronellic acid (major component), β-eudesmol, and five sesquiterpene lactones.[956]

Callitris glauca H. Brown, Murray River pine: The wood is repellent and resistant to *Cryptotermes brevis* termites.[86,118,119,188] The wood oil is somewhat effective as a repellent for *Aedes aegypti*, but the leaf oil was not effective against *Aedes* mosquitoes or *Lucilia cuprina*.[322,342] Guiol, present in the heartwood, is repellent to white ants.[957]

Callitris robusta R. Br., common cypress pine: The wood is resistant to *Anacanthotermes ochraceus*, *Psammotermes fuscofemoralis*, and *P. assuarensis* termites.[119,187]

Cedrus deodara (Roxb.) Loud., deodar, Himalayan cedarwood: The wood is resistant to termites.[207] The ethanol extract of the fresh leaves did not deter feeding by *Lymantria dispar* larvae.[84] The wood oil is toxic to *Callosobruchus analis* and *Musca domestica*. The active compounds are himachalol and β-himachalene.[958]

Larix laricina (Du Roi) K. Koch, tamarack: The needles are not fed upon by *Pristiphora erichsonii*[959] due to their content of resin acids (abietic, neoabietic, dehydroabietic, and isopimaric).[960] Sandarocopimaric acid does not influence feeding but it reduces growth and efficiency of the insect.

Larix occidentalis Nutt., western larch: The wood is very susceptible to attack by *Cryptotermes brevis*[86,207] and *Reticulitermes flavipes* termites.[461]

Picea abies (L.) Karst., Norway spruce: The needle oil was repellent to *Hylobius abietis*[961] but not to *Pissodes strobi* adults.[962] The tree is severely attacked by *Reticulitermes flavipes*,[188,610] *Anacanthotermes*, and *Microcerotermes* termites,[907] and *Elatobium abietinum*.[119] Consecutive extracts prepared from the leaves and twigs with ether and methanol did not deter feeding by *Spodop-*

tera frugiperda larvae.[204] An excellent review of the resistance of conifers of this type to bark beetle attack has recently been published.[963]

Picea asperata **Masters:** The bark is not repellent to adult *Pissodes strobi.*[962]

Picea engelmanni **Parry ex Engelm.,** Engelmann spruce: *Reticulitermes flavipes* termites consumed the wood without harmful effects when force-fed.[461]

Picea glauca **(Moench.) Voss,** white spruce
Picea mariana **(Mill) B.S.F.,** black spruce: The bark of these two species was repellent to *Pissodes strobi* larvae.[962] Pine oil, a byproduct of sulfate wood pulping, protected *P. glauca* from attack by *Dendroctonus rufipennis.*[964]

Picea obovata **Ledeb.,** Siberian spruce: The bark is not repellent to adult *Pissodes strobi.*[962]

Picea omarica **Mayr.,** omarika spruce: Sap expressed from the needles and brushed over the needles of *P. abies* arrested feeding by *Gilpinia hercynaie, G. polytoma,* and *G. abieticola.*[965]

Picea orientalis **(L.) Link,** oriental spruce: An acetone extract of parts of the tree did not repel *Popillia japonica* adults.[126]

Picea rubens **Sargent,** red spruce: The bark repelled adult *Pissodes strobi.*[962] *Coptotermes formosanus* termites refused to feed on the wood.[87]

Picea sitchensis **(Bong.) Carr.,** Sitka spruce: The wood is susceptible to attack by *Cryptotermes brevis* termites.[86,188] Varieties of this tree resistant to attack by *Pissodes strobi* had a lower content of β-phellandrene and a higher content of β-pinene and 3-carene than susceptible varieties.[966]

Pinus banksiana **Lam.,** jack pine: The needles are susceptible to feeding by *Rhyacionia buoliana* larvae.[967] The foliage of juvenile trees inhibits larval feeding by *Neodiprion rugifrons* and *N. swainei.* The components responsible were shown to be hydrophilic in nature by preparation of a hexane extract[968] and aqueous extracts.[969] Although the resin acid (*1*-pimaric) did not inhibit feeding, α-pimaric acid and abietic acid were active at 20% but not at 1%. Larvae were deterred by five 1,4-naphthoquinones.[969] Antifeedants

also occur in the mature foliage in lower concentrations.[970] Of numerous commercial resin acids tested for deterrency, palustric acid was the most effective.[971] The major components responsible for the antifeedant activity were identified as 13-keto-8(14)-podocarpen-18-oic acid and dehydroabietic acid,[974,975] and the former has also been synthesized.[974]

Pinus cembra **L.,** Swiss stone pine: The foliage is resistant to attack by *Rhyacionia buoliana.*[967]

Pinus contorta **Dougl. ex Loud.,** lodgepole pine: The foliage is attacked by *Rhyacionia buoliana,*[119] but it can be protected from *Dendroctonus ponderosae* by spraying with Norpinie-65, a mixture of terpene hydrocarbons extracted from pulp waste.[976]

Pinus densiflora **Sieb. & Zucc.:** Ethane present in the needles repels *Monochamus alternatus.*[977]

Pinus echinata **Mill.,** shortleaf pine, southern yellow pine: The wood is susceptible to termite attack.[86,124,462] Limonene, a constituent of the oleoresin, was toxic to *Dendroctonus frontalis* adults.[978]

Pinus elliotii **Engelm.,** slash pine: An ethanol extract of the fresh leaves failed to deter feeding by *Lymantria dispar* larvae.[84] The wood is generally susceptible to termite attack,[188] but a pentane extract of the wood from mature trees deterred feeding and was toxic to *Reticulitermes flavipes*[979] and *Coptotermes formosanus* termites.[388]

Pinus flexilis **James**
Pinus griffithii **Parl.:** These two species are resistant to attack by *Rhyacionia buoliana.*[967]

Pinus halepensis **Mill.,** Aleppo pine: This tree is comparatively resistant to termites.[119]

Pinus jeffreyi **Grev. & Balf.,** Jeffrey pine: *Dendroctonus jeffreyi* feeds on this species.[980,981]

Pinus lambertiana **Dougl.:** *Coptotermes formosanus* termites fed little on the sapwood,[87] and wood impregnated with a 0.5% solution of chrysin (a constituent of the plant) repelled attack by *Cryptotermes brevis* termites for 55 days.[940] The amount of wood consumed by *Incisitermes minor* termites was inversely proportional to its specific gravity;[361] crude extracts were deterrent at 0.5 mg/cm². Long-chain saturated fatty acids deterred feeding at 0.25 and 0.05% g/cm², and related alpha-halogenated acids were effective at 0.05 mg/cm².[982]

Pinus leiophylla **Schlechtend. & Cham.,** smoothleaved pine: The tree is severely defoliated by *Euproctis terminalis* larvae.[119]

Pinus monticola **Dougl. ex D. Don,** western white pine: The wood is moderately deterrent to *Cryptotermes brevis* termites.[86] *Pissodes strobi* weevils fed upon lateral branches of the tree, with the releasing stimulus for feeding present in the bark.[983]

Pinus mugo **Turra,** Swiss mountain pine: This species is susceptible to feeding by *Rhyacionia buoliana*[967] and *Sirex noctilio*.[119]

Pinus occidentalis **Sw.,** West Indian pine, Haitian pine: The gummy heartwood is resistant to *Cryptotermes brevis* termites.[86,115]

Pinus palustris **Mill.,** longleaf pine: The wood is susceptible to *Cryptotermes brevis* termite attack,[86,124] and resistant to *Anacanthotermes ochraceus, Psammotermes fuscofemoralis, P. assuarensis,*[387] and *Reticulitermes flavipes* termites.[462] The essential oil is moderately repellent to *Apis florea* bees.[149,475] An ethanol extract of the fresh leaves did not deter feeding by *Lymantria dispar* larvae.[84]

Pinus pinaster **Ait.,** maritime pine: Volatile phenolic substances in the branches repelled *Pissodes notatus* weevils.[984]

Pinus ponderosa **Dougl. ex Kawson,** ponderosa pine: The wood is susceptible to *Cryptotermes brevis* and *Incisitermes minor* termites, and the foliage is susceptible to *Rhyacionia buoliana* larvae.[967] *Reticulitermes flavipes* termites force-fed on the wood did not survive,[461] and *Cryptotermes formosanus* termites fed little on the wood.[87]

Pinus pumila **Hort.:** The tree is susceptible to attack by *Rhyacionia buoliana* larvae.[967]

Pinus radiata **D. Don.,** Monterey pine: The wood is susceptible to *Nasutitermes exitiosus, Coptotermes exitiosus,* and *C. acinaceformis* termites.[467]

Pinus resinosa **Ait.,** red pine, Norway pine: 1,8-Cineole and terpineol acetate from this tree repel pine sawyers.[985,986] Larvae of *Anopheles claviger* and *Aedes cantator*

exposed to the oil, which contains limonene and bornyl acetate, did not survive.[182]

Pinus rigida **Mill.,** pitch pine: An acetone extract of the whole plant did not repel *Popillia japonica* adults.[126] The tree is resistant to attack by *Rhyacionia buoliana.* Ethereal and ethanol extracts of the leaves and twigs were not toxic to *Spodoptera frugiperda* larvae.[204]

Pinus spp.: β-Terpineol, α-pinene, Δ-3-carene, and benzoic acid, found in a number of pine oils, were repellent to *Blastophagus piniperda.* Limonene was at first repellent but soon became neutral.[987] High concentrations of α-pinene, β-pinene, terpineol, camphene, borneol, and limonene were repellent to *Scolytus* beetles.[988,989]

Pinus strobus **L.,** eastern white pine: The wood is susceptible to termite attack.[124] The ground bark was repellent to *Pissodes strobi* in the laboratory but not in the field.[962] A methanol extract of the pine cones failed to deter feeding by *Acalymma vittatum* adults.[52] Consecutively prepared ethereal and methanolic extracts of the cones deterred feeding by *Spodoptera frugiperda* larvae.[204] The sapwood was repellent to *Coptotermes formosanus* termites.[87]

Pinus sylvestris **L.,** Scotch pine: The sapwood is very susceptible to attack by *Reticulitermes lucifugus,*[484,610] *R. flavipes,* and *Cryptotermes brevis* termites.[86] Immersion for 10 min in 0.01% solution of pinosylvin monomethyl ether or dihydropinosylvin monomethyl ether, constituents of the heartwood, protected samples of the very susceptible flamboyan wood (*Delonix regia*) from attack for at least 3 months.[86] Larvae of the ticks *Ixodes redikorzevi, Haemaphysalis punctata, Rhipicephalus rossicus,* and *Dermacentor marginatus* exposed to the powdered leaves were killed in 15, 20, 20, and 30 min, respectively.[153] The mechanism of resistance in this species is discussed in detail by Smelyanetz.[990,991]

Pinus taeda **L.,** loblolly pine: The wood is very susceptible to termite attack,[124] especially from *Cryptotermes brevis.*[86] *Reticulitermes flavipes*[462] and *Coptotermes formosanus* termites exposed to the sawdust did not survive.[992,993] An ethanol extract of the fresh leaves did not deter feeding by *Lymantria dispar* larvae.[84] Limonene present in the cones is toxic to *Dendroctonus frontalis* adults following topical application.[978] Feeding on the conelets by nymphs and adults of *Leptoglossus corculus* and *Tetyra bipunctata* caused the conelets to abort.[994] The content (%) of α-pinene, β-pinene, myrcene, limonene, and β-phellandrene in resistant and susceptible trees, respectively, was 80.15, 72.89; 10.78, 23.87; 9.07, 1.55; 0, 1.22; and 0, 0.47.[995]

Pseudotsuga taxifolia **(Poir.) Britt.**, Douglas fir, Oregon pine: The wood is susceptible to attack by *Cryptotermes brevis*,[86] *Reticulitermes lucifugus*,[441] and *Coptotermes formosanus* termites.[87] The variety *glauca* of this pine is comparatively resistant to attack by *Adelges cooleyi*.[119]

Tsuga canadensis **(L.) Carr.**, eastern hemlock: The wood is very susceptible to termite attack.[86,124] Hemlock oil exhibited good repellent action against *Cochliomyia hominivorax* for only 1 or 2 days,[155] and extracts of the tree did not repel *Popillia japonica* adults.[126]

Tsuga heterophylla **(Raf.) Sarg.**, western hemlock: The wood is susceptible to attack by *Reticulitermes flavipes* termites.[188,461]

Tsuga sieboldii **Carr.:** A methanol extract of the leaves killed all *Bombyx mori* larvae exposed to it.[82]

FAMILY PIPERACEAE

Piper aduncum **L.**, spiked pepper: The plant is used in Haiti to repel ants.[996]

Piper cubeba **L. f.**, cubeb: Cubeb oil repelled *Blattella germanica* adults[61] and killed 100% of mosquito larvae tested.[164]

Piper elongatum **Vahl.**, matico: Extracts of the plant did not repel *Popillia japonica* adults.[126]

Piper futokadsura **Sieb. & Zucc.:** A benzene extract of the leaves deterred feeding by *Spodoptera litura* larvae. The active compounds were isolated and identified as piperenone, eudesmin, elemicin, and epieudesmin.[672,997]

Piper guineense **Schumach. & Thomn.,** West African pepper: Wheat was protected from *Sitophilus oryzae* and black-eyed peas were protected from *Tribolium confusum* by coating the grains with 3000 ppm of each of two alkaloidal amides isolated from the fruits.[998] The compounds were not toxic to these species of insects or to *Callosobruchus maculatus* or *Lasioderma serricorne*. A number of alkaloidal amides were isolated and identified from the fruits, stems, and roots.[999-1002]

Piper longum **L.**, Indian long pepper, white pepper: The oil from the fruits repelled adult *Blattella germanica*.[61]

Piper nigrum **L.**, black pepper: Ground pepper was a satisfactory repellent for ovipositing *Heliothis zea*.[1003] Oil from the fruit repelled adult *Blattella germanica*.[61] Piperine obtained from the berries did not deter feeding by adult *Acalymma vittatum* and *Diabrotica undecimpunctata*.[52] Admixture of pepper with yellow-eye beans at 1% reduced damage by *Acanthoscelides obtectus* by more than 90%; lower dosages reduced damage by 50% over 4 months.[1004] Mixing wheat with 0.5% of pepper fruits gave complete kill of adult *Sitophilus oryzae,* and topical application of an ethanol extract of pepper to this insect and to *Callosobruchus maculatus* resulted in high mortality.[1005,1006] Adult *C. chinensis* exposed for 48 hr to mungbean seeds coated with pepper seeds succumbed rapidly[150] but pepper leaves were not effective. Three amides were isolated from the seeds and shown to be responsible for the effects caused by topical application to *C. maculatus*.[1007] *Tribolium castaneum* females fail to oviposit or survive for 20 days when they are maintained on pepper supplemented with yeast.[181] An ethanol extract of pepper applied topically to adult *Anthonomus grandis grandis* was highly toxic to this insect.[1008,1009] A methanol extract of black pepper fruits added to the diet killed *Bombyx mori* larvae.[82] Piperine and several amides were isolated from the fruits.[1010-1012] Piperine was more toxic than pyrethrum to *Musca domestica*.[1013]

Piper retrofractum **Vahl.,** Japanese long pepper: The seed oil was moderately repellent to *Blattella germanica* adults.[61] Several new

amides were isolated from this species.[1014-1016]
Piper sylvaticum **Roxb.**: A new amide was isolated from the seeds.[1017]
Pothomorphe umbellata **(L.) Miq.**, cowfoot: The volatile essential oil is repellent to insects, but the effect is shortlived.[1018]
Saururus cernuus **L.**, common lizardtail: An acetone extract of the whole plant was not repellent to *Popillia japonica* adults.[216]

FAMILY PITTOSPORACEAE

Pittosporum tobira **Ait.**: An ether extract of the fresh leaves deterred feeding by *Lymantria dispar* larvae.[84]

FAMILY PLANTAGINACEAE

Plantago lanceolata **L.**, buckthorn plantain, English plantain: A methanol extract of the whole plant did not repel *Anthonomus grandis grandis*,[487] and an acetone extract did not repel *Popillia japonica* adults.[126]
Plantago major **L.**, common plantain: An acetone extract of the whole plant failed to repel *Popillia japonica* adults.[126]

FAMILY PLATANACEAE

Platanus occidentalis **L.**, American sycamore: This tree was not bothered by an outbreak of *Malacosoma disstria*.[88] An ethanol extract of the fresh leaves did not deter feeding by *Lymantria dispar* larvae.[84] An ether extract and a methanol extract of the twigs failed to deter feeding by adult *Acalymma vittatum*[52] and by *Spodoptera frugiperda* larvae.[204]
Platanus orientalis **L.**, oriental planetree: An ether extract of the seeds deterred feeding by adult *Acalymma vittatum* at 0.5% but not at 0.1%.[89]

FAMILY PLUMBAGINACEAE

Limonium carolinianum **(Walt.) Britton**, sea-lavender: An acetone extract of the whole plant did not repel *Popillia japonica* adults.[126]
Plumbago capensis **Thunb.**: Plumbagin, a naphthoquinone isolated from the root, shows potent antifeedant activity against *Spodop-*

tera exempta and *S. littoralis* larvae,[1019] as well as antimicrobial activity. It also inhibits ecdysis of larval *Pectinophora gossypiella*, *Heliothis virescens*, *H. zea*, and *Trichoplusia ni*.[1020,1021]

FAMILY POACEAE

Agropyron repens **(L.) Beauv.**, quackgrass, couchgrass: An acetone extract of the whole plant was not repellent to *Popillia japonica* adults.[126]
Avena sativa **L.**, oat: The resistance of certain oat cultivars to larvae of *Oscinella frit* increased with the age of the crop.[1022] Large amounts of 18-carbon hydroxydienoic acids identified in triglyceride fractions of a pentane extract induced an avoidance response in *Oryzaephilus surinamemsis*.[1023] Avenacosides A and B, obtained from the seeds and leaves, did not prevent feeding by *Reticulitermes flavipes* termites.[229]
Bambusa tuldoides **Munro**, puntingpole bamboo
Bambusa vulgaris **Schrad. ex Wendl.**, bamboo: The wood of these species of bamboo is susceptible to attack by *Cryptotermes brevis* termites.[86]
Coix lachryma-jobi **L.**, millet: The resistance of millet varieties to *Spodoptera frugiperda* larvae is reviewed.[11]
Cymbopogon citratus **(DC.) Stapf.**, West Indian lemongrass: Oil of lemongrass did not repel *Cochliomyia hominivorax*[155] but the grass was recommended as a repellent for the tsetse fly.[596] As a fumigant, the oil killed 100% of *Culex fatigans* and *Aedes aegypti* mosquitoes in 20 min and 100% of *Musca domestica* in 60 min; it was effective for 30 to 40 min as a repellent.[1024]
Cymbopogon marginatus **Stapf.**: The rootstock repels moths (unidentified).[116]
Cymbopogon nardus **(L.) Rendle**, citronella: Oil of citronella is a repellent for mosquitoes, *Cochliomyia hominivorax*,[155] *Blatta orientalis*,[162] and *Lucilia cuprina*.[668] The major repellent in the oil appears to be borneol.[668]
Cynodon dactylon **(L.) Pers.**, Bermuda grass, devilgrass: The resistance of this grass to *Spodoptera frugiperda* larvae is reviewed.[11]

Echinochloa crus-galli (L.) Beauv. var. crusgalli, barnyardgrass: *Trans*-Aconitic acid was isolated from the grass and shown to be an antifeedant for *Nilaparvata lugens*.[1025]

Elymus canadensis L.: *Melanoplus femurrubrum* grasshoppers reared on a diet containing extracts of this plant failed to reach adulthood.[275] It was shown that at least 0.1 to 0.2 g of this grass/g of diet was needed to produce adults.[244]

Guadua latifolia (Humboldt. & Bonpl.) Kunth., Ecuadorean giant bamboo: The wood is very susceptible to termite attack, especially by *Cryptotermes brevis*.[86]

Holcus lanatus L., velvetgrass, Yorkshire fog: An acetone extract of the whole plant did not repel *Popillia japonica* adults.[126]

Hordeum vulgare L., barley: Barley was relatively susceptible to *Sipha flava* and *Schizaphis graminum* as a host plant.[1026] Although several varieties of barley are resistant to *Rhopalosiphum maidis*, others are susceptible to attack by this insect.[1027] The compound responsible for the resistance is apparently DIMBOA.[1028] About 61% of newly hatched nymphs of *Melanotus sanguinipes* survived to adulthood on this plant.[684]

Imperata cylindrica (L.) Beauv., cogongrass: In Nigeria, cowpeas covered with the grass and placed in a bin over a smoky fire prior to storage are protected from insect attack for 5 months.[28]

Melinis minutiflora Beauv., molasses grass: The grass repels *Boophilus annulatus australis* ticks in Puerto Rico.[1029-1031] The grass has the odor of cumin and is said to repel mosquitoes and tsetse flies,[1032] in addition to ants, chiggers, and snakes;[1033] however, neither a petroleum ether extract nor a water-soluble fraction that separated therefrom were repellent to *Aedes aegypti*.[620]

Oryza sativa L., rice: The seeds of greengram (*Vigna radiata*) can be protected against *Callosobruchus maculatus* for 5 to 6 months by treatment with 0.3 or 0.5% of rice bran oil.[482]

Pennisetum americanum (L.) Leeke, pearl millet: Pearl millet was not fed upon by *Sipha flava*.[1026]

Poa annua L., meadowgrass: The odor of this is repellent to adult *Leptinotarsa decemlineata*.[360]

Poa pratensis L., Kentucky bluegrass: Only 36% of *Melanoplus femurrubrum* reached adulthood when reared on a diet containing extracts of this grass.[275]

Secale cereale L., rye: Rye is resistant to feeding by *Sipha flava*.[1026]

Sorghum bicolor (L.) Moench., grain sorghum, broom-corn, guinea corn: Sorghum is susceptible to attack by *Sipha flava*.[1026] High concentrations of cyanide in some sorghum cultivars are associated with reduced feeding by *Chilo partellus* larvae, and high concentrations of phenolic acids are correlated with reduced feeding by *Locusta migratoria migratorioides, Peregrinus maidis, Rhopalosiphum maidis, Acrida exaltata*, and larvae of *Mythimna separata*.[1034] The epicuticular wax from two varieties of young sorghum at four stages of growth was more deterrent to feeding by *Locusta migratoria migratoroides* when the plants were young. Fractionation of the wax gave a hydrocarbon and an ester fraction; only the latter was deterrent.[1035] A mixture of 14 different phenolic compounds was highly deterrent.[1036] Combinations of phenolic deterrents were additive in their effects.[1037] *p*-Hydroxybenzaldehyde present in the surface wax of sorghum cultivars deters feeding by *Locusta migratoria migratorioides* and the amount of this compound decreases with the age of the plant.[1038]

Triticum aestivum L., common wheat: Feeding studies conducted on newly hatched larvae of *Agrotis orthogonia* to determine why diets prepared from some lyophilized tissues of wheat stems allow only poor growth and development showed that an unidentified toxic substance is present.[1039] Wheat germ oil and the phospholipid fraction of wheat germ evoked feeding activity from older nymphs and adults of *Melanoplus bivittatus* and *Cannula pellucida*[1040] and from *L. migratoria migratorioides* and *Schistocerca gregaria*.[238] The plants are susceptible to feeding by *Sipha flava* and *Schizaphis grami-num.*[1026] However, the resistance of some cultivars to aphids has been traced to their content of DIMBOA.[1028] Ether extracts of the stems of nine varieties of wheat did not deter feeding by adult *Acalymma vittatum*[52,89] and *Diabrotica undecimpunctata.*[52] Of 36 varieties of wheat tested, 27 varieties were completely free of *Rhopalosiphum maidis.*[1027] *Triticum turgidum* **L.,** poulard wheat: This plant is resistant to feeding by *Mayetiola destructor.*[1041]

Vetiveria zizanoides **(L.) Nash ex Small.,** cuscus grass, vetiver: In Africa, the dried roots of this grass, when placed among clothes, prevented insect attack.[287] The roots were used in Haiti to repel insects.[227,996] An aqueous extract of the whole plant deterred feeding by *Attagenus piceus* and *Oncopeltus fasciatus,*[25] but the powdered root failed to deter oviposition by *Callosobruchus chinensis* in mungbean seeds. Topical application of the oil from Java repelled cockroaches and *Musca domestica.* This effect was shown to be due to two aldehydes (zizanal and epizizanal), α-vetivone, β-vetivone, khusimone, and (+)-(1S,10R)-1,10-dimethylbicyclo[4.4.0]-dec-6-en-3-one.[1042]

Zea mays **L.,** corn, maize: In a study of the factors influencing the establishment and survival of *Ostrinia nubilalis* larvae on corn plants, it was found that two or more toxic substances occur in the plant tissues. These toxicants have a deleterious effect on the growth and survival of borer larvae and an inhibitory effect on the growth of other organisms. One of the substances isolated from dent corn was named "resistance factor A",[1043,1044] which was later identified as 2,4-dihydroxy-7-methoxy-(2H)-1,4-benzoxazin-3-(4H)-one (DIMBOA).[1045,1046] 6-Methoxybenzoxazolinone (6-MBOA) was also isolated, but this compound does not confer resistance to insect feeding.[1047-1051] The silks of many corn varieties and cultivars are resistant to *Heliothis zea* larval feeding.[1052,1053] A substance in the silk that severely retards the growth of this insect has been isolated, identified, and designated "maysin."[1054] A

review of plant resistance to *H. zea* has been published.[11] A glycosylflavone inhibitory to the development of the larvae of this insect has been identified from corn silk.[14,1055] MBOA has been shown to deter feeding by *Spodoptera exempta* larvae, and a method has been published for its synthesis.[1056] Nymphs of *Sipha flava* refused to feed on corn in greenhouse tests.[1026] A pollinated cultivar of corn was resistant to the storage pest *Sitophilus zeamais.*[1057] Corn oil repelled adult *Sitophilus granarius* in wheat grain containing 5 ml oil/kg wheat; complete control was obtained with 1 mg/kg.[1058,1059] *Locusta migratoria migratorioides* fed readily on corn oil, but *Schistocerca gregaria* fed reluctantly.[238] A volatile extract of corn stimulated oviposition by female *Sitotroga cerealella.*[106] Only 72% of *Melanoplus femurrubrum* grasshoppers reached adulthood when reared on a diet containing extracts of the corn plant.[319]

FAMILY PODOCARPACEAE

Dacrydium cupressinum **Soland. ex Lamb.,** New Zealand red pine: The wood is moderately resistant to termite attack.[188]

Dacrydium franklinii **Hook. f.,** Huon pine: The wood is resistant to termite attack.[188] Huon pine oil was very effective as a repellent for *Aedes* and *Anopheles* mosquitoes due to its high content of methyleugenol. It also repels March flies, sand flies, and *Lucilia cuprina,* but not bush flies.[322,342] The oil prevents *L. cuprina* from ovipositing on sheep.[1061]

Phyllocladus rhomboidalis **Rich.,** celery-top pine: The wood is resistant to termite attack.[188]

Podocarpus gracilior **Pilg.:** This plant is resistant to attack by *Pectinophora gossypiella, Heliothis zea, Spodoptera frugiperda,* and *Bombyx mori* larvae. The activity is due to four norditerpene dilactones (including nagilactones C, D, and F), podolide, an insecticidal growth inhibitor, several flavones, and the phytoecdysone, ponasterone A.[1062]

Podocarpus macrophyllus **D. Don,** inumaki:

The wood, which is well known for its resistance to termites, yielded the bisnorditerpenoid, inumakilactone, and another compound which is soluble in water. Both compounds are highly toxic to *Coptotermes formosanus* termites.[1063]

***Podocarpus hallii* Kirk**
***Podocarpus neriifolia* D. Don**
***Podocarpus nivalis* Hooker:** Hallactones, podolactones A, B, C, and E, nagilactone C, and podolactone E isolated from the leaves of these species were fed to *Musca domestica* larvae, resulting in complete suppression of the development of these insects.[1064-1066]

***Podocarpus nakai* Hayata:** Ponasterone, isolated from the leaves, interferes with insect development.[1067]

***Podocarpus spicatus* R. B.,** black pine, matai: The wood is susceptible to attack by *Nasutitermes exitiosus*, *Coptotermes lacteus*, and *C. acinacitermis* termites.[467]

***Podocarpus sellowii* Klotsch ex Endl.:** An ether extract of the twigs deterred feeding by adult *Acalymma vittatum* at 0.5% but not at 0.1%. An extract of the stem bark was deterrent at both concentrations.[89]

FAMILY POLEMONIACEAE

***Ipomopsis aggregata* (Pursh.) V. Grant:** *Hylemya* spp. oviposit on leaves of this plant but do not feed.[1068]

***Phlox adsurgens* Torr.:** *Plutella xylostella* larvae feed on this plant.[195]

***Phlox paniculata* L.,** garden phlox

***Phlox subulata* L.,** moss phlox: Acetone extracts of these two species did not repel *Popillia japonica* adults.[126]

***Polemonium foliosissimum* A. Gray:** *Hylemya* spp. oviposit on the leaves of this plant, after which a pheromone is secreted that deters oviposition.[1068]

FAMILY POLYGALACEAE

***Polygala seneca* L.,** Seneca snakeroot: An acetone extract of the whole plant did not repel adult *Popillia japonica*.[126]

FAMILY POLYGONACEAE

***Coccoloba grandiflora* (L.) Lindau,** moralon: The wood is resistant to *Cryptotermes brevis* termites.[86,124]

***Coccoloba rugosa* Desf.,** ortegon: The wood is resistant to *Cryptotermes brevis* termites.[86]

***Coccoloba uvifera* (L.) L.:** An ethanol extract of the fresh leaves did not deter feeding by *Lymantria dispar* larvae.[84] The wood is very susceptible to *Cryptotermes brevis* termites.[86]

***Eriogonum fasiculatum* Benth.:** An ethanol extract of the fresh leaves did not deter feeding by *Lymantria dispar* larvae.[84]

***Fagopyrum cymosum* Meissn.:** A methanol extract of the leaves retarded the growth of *Bombyx mori* larvae fed thereon.[82]

***Fagopyrum esculentum* Moench.,** buckwheat: An acetone extract of the whole plant did not repel *Popillia japonica* adults.[126]

***Polygonum amphibium* L.,** water smartweed: About 81% of *Melanoplus femurrubrum* grasshoppers reached adulthood when reared on a diet containing an aqueous extract of this plant.[275] An aqueous extract of the flower heads did not elicit juvenilization of *Tenebrio molitor* larvae fed thereon.[246]

***Polygonum hydropiper* L.,** water pepper, marsh pepper: It is reported that when wounds or sores on horses are washed with a decoction of this herb, flies avoid them and will not approach the animals even in the heat of summer.[437] Polygodial, a sesquiterpene dialdehyde with pungent taste isolated from the leaves of this plant, is an insect antifeedant.[1069] This compound has been synthesized.[1070,1071] Warburganal, another pungent dialdehyde isolated from the leaves, is an insect antifeedant and cytotoxicant, as well as a molluscicide.[1072]

***Rheum Emodi* Wall:** An acetone extract of the plant showed moderate deterrency toward feeding by *Spilosoma obliqua* larvae.[194]

***Rheum officinale* Baill.,** Chinese rhubarb: Extracts of the rhizomes prepared with ethanol were moderately deterrent to feeding by *Pieris brassicae* larvae,[652] and slightly repelled oviposition by *Cydia pomonella* females.[1073]

Rumex acetosa **L.**, garden sorrel: An acetone extract of the whole plant was not repellent to *Popillia japonica* adults.[126]

Rumex crispus **L.**, curly dock, yellow dock: Only 80% of *Melanoplus differentialis* grasshoppers reached adulthood when reared on a diet containing extracts of this plant.[275,319] A methanol extract of the whole plant tested at 1000 ppm killed 28, 70, and 78% of *Aedes aegypti* larvae in 1, 3, and 7 d, respectively.[245]

Rumex hymenosepalus **Torr.**, canaigre, tanner's dock: The oil of this plant did not repel *Blattella germanica* adults.[61]

Rumex obtusifolius **L.**, broadleaf dock: Ethereal and ethanolic extracts of the combined leaves and stems significantly deterred feeding by *Spodoptera frugiperda* larvae.[204]

FAMILY PONTEDERIACEAE

Pontederia cordata **L. var. cordata**, pickerelweed: An acetone extract of the whole plant did not repel *Popillia japonica* adults.[126]

FAMILY PORTULACCACEAE

Claytonia virginica **L.**, spring beauty: An acetone extract of the whole plant repelled *Popillia japonica* adults.[126]

FAMILY PRIMULACEAE

Anagallis arvensis **L.**, scarlet pimpernel: The plant is known as an insect repellent in India.[116,227]

Cyclamen europaeum **L.**, alpine violet: Filter papers impregnated with 0.5 or 3% solutions of cyclamen, a triterpene saponin from the tubers of this plant, were toxic to *Reticulitermes flavipes* termites and greatly inhibited their feeding.[229]

Lysimachia nummularia **L.**, moneywort: The oil extracted from this plant is destructive to grain insects.[1074]

Lysimachia terrestris **(L.) B.S.P.**, swamp loosestrife: An acetone extract of the whole plant did not repel *Popillia japonica* adults.[126]

Primula elatior **Jacq.**: Filter papers impregnated with a 3% solution of the saponin obtained from the roots were toxic to *Reticu-*

litermes flavipes termites and inhibited feeding by the insect; 0.05 and 0.5% solutions were not effective.[229]

FAMILY PROTEACEAE

Grevillea robusta **Cunn. ex R. Br.**, Australian silver oak: The wood is susceptible to attack by *Cryptotermes brevis* termites,[86,124] bostrichid borers, and furniture beetles.[119] An ethanol extract of the fresh leaves did not deter feeding by *Lymantria dispar* larvae.[84]

FAMILY PSILOTACEAE

Psilotum nudum **(L.) Beam.**: Psilotin, a constituent of the aerial portion of this plant, deterred feeding by larvae of *Ostrinia nubilalis*.[1075]

FAMILY PUNICACEAE

Punica granatum **L.**, pomegranate: Larvae of *Plutella maculipennis* refused to feed on this plant.[195] A 1,1,2-trichloroethylene extract of the bark mixed with soybean oil is reported to be useful as an insecticide.[1076] A pentane extract of the seed did not deter feeding by adult *Anthonomus grandis grandis*.[496]

FAMILY RANUNCULACEAE

Aconitum napellus **L.**, aconite, monkshood friar's cap: An acetone extract of the whole plant did not repel *Popillia japonica* adults.[126]

Adonis vernalis **L.**, spring adonis

Anemone pulsatilla **L.**, European pasqueflower

Anemone quinquefolia **L.**, American wood anemone

Calytha palustris **L.**, marsh marigold: Acetone extracts of these plants did not repel adult *Popillia japonica*.[126]

Cimicifuga racemosa **(L.) Nutt.**, Cohosh bugbane: An acetone extract of the whole plant did not repel adult *Popillia japonica*.[126]

Clematis vitalba **L.**: This plant was reported to repel weevils from stored grain in France. Twigs with leaves and flowers were placed on the bags of grain and kept them free of infestation.[1077]

Helleborus niger L., Christmas-rose: A methanol extract of the leaves strongly retarded the growth of *Bombyx mori* larvae.[82] An acetone extract of the dry rhizomes repelled *Popillia japonica*.[126]

Hepatica americana (DC.) Ker., hepatica: An acetone extract of the whole plant did not repel adult *Popillia japonica*.[126]

Hydrastis canadensis L., goldenseal: An acetone extract of the whole plant failed to repel adult *Popillia japonica*.[126]

Nigella sativa L., black cumin: Oleic and linoleic acids from the seeds were tested against *Sitophilus oryzae*, *Stegobium paniceum*, *Tribolium castaneum*, and *Callosobruchus chinensis*. The highest degree of repellency was obtained against *C. chinensis*, with lesser activity against the other species. The acids were weakly repellent to *T. castaneum*.[1078] The natives of Hindustan sprinkled the seeds among woolen clothes as a preservative against insects.[1079] The plant was also used to protect linen.[81,1080] An acetone extract of the seeds was more or less repellent to *Popillia japonica* adults.[126] An acetone extract of the leaves deterred feeding by adult *Schistocerca gregaria*.[99]

Ranunculus septentrionalis Poir., buttercup
Thalictrum polygonum Muhl., meadow rue: Acetone extracts of these plants did not repel adult *Popillia japonica*.[126]

Xanthorhiza simplicissima Marshall: An ether extract of the combined stems, leaves, and fruits did not deter feeding by adult *Acalymma vittatum*.[89]

FAMILY RESEDACEAE

Reseda odorata L., common mignonette: An acetone extract of the whole plant did not repel *Popillia japonica* adults.[126]

FAMILY RHAMNACEAE

Ceanothus americanus L., Jersey tea: An acetone extract of the leaves and flowers repelled *Popillia japonica* adults.[126]
Ceanothus cuneatus Nutt.

Ceanothus leucodermis Greene: An ethanol extract of the fresh leaves of *C. cuneatus* deterred feeding by *Lymantria dispar* larvae, but an extract of *C. leucodermis* leaves did not.[84]

Colubrina viridis (M. E. Jones) M. C. Johnst.: An ether extract of the combined twigs and leaves failed to deter feeding by adult *Acalymma vittatum*.[89]

Krugiodendron ferreum (Vahl.) Urban, palo, black ironwood: The wood is highly resistant to attack by *Cryptotermes brevis* termites.[86,124]

Maesopsis femini Engl., musizi: The wood is susceptible to attack by termites in Nigeria and Uganda.[370]

Rhamnus alnifolia Pursh., buckthorn: Emodin, a mixture of anthraquinones isolated from the foliage, deterred feeding by *Lymantria dispar* larvae.[1081]

Rhamnus crenata Sieb. & Zucc.: *Plutella maculipennis* larvae refused to feed on this plant.[195]

Zizyphus rugosa Lam.: Wood about 4000 years old, found to contain anthraquinone derivatives, was very resistant to termites.[368] An acetone extract of the whole plant was somewhat deterrent to feeding by *Spilosoma obliqua* larvae.[194]

Zizyphus sativa Gaertn. var. *imermis*: A methanol extract of the leaves was a weak growth retardant for *Bombyx mori* larvae fed thereon.[82]

FAMILY RHIZOPHORACEAE

Anisophyllea cinnamoides Alston: An ether extract of the stem bark did not deter feeding by *Acalymma vittatum* adults.[89]

Anopyxis blainesne Engl., bodica: The wood is resistant to *Reticulitermes lucifugus* termites.[441]

Poga olease Pierre: The wood is susceptible to attack by *Reticulitermes lucifugus* termites.[441]

Rhizophora mangle L., mangrove, red mangrove: The wood is somewhat resistant to attack by *Cryptotermes brevis* termites.[86] An ethanol extract of the fresh leaves deterred feeding by *Lymantria dispar* larvae.[84]

FAMILY ROSACEAE

Adenostoma fasciculatum **Hook. & Arn.,** chamis: An ethanol extract of the fresh leaves did not deter feeding by *Lymantria dispar* larvae.[84]

Agrimonia eupatoria **L. var. pilosa:** A benzene extract of the whole plant tested at 5% showed 80 to 100% deterrency of feeding by *Spodoptera litura* larvae.[258]

Eriobotrya senegalensis **Hkf.:** An acetone extract of the aerial portion deterred feeding by *Spilosoma obliqua* larvae.[194]

Hagenia abyssinica **J. F. Gmel.,** kousso: An acetone extract of the whole plant did not repel *Popillia japonica* adults.[126]

Heteromeles arbutifolia **Roem.:** An ethanol extract of the fresh leaves did not deter feeding by *Lymantria dispar* larvae.[84] The shredded leaves release a volatile toxicant that kills larvae of mites and whiteflies; the toxicant is believed to be hydrogen cyanide.[1082]

Horkelia fusca **Lindl. ssp.** *pseudocapitata* **Keck:** An ether extract of the whole plant did not deter feeding by *Acalymma vittatum* adults.[89]

Licania densiflora **Kelink.,** marishiballi: The wood is resistant to attack by *Cryptotermes brevis* termites.[86]

Malus sikkimensis **Bailey,** crabapple

Malus toringoides **Hughes,** crabapple: Six varieties of crabapples were found to be resistant to the development of larvae of the apple maggot, *Rhagoletis pomonella*. Resistance was correlated with the total phenol content.[1083]

Malus sylvestris **Mill.,** apple: Apple trees are rarely attacked by *Lymantria dispar* larvae[422] and are reported to be resistant to attack by *Eriosoma lanigerum* and *Rhopalosiphum fitchii*.[1084]

Potentilla argentea **L.,** silver cinquefoil: An acetone extract of the whole plant did not repel *Popillia japonica* adults.[126]

Prunus armeniaca **L.,** apricot

Prunus caroliniana **Ait.**

Prunus cerasifera **Ehrh.:** Ethanol extracts of the fresh leaves of these three species did not deter feeding by *Lymantria dispar* larvae.[84]

Prunus dulcis **(Mill.) D. A. Webb,** bitter almond: An ethanol extract of the fresh leaves did not deter feeding by *Lymantria dispar* larvae.[84] At least 44 components were identified in a steam distillate of the almond hulls, which appear to attract and inhibit the growth of *Amyelois transitella*.[1085] Almond oil was slightly repellent to *Lucilia cuprina*.[668]

Prunus laurocerasus **L.,** cherry laurel: A methanol extract of the leaves fed to *Bombyx mori* larvae caused growth retardation.

Prunus occidentale **Sw.,** almendron: The wood is highly resistant to attack by *Cryptotermes brevis* termites.[86,1088]

Prunus padus **L.,** European bird cherry: Applications of hot water extracts combined with the wearing of collars woven from the steamed branches resulted in the delousing of cattle within 3-4 days, with the effect lasting for several months.[1087]

Prunus persica **(L.) Batsch.,** peach, nectarine: A methanol extract of the leaves did not deter feeding by *Lymantria dispar* larvae.[84] Lyophilized leaves or an ethanol extract of the leaves deterred feeding by larvae of *Choristoneura rosaceana* and inhibited their development due to the formation of hydrogen cyanide.[1086]

Prunus puddum **Roxb.:** An ether extract of the seed did not deter feeding by adult *Acalymma vittatum*.[89]

Prunus salicina **Lindl.,** Japanese plum

Prunus serotina **Ehrh.,** black cherry: Ethanol extracts of the fresh leaves of these two species failed to deter *Lymantria dispar* larvae from feeding.[84] An aqueous extract of *P. serotina* leaves did not deter feeding by *Melanotus communis* larvae.[247]

Prunus virginiana **L.,** chokecherry: An ethanol extract of the fresh leaves did not deter feeding by *Lymantria dispar* larvae.[84]

Purshia tridentata **(Pursh.) DC:** An acetone extract of the leaves and stems significantly deterred feeding by *Leptinotarsa decemlineata* larvae. A polyphenolic fraction isolated from the extract was responsible for the activity.[281]

Pyrus ussuriensis **Maxim.:** The polar fraction of the leaf extract had a strong deterrent effect on feeding by *Psylla pyricola* when applied to *Pyrus* leaves.[1089]

Quillaja saponaria Molina, soapbark: An acetone extract of the stem bark did not repel adult *Popillia japonica*.[126] Filter papers impregnated with a 3% solution of a saponin obtained from the bark was toxic to *Reticulitermes flavipes* termites and inhibited their feeding; 0.05 and 0.5% solutions were not effective.[229]

Raphiolepis umbellata C. K. Schneider var. mertensii: A methanol extract of the leaves was moderately retardant to the growth of *Bomyx mori* that fed on the extract.[82]

Rosa arkansana Porter, Arkansas rose: *Melanoplus differentialis, M. sanguinipes,* and *M. femurrubrum* grasshoppers failed to reach adulthood when reared on a diet containing aqueous extracts of the plant.[19,275]

Rosa canina L., dog rose: An aqueous extract of the leaves failed to deter feeding by *Pieris brassicae* larvae[652] and did not prevent oviposition by *Cydia pomonella*.[1073]

Rosa gallica L., French rose: Rose oil was a strong repellent for adult *Blattella germanica*.[61]

Rosa multiflora Thunb. ex Murr., multiflora rose: Extracts of the stems prepared first with ether and then with methanol failed to deter feeding by *Spodoptera frugiperda* larvae.[204] Ethereal and ethanolic extracts of the leaves did not deter feeding by adult *Acalymma vittatum*.[52]

Rubus allegheniensis Porter, blackberry: An ethanol extract of the fresh leaves did not deter feeding by *Lymantria dispar* larvae.[84]

FAMILY RUBIACEAE

Adina cordifolia Benth. & Hook. f.: A 5% ethanol extract of the leaves gave complete protection against *Amphocerus cornutus* adults for 12 months when brushed on timber.[120]

Anthocephalus chinensis Lam.: An ether extract of the leaves did not deter feeding by *Acalymma vittatum* adults.[89]

Calycophyllum candidissimum (Vahl) DC, degame, dagame: Acetone and pentane ex-

tracts of the wood did not deter feeding by *Coptotermes formosanus* termites, but 92% of those that fed on the pentane extract did not survive.[369]

Cephalanthus occidentalis L., buttonbush: An acetone extract of the whole plant did not repel *Popillia japonica* adults.[126]

Cinchona officinalis L., cinchona, quinine Cinchona pubescens Vahl., quinine: Quinidine sulfate showed marked mothproofing properties when used with woolen clothing.[456] An acetone extract of the dried bark did not repel *Popillia japonica* adults.[126]

Coffea arabica L., coffee: The wood is very susceptible to attack by *Cryptotermes brevis* termites.[86,124] Caffeine had some mothproofing value but it was not sufficient for practical use.[1090] The compound, 1,3,7-trimethylxanthine, was isolated from the seeds and shown to be a chemosterilant for *Callosobruchus chinensis* at a concentration of 1.5%.[1091] Caffeine and other methylxanthines have been shown to be insecticidal as well as deterrent to feeding.[1092]

Coffea liberica Bull ex Hiern., liberica coffee: An ether extract of the combined leaves and fruit failed to deter feeding by *Acalymma vittatum* adults at 0.1% but completely prevented feeding at 0.5%.[89]

Cunninghamia sinensis R. Brown: The wood is susceptible to attack by *Cryptotermes brevis* termites.[86]

Galium aparine L., bedstraw, catchweed, cleavers: An acetone extract of the whole plant did not repel *Popillia japonica* adults.[126] A methanol extract of the plant did not kill *Aedes* and *Culex* mosquito larvae or inhibit their development.[245]

Galium triflorum Michx. fragrant bedstraw An acetone extract of the whole plant did not repel *Popillia japonica* adults.[126]

Gardenia jasminoides Ellis, Cape jasmine: A methanol extract of the leaves did not prevent feeding by *Bombyx mori* larvae, but it weakly retarded the growth of those that fed thereon.[82]

Genipa americana L., marmelade box, jagua: The wood is very susceptible to feeding by *Cryptotermes brevis* termites.[86,124]

Genipa clusiifolia **Griseb.:** An ethanol extract of the fresh leaves did not deter feeding by *Lymantria dispar* larvae.[84]

Guettarda elliptica **Sw.:** An ethanol extract of the fresh leaves did not deter feeding by *Lymantria dispar* larvae.[84]

Guettarda laevis **Urban,** cucubano: The wood is highly susceptible to attack by *Cryptotermes brevis* termites.[86]

Ixora macrothyrsa **(Teijem. & Binn.) Moore:** An ethanol extract of the whole plant did not repel *Anthonomus grandis grandis* larvae or adults.[487]

Laugeria resinosa **Vahl.,** aquelon: The wood was somewhat resistant to attack by *Cryptotermes brevis* termites.[86,124]

Luculia poinciana **Hook.:** An acetone extract of the whole plant did not prevent feeding by *Spilosoma obliqua* larvae.[194]

Mitchelia repens **L.,** partridgeberry: An acetone extract of the whole plant did not repel *Popillia japonica* adults.[126]

Mitragyna macrophylla **Hiern.:** Shavings of the wood were neither repellent nor resistant to *Reticulitermes lucifugus* termites.[187]

Mitragyna stipulosa **(DC) O. Kuntze,** abura: The wood is susceptible to attack by *Coptotermes formosanus* termites.[388]

Moringa oleifera **Lam.,** moringa: Mixing the powdered bark with stored wheat seeds did not prevent *Sitophilus oryzae* larvae from feeding on the seeds.[761]

Mussaenda frondosa **L.:** An ether extract of the twigs deterred feeding by *Acalymma vittatum* at 0.5% but not at 0.1%. An extract of the leaves was not effective at either concentration.[89]

Nauclea diderrichii **Merrill.,** kusia: The wood is resistant to *Reticulitermes lucifugus* termites[441] but not to *Coptotermes formosanus.*[388]

Pavetta siphonantha **Dalz.:** An acetone extract of the whole plant provided 50% protection from attack by *Spilosoma obliqua* larvae.[194]

Psychotria capensis **(Eckl.) Vatke:** Ether extracts of the combined twigs and leaves and of the combined woody stems with bark strongly deterred feeding by *Acalymma vittatum* adults at 0.5% but not at 0.1%.[89]

Psychotria undata **Jacq.:** An ethanol extract of the fresh leaves did not deter feeding by *Lymantria dispar* larvae.[84]

Randia dumetorum **Lam.:** In India, the bruised roots and unripe fruit were mixed with grain to preserve it from insect attack.[191] A petroleum ether extract of the fruit, used as a spray, showed little toxicity to *Musca domestica* but was repellent to *Aedes aegypti* mosquitoes.[79] An ether extract of the combined woody stems and bark did not deter feeding by *Acalymma vittatum* adults.[89]

Remijia peruviana **Standl.:** An ether extract of the leaves deterred feeding by *Acalymma vittatum* adults at 0.5% but not at 0.1%.[89]

Rothmannia urcelliformis **(Schweinf. ex Hirn.) Bullock:** An ether extract of the combined twigs and stem bark deterred feeding by adult *Acalymma vittatum* at 0.5% but not at 0.1%.[89]

Rubia tinctorum **L.,** madder: Alizarin, obtained from this plant, is quite repellent to termites.[118]

Sarcocephalus diderichii **De Wild.,** opepe: The wood is susceptible to attack by *Cryptotermes brevis* termites.[86]

Sarcocephalus trillesii **Piene:** Shavings of the wood were neither repellent nor resistant to *Reticulitermes lucifugus* termites.[187]

Tricalysia pallens **Hiern.:** Ether extracts of the combined twigs and bark and of the combined twigs and leaves deterred feeding by *Acalymma vittatum* adults at 0.5% but not at 0.1%.[189]

Uncaria gambir **(Hunter) Roxb.,** gambier: An acetone extract of the whole plant was repellent to *Popillia japonica* adults.[126]

Xeromphis nilotica **Keay.:** It has been said[116] that the plant was used as an insecticide, but this has not been confirmed.

Xeromphis spinosa **Keay.:** In India, the bruised fruit is mixed with grain to preserve the latter from attack by insects. A 10% aqueous extract of the root is effective against green scale of coffee.[116]

FAMILY RUTACEAE

Agathosma betulina **(Berg.) Pillans,** buchu: An acetone extract of the whole plant did not repel *Popillia japonica* adults.[126]

Amyris balsamifera L., balsam amyris: Caryophyllene, present in the wood, offers little protection against termite attack at low concentrations, but 5% caryophyllene solution did prevent termites from eating treated wood for almost 4 weeks.[118] The wood oil did not repel adult *Blattella germanica*.[61]

Amyris elemifera L., torchwood: The wood is very resistant to *Cryptotermes brevis* attack.[86] Termites will not even come to rest on the heartwood.[118]

Araliopsis tabouensis Aubrev. & Pellegr.: An ether extract of the stem bark was fairly repellent to adult *Acalymma vittatum* at 0.1% and 0.5%.[89]

Boenninghausenia albiflora (Hook.) Reichb.: A benzene extract of the whole plant at 10% concentration was a strong feeding deterrent for *Spodoptera litura* larvae,[258] but an alcohol extract of the leaves and the resin from this extract gave no protection from termites on the wood.[120] The essential oil extracted from the leaves, which have a disagreeable odor and are used in India as a flea repellent, strongly repelled *Ctenocephalides canis* for almost 1.5 h when spread on the skin; it was not toxic to the cat.[1093] The leaves are also scattered underneath the bed linen in India to ward off bugs and fleas at night.[79] The fraction of the distilled leaf oil free of hydrocarbons was an effective flea repellent.[1094]

Boronia megastigma Nees ex Bartl., sweet boronia: The leaf essential oil did not repel adult *Blattella germanica*.[61]

Calodendrum capensis Thunb.: An ether extract of the woody stems with bark deterred feeding by adult *Acalymma vittatum* at 0.5% but not at 0.1%.[89]

Citrus aurantiifolia (Christm.) Swingle, lime: Lime oil did not repel *Blattella germanica* adults.[61] When black-eyed peas treated with 0.25 to 1.0% of the oil were exposed to adult *Callosobruchus maculatus*, delayed emergence of progeny from eggs, without toxicity, resulted.[1095] The lyophilized oil from the peel was applied topically at 25 and 50 μg to *Attagenus megatoma* larvae, *Plodia interpunctella* larvae, *Lasioderma serricorne*

adults, *Sitophilus oryzae*, and *C. maculatus* adults. High toxicity was shown by *C. maculatus* and moderate toxicity by *S. oryzae*.[1096] An ethanol extract of the fresh leaves deterred feeding by *Lymantria dispar* larvae.[84]

Citrus aurantium L., sour orange: *Plutella xylostella* larvae refused to feed on this plant.[195] A mixture of the dried powdered peel at 3 and 5% with stored wheat for 75 d resulted in 12.53 and 9.96% grain damage, respectively.[61] The essential oil did not repel adult *Blattella germanica*,[61] but it was highly repellent to *Apis florea* bees. The major active components are linalool[149,475] and limonene.[1097] A petroleum ether extract of the leaves applied topically to the abdominal tergites of *Dysdercus cingulatus* nymphs resulted in deformed adultoids and dead adults.[109]

Citrus deliciosa Tenore var. Dancy, tangerine: An ethanol extract of the fresh leaves was highly deterrent to feeding by *Lymantria dispar* larvae.[84]

Citrus limon (L.) Burm. f., lemon: An ethanol extract of the fresh leaves did not deter feeding by *Lymantria dispar* larvae.[84] Lemon oil did not repel adult *Blattella germanica* adults.[61] Black-eyed peas treated with 0.25 to 1.0% of the oil were exposed to adult *Callosobruchus maculatus*, resulting in delayed emergence of progeny without toxicity.[1095] Oil extracted from the peel was highly toxic when applied topically to *C. maculatus* and moderately toxic when applied to *Sitophilus oryzae*.[1096] The toxicity could be ascribed to several fractions isolated from the oil.[1098] Su and Horvat[1079] isolated four limonoids from lemon peel, several of which were toxic to *Callosobruchus maculatus* and *Sitophilus oryzae* adults when applied topically. Limonin has been shown to be an antifeedant for *Leptinotarsa decemlineata* larvae and adults,[1100] and other limonoid byproducts such as nomilin and obacunone are feeding deterrents for larvae of *Spodoptera frugiperda, S. exempta, Eldana saccharina, Heliothis zea,* and *Maruca testulalis*.[824,1101] Powdered lemon rind killed *Ixodes* ticks in 17 min when the insects were exposed to the vapors.[1102] The wood of the lemon tree is susceptible to attack by *Cryptotermes brevis* termites.[86]

***Citrus paradisi* Macfad.**, grapefruit: Grapefruit oil applied to black-eyed peas at 0.5 to 1.0% prevented oviposition by *Callosobruchus maculatus*,[1095] and topical application to this insect and to *Attagenus megatoma*, *Tribolium castaneum*, *Lasioderma serricorne*, *Tribolium confusum*, and *Sitophilus oryzae* was highly toxic.[1096] However, the oil did not repel adult *Blattella germanica*.[61] An ethanol extract of the fresh grapefruit leaves deterred feeding by *Lymantria dispar* larvae.[84] The limonoid, isolimonic acid, was isolated from the seeds as the methyl ester.[1103] Limonin was also isolated from grapefruit seeds.[1104]

***Citrus reticulata* Blanco,** tangerine: Tangerine oil was repellent to *Apis florea* bees.[199,475] The active compound is probably linalool. It is also repellent to *Acromyrmex octospinosus*, and two lipid fractions responsible for this activity were isolated from the leaves.[1105] Four limonoids were isolated from seeds of the variety *austera*.[1106] Tangerine oil applied to black-eyed peas at 0.5 to 1.0% prevented oviposition by *Callosobruchus maculatus*.[1025] Topical application of 50 µg of the oil to the abdominal tergites of *C. maculatus* and *Sitophilus oryzae* was toxic.[1096]

***Citrus sinensis* (L.) Osb.,** sweet orange: The wood is highly susceptible to attack by *Cryptotermes brevis* termites.[86] The oil was repellent to *Apis florea* bees.[149,475] The peel is toxic to *Musca domestica*[1107] and the Caribbean fruit fly.[1108] The limonoid, nomilin, is an excellent antifeedant for *Spodoptera frugiperda* and *Trichoplusia ni* larvae.[1109] Limonin, deoxylimonin, citrolin, and acetoxyharrisonin did not deter feeding by *S. exempta* larvae, but at high concentrations (100 µg) obacunone and harrisonin were mildly effective. Harrisonin and obacunone were highly effective against *Eldana saccharina* and *Maruca testulalis*.[1110]

***Clausena anisata* Hook f.,** samanobere: The plant was commonly hung in houses to repel mosquitoes in the Gold Coast.[121,1111] An ether extract of the root completely prevented feeding by *Acalymma vittatum* adults.[89] Several coumarins were isolated from the root bark.[1112] The volatile oil from the leaves yielded estragole, which is probably responsible for the repellency to mosquitoes and which was toxic to *Zonocerus variegatus* nymphs when applied topically.[1112,1113]

***Euxylophora paraensis* Huber,** Brazilian satinwood: The wood is resistant to *Cryptotermes brevis* termites.[86,124]

***Fagara chalybea* Engl.**

***Fagara holstii* Albuq.:** The alkaloid, *N*-methylflindersine, isolated from the root bark of these species, deterred feeding by *Spodoptera exempta*, *S. littoralis*, and *Epilachna varivestis*.[1114]

***Fagara flavum* Vahl.,** satinwood

***Fagara macrophylla* (Oliv.) Engler**

***Fagara trinitensis* (Williams) J. S. Beard,** bosoo: The wood of these three species is highly resistant to attack by *Cryptotermes brevis* termites.[86]

***Flindersiana ifflaiana* F. v. M.,** hickory ash: The wood is resistant to termites.[188]

***Fortunella hindsii* (Champ) Swingle,** Hong-kong kumquat: Oil extracted from the peel was highly toxic when applied topically to *Callosobruchus maculatus* nymphs.[1096] The furanocoumarin, xanthotoxin, found in this plant, incorporated into the diet of *Spodoptera eridania* larvae at 0.1 and 1.0% was highly toxic to this insect.[1114]

***Oryxa japonica* Thunb.:** The components isopimpinellin, bergaptene, xanthotoxin, kokusagine, evoxine, and japonin were isolated from the leaves and shown to inhibit feeding by *Spodoptera litura* larvae. Isopimpinellin was the most active compound.[1115,1116]

***Pilocarpus jaborandi* Holmes:** The alkaloid, pilocarpine, is a strong muscarinic agonist toward *Boophilus microplus* (Yeerongpilly strain) ticks.[231] An acetone extract of the plant did not repel *Popillia japonica* adults.[126]

***Ptelea trifoliata* L.:** An ethanol extract of the fresh leaves deterred feeding by *Lymantria dispar* larvae.[84]

Ruta graveolens L., common rue: An acetone extract of the whole plant was repellent to *Popillia japonica* adults.[126] The essential oil did not repel adult *Blattella germanica*.[61]

Teclea grandifolia Engl.: Tecleanin, a possible precursor of the insect antifeedant, limonin, and two other tetranortriterpenoids were isolated from this plant.[1117]

Teclea trichocarpa Engl.

Thamnosma africana Engl.: These two species have been used as flea and ant repellents.[116] The 9-acridone alkaloids, melicopicine and tecleanthine, isolated from the bark exhibited mild antifeedant activity against *Spodoptera exempta* larvae. These alkaloids, as well as 6-methoxytecleanthine, showed antimicrobial activity against the fungus, *Cladosporium cucumerinum,* and against *Bacillus subtilis*.[1118]

Zanthoxylum alatum Roxb.: The dried seeds or leaves mixed with maize gave protection from insects in Nepal.[28]

Zanthoxylum americanum Mill.: An ethanol extract of the fresh leaves deterred feeding by *Lymantria dispar* larvae.[84] An acetone extract of the dried bark repelled *Popillia japonica* adults.[126]

Zanthoxylum arborescens Rose: An ether extract of the combined stems, leaves, and flowers did not deter feeding by *Acalymma vittatum* adults.[89]

Zanthoxylum clava-herculis L., southern prickly ash, Hercules-club: The powdered leaves repelled cotton caterpillars.[628] Herculin, a pungent isobutylamide isolated from a petroleum ether extract of the bark, was as toxic as the pyrethrins to *Musca domestica,* including rapid knockdown.[1119]

Zanthoxylum flavum Vahl., satinwood: The wood is resistant to termites.[124]

Zanthoxylum martinicense (Lam.) DC.: The wood is susceptible to termite attack.[86]

Zanthoxylum monophyllum Lam.: Zanthophylline, an alkaloid isolated from this plant,[1120] deterred feeding by *Hemileuca oliviae* larvae, adult *Melanoplus sanguinipes,* and adult *Hypera postica* in both choice and no-choice tests. The compound did not deter feeding by *Schizaphis graminum*.[1121]

Zieria smithii Andr.: The essential oil was highly repellent to *Aedes* and *Anopheles* mosquitoes and to *Lucilia cuprina* but not to *Musca vetustissima*. The active constituents are elemicin and methyleugenol.[342]

FAMILY SABIACEAE

Meliosma herberti Rolfe, aguacatillo: The wood is very susceptible to attack by *Cryptotermes brevis* termites.[86]

FAMILY SALICACEAE

Populus alba L., white poplar: Poplar stems buried in Egyptian soil withstood attack by *Anacanthotermes ochraceus, Psammotermes fuscofemoralis,* and *P. assuarensis* for 6 months.[387]

Populus deltoides Bartr. ex Marsh, eastern cottonwood: An ethanol extract of the fresh leaves did not deter feeding by *Lymantria dispar* larvae.[84]

Populus grandidentata Michx., bigtooth aspen, large-toothed aspen: The sapwood is susceptible to attack by *Reticulitermes flavipes* termites.[462]

Populus nigra L., black poplar: An acetone extract of the whole plant did not repel *Popillia japonica* adults.[126]

Populus spp.: Several species of *Populus* contain salicin, a feeding deterrent for *Spodoptera exempta* larvae.[1122]

Salix chilensis Mol., Humboldt's willow: The wood is very susceptible to termite attack.[124]

Salix humboldtiana Willd., Humboldt's willow: The wood is susceptible to attack by *Cryptotermes brevis* termites.[86]

Salix spp.: Some willows are resistant to *Earias chlorana* larvae, apparently due to the presence of high concentrations of tannins and organic acids, catechins, and substances similar to quercitrin[1123] and salicin.[1122] Only 20% of *Melanoplus femurrubrum* grasshoppers reached adulthood when reared on a diet containing extracts of *Salix* spp.[319]

Salix tetrasperma Roxb., Egyptian willow: Willow stakes buried in Egyptian soil withstood attack by *Anacanthotermes ochraceus,*

Psammotermes fuscofemoralis, and *P. assuarensis* termites for four months.[387] The content of chlorogenic acid is responsible for the resistance of the leaves to feeding by *Lochmaeae capreae cribrata.* However, this compound did not prevent feeding by *Plagiodera versicolora.*[1124]

FAMILY SANTALACEAE

Comandra umbellata **(L.) Nutt.,** comandra: An acetone extract of the whole plant repelled *Popillia japonica* adults.[126]

Eucarya spicata **(R. Br.) Sprague & Somm.,** sandalwood: The essential oil and the resin oil are not repellent to *Aedes* mosquitoes.[342] A proprietary dressing containing sandalwood oil was repellent to *Lucilia cuprina.*[668]

Santalum album **L.,** sandalwood: The wood is resistant to attack by *Cryptotermes brevis* termites.[86] Sandalwood oil repels *Musca nebulo* adults,[623] *Culex fatigans* adults,[623] and *Anthonomus grandis grandis* adults.[355] An ether extract of the leaves did not deter feeding by *Acalymma vittatum* adults.[89] Topical application of the palmitic acid ester of a triterpene alcohol with urs-12-ene skeleton, isolated from sandalwood bark to *Eligma narcissus,Atteva fabriciella,Eupterote geminata,* and *Pyrausta machaeralis* larvae produced morphologically defective pupae and adults. Topical application to *Tribolium castaneum* pupae resulted in sterile adults.[1125,1126] Allatectomized female *Dysdercus cingulatus* treated topically with an acetone extract of the stems showed juvenilization.[123]

FAMILY SAPINDACEAE

Cupania americana **L.,** guara: The wood is susceptible to attack by *Cryptotermes brevis* termites.[86]

Dodonea viscosa **Jacq.**

Filicium decipiens **Thw.**

Koelreuteria paniculata **Lamm.,** goldenrain tree: Ether extracts of the combined stems, bark, and seeds of each of these three species did not deter feeding by *Acalymma vittatum* adults.[89]

Matayba domingensis **(DC.) Radlk.:** The wood is very susceptible to attack by *Cryptotermes brevis* termites.[86]

Sapindus marginatus **Willd.,** soapberry: Three berries of this plant were sufficient to preserve a bushel of wheat from insect infestation, and the powdered or liquid form of the berries mixed with dried foodstuffs repelled weevils and other insects.[1127]

Sapindus mukurossi **Gaertn.:** *Bombyx mori* larvae exposed to the vapors of a methanol extract of the leaves quickly succumbed.[82]

Sapindus trifoliata **L.,** soapnut: Extracts of the seeds placed with wheat repelled the larvae of *Sitophilus oryzae* and adult *Tribolium castaneum.*[512]

Schleichera trijuga **Willd.:** A petroleum ether extract of the plant tested at 0.1 and 0.5% failed to deter feeding by *Acalymma vittatum* adults.[52]

FAMILY SAPOTACEAE

Butyrospermum paradoxum **(Gaertn. f. Hepper subsp. parkii,** shea butter tree, butterseed: The plant is mixed with stored cereals to repel insect pests.[458]

Chrysophyllum pruniforme **(Pierre) Engl.,** dustadwe: The wood is very resistant to attack by *Coptotermes formosanus* termites.[388]

Lucuma multiflora **A. DC.:** The wood is susceptible to termite attack.[388]

Madhuca indica **Gmel.,** moa tree: The seed oil is used as a surface protectant of redgram seeds to check infestations of *Callosobruchus chinensis.*[291] Stored grain is coated with the oil and kept in closed receptacles.[97]

Madhuca longifolia **(L.) Macbr.,** mahua: The wood is resistant to attack by *Anacanthotermes ochraceus, Psammotermes fuscofemoralis,* and *P. assuarensis.*[387] An ethanol extract of the whole plant failed to deter feeding by *Acalymma vittatum* adults at 0.1 and 0.5%.[52] The seed oil, mixed with green gram seeds at 1 ml/100 g, killed all *Callosobruchus chinensis* eggs and adults in 3 d.[234]

Manilkara excelsa **(Ducke) Standl.,** massaranduba

Manilkara nitida (Sesse. & Moc.) Dubard, bulletwood: *M. excelsa* is much more resistant than *M. nitida* to termite attack.[118]

Manilkara multinervis Dubard: The wood is resistant to *Coptotermes formosanus* termites.[388]

Manilkara vapota (L.) van Royen, sapodilla, chicle: Wood about 1000 years old and its ethanol extract were very toxic to termites.[369,1128,1129]

Micropholis chrysophylloides Pierre, leche prieto

Micropholis curvata Pierre

Micropholis garcinifolia Pierre, calmitillo verde: The wood of these three species is very susceptible to attack by *Cryptotermes brevis* termites.[86,124]

Mimusops emarginata (L.) Britton: An ethanol extract of the fresh leaves did not deter feeding by *Lymantria dispar* larvae.[84]

Mimusops heckelii (A. Chev.) Hutch. & Dalz., makore, baku: The wood is very resistant to termite attack.[370]

Mimusops hexandra Roxb.

Mimusops schimperi Hochst., persee: The wood of these two species is very resistant to attack by *Anacanthotermes ochraceus*, *Psammotermes fuscofemoralis*, and *P. assuarensis* termites.[387]

Palaquium ridleyi King & Gamble

Palaquium tellatum King & Gamble: The wood of these trees is sufficiently durable to be safe for use without preservative treatment even when exposed to termites.[480]

Pouterium chiricana L.: An acetone extract of the wood was highly toxic to *Coptotermes formosanus* termites.[369]

Pouterium demerarae Sandw., asipoko: The wood is susceptible to attack by *Cryptotermes brevis* termites.[86,124]

Pouterium lasiocarpa (Mart.) Radlk.: An ether extract of the twigs was moderately deterrent to feeding by *Acalymma vittatum* adults at 0.5% but not at 0.1%.[89]

Pouterium multiflora A. DC., jacana: The wood is susceptible to attack by *Cryptotermes brevis* termites.[86]

Pouterium torta (Mart.) Radlk.: An ether extract of the twigs was moderately deterrent

to feeding by *Acalymma vittatum* adults at 0.5% but not at 0.1%.[89]

FAMILY SAXIFRAGACEAE

Hydrangea macrophylla Seringe var. *Acuminata:* A benzene extract of the whole plant failed to deter feeding by *Spodoptera litura* larvae.[258]

Hydrangea macrophylla Seringe var. *Macrophylla:* A methanol extract of the leaves strongly retarded the growth of *Bombyx mori* larvae feeding thereon.[82]

Ribes vulgare Lam.: An ethanol extract of the leaves did not deter feeding by *Lymantria dispar* larvae.[84]

FAMILY SCROPHULARIACEAE

Aureolaria pedicularia (L.) Raf., gerardia: An acetone extract of the leaves and flowers was repellent to *Popillia japonica* adults.[126]

Aureolaria virginica (L.) Pennell: This plant is reported to prevent attacks of flies on horses.[303]

Chelone glabra L., turtlehead: An acetone extract of the dry leaves was more or less repellent to *Popillia japonica* adults.[126]

Digitalis lanata Ehrh., digitalis: Lanatasaponin, isolated from the seeds, was toxic to *Reticulitermes flavipes* termites and prevented feeding when tested as a 3% solution on filter papers.[229]

Digitalis purpurea L., foxglove: Digitonin, isolated from the seeds, was toxic to *Reticulitermes flavipes* termites and prevented feeding when tested as a 5% solution on filter papers.[229] The juice pressed from the fresh leaves did not deter feeding by *Melanoplus communis* larvae.[247] A methanol extract of the leaves strongly retarded the growth of *Bombyx mori* larvae feeding thereon;[82] digitalin and digitoxin are presumed to be the active principles.[83] An acetone extract of the whole plant did not repel *Popillia japonica* adults.[126]

Euphrasia officinalis L., eyebright: An acetone extract of the whole plant did not repel *Popillia japonica* adults.[126]

Linaria canadensis L., toadflax: The juice pressed from the fresh leaves did not deter feeding by *Melanotus communis* larvae.[247]
Linaria vulgaris Mill., toadflax, butter-and-eggs: An acetone extract of the whole plant did not repel *Popillia japonica* adults.[126]
Lindenbergia grandiflora Benth.: An acetone extract of the whole plant prevented feeding by *Spilosoma obliqua* larvae.[194]
Melampyrum lineare Desr., cow-wheat: An acetone extract of the whole plant did not repel *Popillia japonica* adults.[126]
Penstemon grandiflorus Fras.: Of *Melanoplus femurrubrum* grasshoppers reared on a diet containing extracts of this plant, 63% reached adulthood.[319] Of *M. differentialis* and *M. sanguinipes* grasshoppers reared on the diet, 50% and 25% respectively, reached adulthood.[275]
Scrophularia marilandica L., figwort: An acetone extract of the whole plant did not repel *Popillia japonica* adults.[126]
Scrophularia ningpoensis Hemsl.: A methanol extract of the leaves caused moderate growth retardation of *Bombyx morii* larvae feeding thereon.[82]
Silvia itauba (Meissn.) Ducke: The wood is resistant to termite attack.[124]
Verbascum blattaria L., moth mullein: A methanol extract of the whole plant tested at 1000 ppm killed 100% of *Aedes aegypti* larvae.[245]
Verbascum thapsus L., mullein: An acetone extract of the whole plant did not repel *Popillia japonica* adults,[126] and an ethanol extract failed to repel larvae and adults of *Anthonomus grandis grandis*.[487] Consecutively prepared ether and methanol extracts of the leaves tested at 15,000 ppm in the diet did not deter feeding by *Spodoptera frugiperda* larvae.[204] When maxillectomized *Manduca sexta* larvae were reared on this plant, growth and reproduction were poor, probably because of the low rate of feeding.[1130]

FAMILY SIMAROUBACEAE

Ailanthus altissima (Mill.) Swingle, ailanthus, tree-of-heaven: An acetone extract of

the whole plant did not repel *Popillia japonica* adults.[126]
Hannoa undulata (Guillerm. & Perr.) Planch: An ether extract of the root bark deterred feeding by *Acalymma vittatum* adults.[89]
Harrisonia abyssinica Oliv.: The root bark contains a hydroperoxychromane and pedonin, a spiro tetranortriterpenoid, both of which show antifeedant activity against *Spodoptera eridania* and *S. exempta* larvae.[1131,1132]
Simaba multiflora A. Juss.:
Scilamea soulameoides (Gray) Nooteboom: Six quassinoids have been isolated from these plants that deter feeding by larvae of *Heliothis virescens* and *Spodoptera frugiperda*.[1133] Numerous quassinoids have been isolated from plants of the genera *Picrasma* and *Quassia*, and a goodly number of the compounds have been shown to be insect antifeedants. The subject is well reviewed by Leskinen et al.[1134] and Lidert et al;[1135] the latter scientists tested 46 quassinoids as antifeedants for *Heliothis virescens* and *Agrotis ipsilon* larvae.

FAMILY SOLANACEAE

Acnistus arborescens Schlecht.: This plant was readily accepted as food by *Manduca sexta* larvae, and growth was normal.[1136]
Atropa bella-donna L., belladonna, deadly nightshade: Atropine, the major alkaloid of this plant, fed to larvae and adults of *Leptinotarsa decemlineata* at 0.4 to 0.9% (concentrations normally found in the leaves) mixed with potato leaves caused intestinal epithelial tumors or was extremely toxic.[1137] However, the plant was readily accepted by *Manduca sexta* larvae.[1136] Larvae of *Euxoa messoria* were tolerant of levels up to 0.1% atropine in the diet.[1138] Larvae of *Leptinotarsa decemlineata, Epilachna vigintioctomaculata,* and *E. vigintioctopunctata* survived well on this plant, with 70 – 100% reaching adulthood.[583]
Browallia americana L.: This plant was a good host for *Manduca sexta*.[327]

Browallia demissa L.: Larvae of *Leptinotarsa decemlineata* and *Epilachna vigintioctomaculata* could not survive on this plant, but about 50% of *E. vigintioctopunctata* survived to the adult stage.[583]
Brunfelsia americana L.: This plant was a good host for *Manduca sexta*.[327]
Capsicum annuum L. var. annuum, bell pepper, green pepper: An ether extract of the fruits did not deter feeding by *Acalymma vittatum* adults.[89] Leaves and fruits were accepted as food by *Manduca sexta* larvae, but growth was slow on the leaves and a high rate of mortality was caused by the fruits.[1130,1136] With the exception of the youngest seedlings, the plant is resistant to attack by *Leptinotarsa decemlineata, Epilachna vigintioctomaculata, and E. vigintioctopunctata* larvae.[583]
Capsicum frutescens L., tabasco pepper, chili pepper, red pepper: The plant is used in China as a fumigant for stored grain.[283] The ground dried roots are sprinkled among materials to be preserved in El Salvador.[295] The juice pressed from the fresh leaves did not deter feeding by *Melanotus communis* larvae.[247] The crushed plant did not kill four species of stored product insects.[1139] Broken or powdered red peppers are used in several countries of Asia and Africa to preserve stored rice, cowpeas, beans, and maize from insect attack for periods up to 6 to 10 months.[28] The fruits repel larvae of *Diacrisia obliqua* and deter feeding by this insect.[1140] A methanol extract of the fruits, prepared following extraction with petroleum ether, was quite toxic to *Sitophilus oryzae*.[1141] The alkaloid, capsaicin, is probably responsible for many of the properties detrimental to insects.[1142]
Cestrum nocturnum L., night jessamine: The powdered leaves were detrimental to the emergence and development of *Corcyra cephalonica* larvae in wheat.[220]
Cestrum parqui L'Her.: An ether extract of the roots failed to deter feeding by *Acalymma vittatum* adults.[89]
Datura arborea L.: The plant is used in Cuba to repel giant ants,[116,996] but it is accepted as food by *Manduca sexta* larvae.[1136]

Datura ferox L.
Datura innoxia Mill.
Datura meteloides DC.
Datura quercifolia H.B.K.: These four species of *Datura* are accepted as food by *Manduca sexta* larvae.[1136]
Datura stramonium L., jimson weed, thorn apple: Treating groundnuts with the seeds or leaves of this plant reduced damage by *Pachmerus longus* in clay pots for 45 d from 33 to 25%.[1143] The leaves are a satisfactory food for *Manduca sexta* larvae,[1130] but when used as a fumigant the leaves were not effective against bedbugs, cockroaches, flies, clothes moths, and mosquitoes.[679] The juice expressed from the fresh leaves did not deter feeding by *Melanotus communis* larvae.[247] A petroleum ether extract of the fruits applied topically to *Dysdercus cingulatus* nymphs caused death.[108,109]
Datura spp.: Withanolides have been isolated and identified by Ascher et al[1144] from several species of *Datura* that are antifeedants for the larvae of *Spodoptera littoralis, Epilachna varivestis,* and *Tribolium castaneum.*
Dunalia arborescens (L.) Sleum.: The first withanolide glycosides reported were isolated and identified from this plant; they have been designated as dunawithanine A and B.[1145]
Hyoscyamus niger L., henbane, black henbane: The plant was readily accepted as food by *Manduca sexta* larvae,[1136] but the growth of the insect was strongly retarded.[82]
Juanulloa aurantiaca DC.: Larvae of *Leptinotarsa decemlineata, Epilachna vigintioctomaculata, and E. vigintioctopunctata* could not survive to adulthood on this plant.[583]
Lycium chinense Mill., gow-kee: A methanol extract of the leaves caused growth retardation in larvae of *Bombyx mori* feeding thereon.[82]
Lycium halimifolium Mill., matrimony vine: The leaves are a satisfactory host for *Manduca sexta* larvae.[327,1130]
Lycopersicon esculentum Mill., tomato: The juice expressed from the fresh leaves deterred feeding by *Melanoplus communis* larvae.[247] The alkaloids, tomatine and tomatid-

ine, obtained from the plant were tested for feeding response by *Manduca sexta* larvae; tomatine stimulated feeding slightly.[1146] Compounds isolated from the leaves that retard the larval growth of *Heliothis zea* have been isolated and identified as α-tomatine, chlorogenic acid, and rutin.[1147] Phenolic compounds also influence the growth of this insect.[1148] α-Tomatine is also toxic to an endoparasite, *Hyposoter exigua*, of the larvae.[1149] Tomatine inhibits feeding by *Leptinotarsa decemlineata*.[327] Those larvae that did feed required 22.3 d for development to the pupal stage, with 77.2% mortality.[1150] *Plutella xylostella* larvae refused to feed on tomato leaves.[195] 2-Tridecanone is a good repellent and feeding deterrent for stored product insects. It is effective against granary, rice, and maize weevils, but not against *Tribolium confusum, T. castaneum, Rhyzopertha dominica, Plodia interpunctella,* and *Cadra cautella*, which lay their eggs on the surface of the grain.[1151]

Lycopersicon hirsutum f. *glabratum* **Humb. & Bonpl.**: The resistance of the tomato plant to *Manduca sexta* and *Heliothis zea* larvae is probably due to 2-tridecanone.[1152,1153] This compound also plays an important role in the resistance to *Leptinotarsa decemlineata*.[1154] Dymock et al.[1155] tested 2-tridecanone and a series of 15 related compounds for toxicity to *Heliothis zea* larvae. Chain length had a strong influence on toxicity.[1155] The resistance of a number of wild tomato and domestic tomato accessions to *Spodoptera littoralis, Plusia chalcites, Heliothis armigera,* and *Phthorimaea operculella* was studied in the laboratory, greenhouse, and field. Of these, *L. hirsutum* f. *glabratum* was highly resistant to all four species of insects.[1156]

Nicandra physalodes **(L.) Gaertn.**, apple-of-Peru, shoofly: An ether extract of the leaves deterred feeding by *Acalymma vittatum* adults.[89] Withanolides have been isolated and identified for several species of *Nicandra* that are antifeedants for the larvae of *Spodoptera littoralis, Epilachna varivestis,* and *Tribolium castaneum*.[1144] Larvae of *Leptinotarsa decemlineata* did not survive on this

plant, but about 50% of *Epilachna varivestis* did survive to adulthood.[583] The leaves were acceptable to *Manduca sexta* larvae after 8 hr of contact, but development was slow with high mortality.[327] An insecticide highly toxic to *Musca domestica* and *Manduca sexta* has been isolated from the fresh foliage; it is a conjugated ketone called "nicandrenone".[1157-1159] A few dozen of the plants distributed in a greenhouse will eliminate whiteflies.[1160]

Nicotiana tabacum **L.**, tobacco: The juice expressed from the fresh leaves deterred feeding by *Melanotus communis* larvae.[247] The chopped dry leaves mixed with stored groundnuts in Zambia gave some protection from *Caryedon serratus*.[1143]

Nierembergia hippomanica **Miers.**: Larvae of *Leptinotarsa decemlineata, Epilachna vigintioctomaculata,* and *E. vigintioctopunctata* did not survive to adulthood on this plant.[583]

Petunia hybrida **Vilm.**, petunia: This plant was readily accepted as food by *Manduca sexta*, but the larvae did not develop and suffered premature death.[327] An acetone extract of the fresh leaves did not repel *Popillia japonica* adults.[126]

Physalis alkekengi **L.**, Chinese lanternplant: This plant is a good host for *Manduca sexta* larvae.[327]

Physalis heterophylla **Nees.**, clammy ground-cherry: The leaves are readily accepted by *Manduca sexta* larvae.[1130,1136]

Physalis ixocarpa **Brot. ex Hornem.**, tomatillo, husk tomato: The leaves are a satisfactory host for *Manduca sexta* larvae.[327] An ether extract of the fresh leaves at 0.5% prevented feeding by *Acalymma vittatum* larvae, but 0.1% was not effective.[89]

Physalis longifolia **Nutt.**: An ether extract of the leaves prevented feeding by *Acalymma vittatum* adults at 0.5% but not at 0.1%.[89]

Physalis peruviana **L.**, Peruvian groundcherry: Extracts of this plant were tested for feeding deterrency against *Spodoptera littoralis, S. frugiperda, Boarmia selenaria, Epilachna varivestis, Acalymma vittatum, Diabrotica undecimpunctata howardi, Tri-*

bolium castaneum, T. confusum, Sitophilus oryzae, Callosobruchus maculatus, and *Oryzaephilus surinamensis,* and a number of active withanolides were isolated and identified.[1161] Several withanolides and related ergostane-type steroids were deterrent to feeding by *Epilachna varivestis* larvae.[1162]

***Physalis pubescens* L.,** downy groundcherry: An ether extract of the leaves prevented feeding by *Acalymma vittatum* adults at 0.1 and 0.5%.[89]

***Solanum acaule* Bitter:** The glycoalkaloid, solacauline, in the leaves of this plant inhibited feeding by larvae of *Leptinotarsa decemlineata.*[327] Acauline, another glycoalkaloid present in the leaves, is toxic to this insect, killing some larvae and causing delayed development in others.[1163]

***Solanum auriculatum* Ait.,** fuma bravo: The glycoalkaloids, solamargine and solasonine, in this plant do not inhibit feeding by *Leptinotarsa decemlineata* larvae.[327] The wood is susceptible to attack by *Cryptotermes brevis* termites.[86]

***Solanum aviculare* G. Forst.:** The glycoalkaloids, solamargine and solasonine, do not inhibit feeding by *Leptinotarsa decemlineata* larvae.[327]

***Solanum berthaultii* Hawkes,** wild potato: The leaves repel *Myzus persicae* aphids from a distance of 1 to 3 mm, due to their volatilization of *trans*-β-farnesene.[1164]

***Solanum brachystachys* Dun.:** An ether extract of the combined stems, leaves, flowers, and fruit failed to deter feeding by *Acalymma vittatum* adults.[89]

***Solanum capsicoides* Hort.:** An acetone extract of the whole plant did not inhibit feeding by *Spilosoma obliqua* larvae.[194]

***Solanum carolinense* L.,** horse nettle: An acetone extract of the whole plant did not repel adult *Popillia japonica.*[126] The plant is also a good host for *Manduca sexta* larvae.[327]

***Solanum chacoense* Bitter:** An aqueous extract of the leaves inhibited feeding by *Leptinotarsa decemlineata* larvae. The leaves contain considerable amounts of water-soluble glycoalkaloids, which hydrolyze to form leptines. The leptine content is sufficient to explain the resistance to both the larvae and

adults of *L. decemlineata.*[1165,1166] α-Chaconine does not inhibit feeding by *L. decemlineata,*[327] but when various *Solanum* alkaloids were incorporated into the diet feeding was discouraged by all these compounds.[1167] Leaf extract preparations of impure leptine I reduced feeding and caused increased mortality of *Empoasca fabae* larvae.[1168]

***Solanum demissum* Lindl.:** The plant is resistant to feeding by *Leptinotarsa decemlineata* larvae[29,1169] due to its content of the glycoalkaloid, demissine.[327]

***Solanum dulcamara* L.,** bitter nightshade: The leaves are readily accepted by *Manduca sexta* larvae.[1130,1136,1150]

***Solanum jamesii* Torr.**

***Solanum luteum* Mill**

***Solanum macole* Sukasov:** These three species are resistant to *Leptinotarsa decemlineata* larvae.[29,1063]

***Solanum melongena* L.,** eggplant: The juice expressed from the fresh leaves did not deter feeding by *Melanotus communis* larvae.[247] The plant is an acceptable host for *Manduca sexta* larvae,[327] but the fruits do not permit normal growth.[1130]

***Solanum nigrum* L.,** black nightshade: The odor of the plant is attractive to *Leptinotarsa decemlineata* adults.[1150] Water-soluble substances from the fresh leaves elicit feeding responses from *Manduca sexta* larvae.[1130,1136]

***Solanum pampasense* Hawkes:** This diploid species is resistant to *Empoasca fabae* larvae after the seedling reaches the age of 2 weeks, when the total glycoalkaloid content is high.[1170]

***Solanum polyadenium* Greenm.:** The glycoalkaloid tetraside in this plant inhibits feeding by *Leptinotarsa decemlineata* larvae.[327]

***Solanum pseudocapsicum* L.,** Jerusalem cherry: Tomatine in the plant slightly deters feeding by *Manduca sexta* larvae.[1146]

***Solanum rostratum* Dunal,** buffalobur: The plant is a good host for *Manduca sexta* larvae.[327]

***Solanum schickii* Juzepczuk & Sukasov:** Although solanine was isolated from this plant, its resistance to *Leptinotarsa decemlineata* is due to other unidentified factors.[196]

Solanum sodomeum **L.:** The glycoalkaloids solamargine and solasonine in this plant did not inhibit feeding by *Leptinotarsa decemlineata* larvae.[327]

Solanum tuberosum **L.**, potato, Irish potato: *Leptinotarsa decemlineata* larvae feed on this plant because the leaf surface waxes are composed mainly of saturated long-chain esters and alcohols.[1171] The juice pressed from the fresh leaves does not deter feeding by *Melanotus communis* larvae,[247] but *Plutella xylostella* refuse to feed on the leaves.[195] Five *Solanum* alkaloids (tomatine, tomatidine, α-chaconine, α-solanine, solanidine) tested at 10^{-6} *M* significantly deterred feeding by larvae of *Choristoneura fumiferana*.[1172] An analysis of feeding behavior by *Leptinotarsa decemlineata* indicated that variable acceptance of host plants among regional populations has evolved independently of adaptations to alkaloids.[1173]

Withania somnifera **(L.)** Dunal.: An ether extract of the roots deterred feeding by *Acalymma vittatum* adults at 0.1 and 0.5%.[89] The powdered root mixed with wheat at 1 to 3% reduced the emergence of *Corcyra cephalonica* for a period of 2 months.[220] The root was slightly toxic to *Callosobruchus chinensis* and reduced oviposition by this insect.[150] The structures of a large number of withanolides and their derivatives were determined by nuclear magnetic resonance (NMR) spectra,[1174] and the steroidal lactone moiety of withanolides was prepared synthetically.[1175] A number of insect antifeedant withanolides were tested by Ascher et al.[1176–1178]

FAMILY SPARGANIACEAE

Sparganium americanum **Nutt.**, bur reed: An acetone extract of the whole plant did not repel adult *Popillia japonica*.[126]

FAMILY STAPHYLIACEAE

Turpinia heterophylla **(R & P)** Thul.: An ether extract of the stem bark tested at 0.1 and 0.5% strongly deterred feeding by adult *Acalymma vittatum*.[89]

Turpinia paniculata **Vent.:** The wood is resistant to *Cryptotermes brevis* termites.[86,124]

FAMILY STEMONACEAE

Stemona japonica **Miq.:** Stemospironine, an alkaloid isolated from the fresh leaves, fed to *Bombyx mori* larvae was highly toxic to this insect.[1179]

Stemona tuberosa **Lour.**, paipu: A methanol extract of the leaves was highly toxic to *Bombyx mori* larvae feeding thereon.[82]

FAMILY STERCULIACEAE

Dombeya rotundifolia **Planch.:** The seasoned wood is termite-proof.[116]

Eriolaena hookeriana **Wight & Arn:** An ether extract of the seeds strongly deterred feeding by adult *Acalymma vittatum.*

Firmiana platanifolia **Schott. & Endl.:** The leaves were accepted as food by *Spodoptera littoralis* larvae.[91]

Guazuma ulmifolia **Lam.**, West Indian elm: The wood is susceptible to attack by *Cryptotermes brevis* termites.[86]

Helicteres isora **L.:** An ether extract of the root strongly deterred feeding by *Acalymma vittatum* adults.[89]

Mansonia altissima **(A.) Chev.**, mansonia: The wood was toxic to *Coptotermes formosanus* termite attempts to feed thereon.[388]

Melochia tomentosa **L.:** An ether extract of the combined stems, leaves, and fruits deterred feeding by *Acalymma vittatum* adults at 0.5% but not at 0.1%.[89]

Pentapetes phoenicea **L.:** Larvae of *Bombyx mori* feeding on a methanol extract of the leaves were somewhat retarded in development.[82]

Pterocota bequaertii **De Wild.**, pterygota
Pterocymbium beccardi **K. Schum.**, amberod: The wood of these plants is susceptible to termite attack.[188,370]

Sterculia campanulata **Wall.**, papita: No insect pests are reported from plantings of this tree in India.[119]

Sterculia caribaea **R. Br.**, mahoe: The wood is susceptible to attack by *Cryptotermes brevis* termites.[86]

Sterculia foetida L., hazel sterculia, Java olive: Both the bark and the leaves are reported to be repellent to insects.[116] *Musca domestica* adults fed a diet containing 2.5 and 5% of the seed oil produced no eggs. The oil is reported to be edible by humans.[1180]

Sterculia oblonga Mast., yellow sterculia: The wood is susceptible to termite attack.[370]

Sterculia pruriens (Aubl.) K Schum., maho yahu

Sterculia rhinopetala K. Schum., brown sterculia: The wood of these two species is resistant to termites.[370]

Tarrietia argyrodendron Benth., brown tulip oak

Tarrietia javanica Blume

Tarrietia utilis Sprague: The wood of these three species is resistant to attack by *Reticulitermes lucifugus* termites.[188,441]

Theobroma cacao L., subsp. cacao, cacao, cocoa: A chocolate manufacturer mentioned an odd experience with the disposal of cacao shells. Some of the shells were used as bedding for dogs. Later, the keeper credited the shells with having caused the disappearance of fleas that had infested the dogs.[1181] *Bombyx mori* larvae fed on a diet containing a methanol extract of the leaves showed no harmful effects.[82]

FAMILY STYRACACEAE

Styrax benzoides Craib., Siam benzoin: Benzoin was one of the mothproofing materials claimed in a German patent.[208]

Styrax camphorum Pohl.: An ether extract of the stem bark deterred feeding by adult *Acalymma vittatum* at 0.1 and 0.5%.[89]

Styrax obassia Sieb. & Zucc.: *Bombyx mori* larvae fed on a diet containing a methanol extract of the leaves did not survive.[82]

FAMILY SYMPLOCACEAE

Symplocos poliosa Wight: An acetone extract of the aerial portion of this plant afforded only 11.5% protection from feeding by *Spilosoma obliqua* larvae.[194]

FAMILY TAMARICACEAE

Tamarix aphylla (L.) Karst., athel tamarisk: The wooden stakes of this tree withstood attack by *Anacanthotermes ochraceus*, *Psammotermes fuscofemoralis*, and *P. assuarensis* termites.[387]

FAMILY TAXACEAE

Taxus baccata L., English yew

Taxus baccata cv. *fastigiata* Loud.: Methanol extracts of the leaves fed to *Epilachna varivestis* larvae caused mortality or prevented molting. The component responsible is a taxane derivative.[1182]

Taxus brevifolia Nutt., Pacific yew: An ether extract of the stem bark completely prevented feeding by *Acalymma vittatum* adults at 0.1 and 0.5%.[89]

Taxus canadensis Marsh, American yew: An aqueous extract of the plant incorporated in the diet fed to *Macrosiphum euphorbiae* aphids and *Tribolium confusum* affected the survival and reduced the fecundity of both species of insects.[1183]

Taxus cuspidata Sieb. & Zucc., Japanese yew: An acetone extract of the plant did not repel *Popillia japonica* adults.[126] A methanol extract of the leaves incorporated into a diet fed to *Bombyx mori* larvae caused high mortality.[82]

FAMILY TAXODIACEAE

Sequoia sempervirens (D. Don) Endl., redwood: The sapwood was rather susceptible to feeding by *Reticulitermes flavipes* termites,[462] but it was also toxic to this species[461] as well as to *Incisitermes flavipes*.[361] The heartwood was resistant to attack by *Coptotermes formosanus* and *Reticulitermes virginicus* termites.[87] An ethanol extract of the fresh leaves failed to deter feeding by *Lymantria dispar* larvae.[84]

Taiwania chrystomericides Hayata: Caryophyllene, present in the wood, offers little protection against termite attack at low concentrations, but 5% caryophyllene did prevent termites from eating treated wood for almost 4 weeks.[115]

***Taxodium distichum* (L.) L. C. Rich.,** bald cypress: The gummy heartwood is very resistant to termites, but the fine-grained heartwood is susceptible to attack.[124,462] *Reticulitermes flavipes* fed on the wood,[461] but the heartwood was resistant to attack by *Coptotermes formosanus*.[1184] The two most active heartwood constituents were identified as ferruginol and manool.[1184] 2-Furaldehyde, which was also isolated from the heartwood, inhibits growth and is toxic to the larvae of *Bombyx mori* at concentrations as low as 1 ppm.[1185] An ethanol extract of the fresh leaves did not deter feeding by *Lymantria dispar* larvae.[84]

FAMILY TERMINALIACEAE

***Bucida buceras* L.,** black olive: The wood is resistant to termites.[124]

***Conocarpus erectus* L.,** button mangrove: The wood is susceptible to termite attack.[124]

FAMILY TERNSTROEMIACEAE

***Schima rhasiana* Dyer:** An acetone extract of the whole plant was fairly deterrent to feeding by *Spilosoma obliqua* larvae.[194]

FAMILY THEACEAE

***Camellia japonica* L.,** common camellia: An ethanol extract of the fresh leaves deterred feeding by *Lymantria dispar* larvae.[84] Camellidin II isolated from the leaves is an antifeedant for the larvae of *Eurema hacabe mandarina*.[1186]

***Camellia sasangua* Thunb.,** sasangua camellia: An ethanol extract of the fresh leaves deterred feeding by *Lymantria dispar* larvae.[84]

***Camellia sinensis* (L.) Ktze.,** tea: Filter papers impregnated with a 3% solution of tea saponin from the fruits were toxic to *Reticulitermes flavipes* termites and inhibited feeding by the insect; 0.05 and 0.5% solutions

were not effective.[229] A butanol extract of the seeds and roots incorporated into the diet at 1000 ppm failed to suppress brood development in *Xyleborus fronicatus*.[1187]

***Gordonia lasianthus* (L.) Ellis,** loblolly bay: An ethanol extract of the fresh leaves deterred feeding by *Lymantria dispar* larvae.[84]

***Nesogordonia papaverifera* (A. Chevalier) Capuron,** danta: The wood is resistant to termite attack.[370]

FAMILY THYMELAEACEAE

***Daphne cannabina* Wall.:** Ether extracts of the roots and of the stems deterred feeding by *Acalymma vittatum* adults at 0.5% but not at 0.1%.[89]

***Daphne odora* Thunb.:** *Bombyx mori* larvae feeding on the leaves did not survive.[82]

***Edgeworthia papyrifera* Sieb. & Zucc.:** *Bombyx mori* larvae feeding on the leaves did not survive.[82]

***Gnidia kraussiana* Meisn.:** An ether extract of the combined stems, leaves, and roots and an extract of the flowers deterred feeding by *Acalymma vittatum* adults.[89]

***Gonystylus banganus* Baill.,** ramin
***Gonystylus warburgianus* Gilg. ex Domke,** ramin: The wood of these species is not resistant to termite attack.[188,370]

***Stellera chamaejasme* L.,** lang-tu: An alcohol extract of the rootstock was repellent and toxic to tent caterpillars but nontoxic to *Tribolium confusum, Sitophilus oryzae, Aphis gossypii,* and *Periplaneta americana*. The noncrystalline portion of a petroleum ether extract of the rootstock was repellent and toxic to all of these insect species.[424]

FAMILY TILIACEAE

***Erinocarpus nimmonii* Grahm.:** An acetone extract of the whole plant provided some degree of protection from feeding by *Spilosoma obliqua* larvae.[194]

***Luehua seemannii* Blanch.:** Pentane and acetone extracts of the wood were fed upon by *Coptotermez formosanus* termites with impunity.[369]

Pentace bermanica **Kurz.**, thitka: The wood is slightly suseptible to attack by subterranean termites.[370]

Sloanea berteriana **Choisy**: The wood is susceptible to termite attack.[86]

Tilia europaea **L.**, European linden: The wood is susceptible to aphid attack.[119]

FAMILY TROPAEOLACEAE

Tropaeolum majus **L., garden nasturtium,** Indian cress: *Bombyx mori* larvae fed on a diet containing a methanol extract of the leaves.[82]

FAMILY TYPHACEAE

Typha angustifolia **L.**, narrowleaf cattail

Typha latifolia **L.**, common cattail: Acetone extracts of these two plants were not repellent to *Popillia japonica* adults.[126]

FAMILY ULMACEAE

Celtis adolphi-frederici **Engl.**, celtis: The wood is not resistant to termites.[370]

Celtis laevigata **Willd.**, suga, hackberry: This tree was not infested during an outbreak of *Malacosoma disstria* in southern Louisiana.[88] In a choice feeding test, *Coptotermes formosanus* termites fed little on the wood, but *Reticulitermes virginicus* fed well.[87]

Celtis mildbraedii **Engl.**, esa: The wood was moderately damaged by *Coptotermes formosanus* termites.[388]

Celtis occidentalis **L.**, hackberry: An ethanol extract of the leaves failed to deter feeding by *Lymantria dispar* larvae.[84]

Celtis sinensis **Perm. var.** *japonica* **Nakai:** A methanol extract of the whole plant did not deter feeding by larvae of *Eurema hecabe mandarina*.[90]

Phyllostylon brasiliensis **Cap.**, San Domingo boxwood: The wood is very susceptible to attack by *Cryptotermes brevis* termites.[86]

Ulmus alata **Michx.**, winged elm: An ethanol extract of the foliage did not deter feeding by *Lymantria dispar* larvae.[84]

Ulmus americana **L.**, American elm: The plant was rejected as food by *Manduca sexta*

larvae.[327] Consecutively prepared ether and methanol extracts of the twigs incorporated in the diet at 0.1 and 0.5% failed to deter feeding by *Acalymma vittatum* adults.[52]

Zelkova serrata **(Thunb.) Makino,** Japanese zelkova: This tree can serve as a host for *Scolytus multistriatus*.[1188]

FAMILY URTICACEAE

Boehmeria macrophylla **D. Don.:** An ether extract of the roots deterred feeding by *Acalymma vittatum* adults at 0.5% but not at 0.1%.[89]

Chaetoptelea mexicana **Liebm.**, cenizo: The heartwood of this tree from Panama was moderately resistant to termites.[371]

Coussapoa ovalifolia **Trecul.:** An ether extract of the leaves deterred feeding by *Acalymma vittatum* adults at 0.5% but not at 0.1%.[89]

Parietaria pennsylvanica **Muhl.**, Pennsylvania pellitory: An acetone extract of the whole plant did not repel *Popillia japonica* adults.[126]

Pellionia pulchra **N. E. Br.:** *Plutella xylostella* larvae would not feed on this plant.[195]

Urtica dioica **L.**, stinging nettle, European nettle: An aqueous extract of the leaves, as well as the powdered leaves, did not deter feeding by *Pieris brassicae* larvae.[652]

Urtica procera **Muhl.:** *Bombyx mori* larvae fed to a degree on this plant but they failed to survive.[327]

Urtica thunbergiana **Sieb. & Zucc.:** A methanol extract of the foliage was fed upon by *Bombyx mori* larvae, but the growth and development of these larvae were retarded.[82]

FAMILY VALERIANACEAE

Centranthus macrosiphon **Poiss.:** The seed oil was highly effective as a feeding deterrent for *Anthonomus grandis grandis* adults.[355]

Patrinia scabiosaefolia **Lk.:** A methanol extract of the leaves was moderately retardant to the growth and development of *Bombyx mori* larvae feeding thereon.[82]

Valeriana officinalis **L.**, common valerian, garden heliotrope: The root of this plant was of no value against the screwworm either as a repellent or an attractant,[155] and root extracts were nontoxic to *Culex pipiens* larvae.[182] An ethanol extract of the roots, as well as the powdered roots, did not deter feeding by *Pieris brassicae* larvae[652] or oviposition by *Cydia pomonella*.[1073]

Valerianella radiata (**L.**) **Dufr.**: Cotton plants imbibing alcoholic extracts of this plant failed to repel larvae or adults of *Anthonomus grandis grandis*.[487]

FAMILY VERBENACEAE

Avicennia germinans (**L.**) **L.**, black mangrove: The wood is susceptible to attack by *Cryptotermes brevis* termites.[86]

Callicarpa japonica **Thunb.**: A benzene extract of the whole plant was highly deterrent to feeding by *Spodoptera litura* larvae.[258]

Caryopteris divaricata **Maxim.**: A benzene extract of the leaves tested at a concentration of 1% showed strong antifeeding activity with *Ostrinia nubilalis* larvae.[1189] Eight diterpenes responsible for this activity have been isolated and identified.[1190-1192] The benzene extract was also highly deterrent to *Spodoptera litura* larvae.[258]

Citharoxylum fruticosum **L.**, old woman's bitter: The wood is susceptible to attack by *Cryptotermes brevis* termites.[86]

Clerodendron calamitosum **L.**: The diterpenoid, 3-epicaryoptin, isolated from the leaves deters feeding by *Spodoptera litura* larvae.[1193]

Clerodendron glabrum **E. Mey.**: An African tribe believes that the odor of the leaves is repellent to beetles (unidentified).[116]

Clerodendron inerme **R. Br.**: A petroleum ether extract of the leaves was evaluated as a surface protectant for cowpea seeds against *Callosobruchus chinensis* adults. The results were not promising.[196] A mixture of the powdered leaves with wheat retarded the development of *Corcyra cephalonica* larvae.[220]

Clerodendron infortuanatum **Gaertn.**: The *trans* decalin unit of clerodendrin is an antifeedant for *Pieris brassicae* larvae but not for the larvae of *Spodoptera exigua* and *S. littoralis*.[1194] Myricoside isolated from the roots is a potent antifeedant at 10 ppm for *Spodoptera exempta* larvae.[1195] Clerodendrins A and B

and clerodin deterred feeding by *Spodoptera litura* larvae.[1196-1198]

Clerodendron trichotomum **Thunb.**: A benzene extract of the leaves deterred feeding by *Spodoptera litura* larvae.[1199] Clerodendrins A and B were isolated as the active compounds.[1189,1199-1202]

Duranta repens **L.**: An ethanol extract of the foliage did not deter feeding by *Lymantria dispar* larvae.[84]

Gmelina arborea **Roxb.**, yemane: An ether extract of the twigs deterred feeding by *Acalymma vittatum* adults at 0.5% but not at 0.1%.[89] Filter papers impregnated with an ethanol extract of the wood killed 60% of *Reticulitermes flavipes* termites.[368]

Gmelina leichardtii **F. v. M.**, white beech, gray teak: The wood is resistant to termites.[188]

Lantana camara **L.**, lantana: Larvae of *Plutella xylostella* would not feed on this plant.[195] A chloroform extract of the flowers prevented feeding by *Schistocerca gregaria*.[99] The oil was an effective repellent for mosquitoes and species of *Tabanus*.[1203] An ether extract of the leaves deterred feeding by *Athalia proxima* larvae.[213] An acetone extract of the stems applied topically to the abdominal tergites of *Dysdercus cingulatus* nymphs caused juvenilization,[595] but petroleum ether extracts and alcohol extracts of the leaves, as well as alcohol extracts of the roots, were not toxic to *Musca domestica* and *Aedes aegypti* when tested as sprays.[79]

Lantana rugosa **Willd. ex Schau.**: The leaves are used in Nigeria to protect stored products from insects.[458]

Lippia dulcis **Trevir,** lippia: An acetone extract of the whole plant did not repel *Popillia japonica* adults.[126]

Lippia oatesii **Rolfe**: The plant has been used in Rhodesia as a mosquito repellent, especially after the leaves are bruised.[116]

Lippia ukambensis **Vatke**: The leaves are said to have a preservative effect on foods in Tanzania,[1204] but the essential oil of the leaves, which contains 36.5% of camphor and eight other identified terpenoids, was not insecticidal, larvicidal, or repellent to insects when tested on *Musca domestica* and *Aedes aegypti*.[1205]

Petitia domingensis **Jacq.**, fiddlewood: The wood is somewhat resistant to attack by *Cryptotermes brevis* termites.[86]

Premna odorata **Blanco**, fragrant premna: The leaves are reported to repel insects.[620]

Stachytarpheta mutabilis **Vahl.**: Iplamide, an iridoid glycoside isolated from the leaves, inhibited feeding by *Spodotera littoralis* larvae and by *Schistocerca gregaria* and *Melanoplus sanguinipes*.[1206]

Tectona grandis **L. f.**, teak: White ants would not touch teakwood,[1080] but the sapwood is susceptible to termites.[124] However, tectoquinone (β-methylanthraquinone), a constituent of the resin, is so repellent to termite attack that , at a dilution of 0.05%, the insect would not even rest on the wood.[118] Lapachol from the wood is repellent to *Reticulitermes lucifugus*[368] and *Cryptotermes brevis*.[86] The Indonesian wood was resistant to 21 species of termites.[481] The wood is also reported to be repellent to the subterranean termites *Anacanthotermes ochraceus*, *Psammotermes fuscofemoralis*, and *P. assuarensis*.[387] There are probably several other minor factors involved in resistance.[951] Immersion for 10 min in a 0.05% solution of tectoquinone protected a sample of susceptible *Delonix regia* wood from attack by *Crytotermes brevis* for more than 1.5 years; treatment with 1% solution provided resistance for 6 years.[86] Various components of teakwood showed the following effectiveness against *R. flavipes* (in percentages): anthraquinone 74, 1,4-dihydroxy-2-methylanthraquinone 81, 2,3-dimethylallylnaphthoquinone 83, lapachol 84, dehydro-α-lapachone 87, tectoquinone 89, deoxylapachol 90, *p*-naphthoquinone 91, lapachol 95.[941] The heartwood is susceptible to attack by *Reticulitermes flavipes*.[462] Nymphs of *Incisitermes minor* termites refused to feed on the wood.[361] Another diterpene, tectograndinol, has been isolated from the leaves of this tree but has not been tested as an antifeedant.[1207]

Verbena hastata **L.**, blue vervain: An acetone extract of the whole plant did not repel adult *Popillia japonica*.[126]

Vitex agnus-castus **L.**, chaste-tree: An alcoholic extract of this shrub did not repel larvae or adults of *Anthonomus grandis grandis*.[487]

Vitex altissima **L.**: The wood is resistant to *Anacanthotermes ochraceus*, *Psammotermes fuscofemoralis*, and *P. assuarensis* termites.[387]

Vitex cannabifolia **Sieb. & Zucc.**: *Bombyx mori* larvae did not survive after feeding on a methanol extract of the leaves.[82]

Vitex copassus **Reinw. ex Blume**, New Guinea teak: The wood is resistant to termites.[188]

Vitex divaricata **Sw.**, lizardwood, black fiddlewood: The wood is susceptible to attack by *Cryptotermes brevis* termites.[86]

Vitex floridula **Duchass. & Walp.**: The wood is resistant to *Coptotermes formosanus* termites.[369]

Vitex madiensis **Oliver**: The root bark is relatively free from insect attack. A methanol extract incorporated in the diet fed to larvae of *Spodoptera frugiperda* and *Pectinophora gossypiella* prevented normal molting and caused death. The active component proved to be 20-hydroxyecdysone.[1208]

Vitex negundo **L.**, Indian privet: The leaves were said to preserve rice and clothes from insect attack. In India the leaves were often placed between the pages of books for the same purpose.[1080] Mixing the leaves with stored grain in closed receptacles did not protect the grain from attack by weevils.[97] However, a 5% alcohol extract of the leaves killed 90% of *Plutella xylostella*, 50 to 75% of *Spodoptera litura*, 20% of *Crocidoloma binotalis* and 100% of *Euproctis fraterna*, *Pericalia ricini*, and *Achaea janata* larvae.[81] The leaves are used to repel fleas in the Phillippines.[94]

Vitex pachyphylla **Baker**: An ethanol extract of the wood exposed to *Reticulitermes flavipes* termites killed 40% of the insects.[368]

Vitex parviflora **Baker**: The wood is reported to be resistant to *Anacanthotermes ochraceus*, *Psammotermes fuscofemoralis*, and *P. assuarensis* termites.[387]

FAMILY VIOLACEAE

Rinorea racemosa **(Hart, & Zucc.) Kuntze**: An ether extract of the stems deterred feeding by *Acalymma vittatum* adults at 0.1 and 0.5%.[89]

Viola papilionacea **Pursh.**, butterfly violet

Viola tricolor **L.**, common pansy: Acetone extracts of these plants did not repel *Popillia japonica* adults.[126]

Viola verecunda **A. Gray.**: Larvae of *Bombyx mori* fed on a diet containing a methanol extract of the leaves showed weak growth retardation.[82]

FAMILY VITACEAE

Ampelopsis brevipedunculata **Maxim. ex Trautv.**: A benzene extract of the whole plant deterred feeding by *Spodoptera litura* larvae.[258]
Cissus rhombifolia **Vahl.**, grape ivy: *Plutella xylostella* larvae refused to feed on the plant.[195]
Parthenocissus inserta **(Kerner) K. Fritsch.**, Virginia creeper
Parthenocissus quinquifolia **(L.) Planch.**, Virginia creeper: Ethanol extracts of the fresh leaves of these two species failed to deter feeding by *Lymantria dispar* larvae.[84]
Vitis labrusca **L.**, fox grape: An ethanol extract of the fresh leaves failed to deter feeding by *Lymantria dispar* larvae.[84]
Vitis vinifera **L.**, wine grape: Ethanol extracts of the fresh leaves of the varieties Carignane, Emperor, and Thompson Seedless did not deter feeding by *Lymantria dispar* larvae.[84]

FAMILY VOCHYSIACEAE

Qualea albiflora **Warm.**
Qualea dinizii **Ducke,** kwalie: The wood of these two species is susceptible to attack by *Cryptotermes brevis* termites.[86]
Vochysia maxima **Ducke,** guaruba: The wood is susceptible to attack by *Cryptotermes brevis* termites.[86]
Vochysia melinonii **Beckman,** wane kwarie: The wood is resistant to attack by *Cryptotermes brevis* termites.[86]
Vochysia mondurensis **Sprague,** yemeri: The wood is susceptible to termites in Honduras.[370]
Vochysia tetraphylla **(C. F. W. Mey.) DC,** sieballi: The wood is susceptible to attack by *Cryptotermes brevis* termites.[86]

FAMILY ZINGIBERACEAE

Alpinia officinarum **Hance,** lesser galanga: An acetone extract of the whole plant did not repel *Popillia japonica* adults.[126]

Amomum melegueta **Rosc.**: A decoction of this plant was rubbed on the skin of domestic animals in Africa to repel tsetse flies.[1209]
Curcuma domestica **Val.**, turmeric: The plant is used as an ant repellent in India.[116] Stored wheat containing 3 or 5% of powdered turmeric was not not protected from damage by *Sitophilus oryzae*,[761] but the powdered rhizomes did repel adults of *S. granarius, Tribolium castaneum*, and *Rhyzopertha dominica*.[606] Few adults of *T. castaneum* settled in rice grain treated with 100, 500, or 1000 ppm of turmeric oil; after two weeks the repellency decreased rapidly.[1210] Powdered turmeric rhizomes and a crude petroleum ether extract of the rhizomes were strongly repellent to adults of *Tribolium confusum, Rhyzopertha dominica*, and *Sitophilus granarius* and to larvae of *Attagenus megatoma*. The responsible compounds were isolated and identified as 2-methyl-6-(4-methylphenyl)-2-hepten-4-one (*ar*-turmerone) and 2-methyl-6-(4-methyl)-1,4-cyclohexadien-1-yl)-2-hepten-4-one (turmerone).[1211]
Curcuma zedoaria **(Christm.) Roscoe,** zedoary shoti: An acetone extract of the whole plant failed to repel adult *Popillia japonica*.[126]
Elettaria cardamomum **(L.) Maton,** cardamon, cardamom: *Tribolium castaneum* beetles failed to oviposit and died within 30 d when they were maintained on a diet supplemented with cardamon oil.[181] Cardamom oil failed to repel *Blattella germanica* nymphs in both choice and no–choice tests.[61]
Elettaria zeylanicum **L.**: Nearly complete repellency of *Apis florea* bees was obtained with the essential oil at 5 g/l air in an arena.[475]
Hedychium gracillimum **Koenig:** An acetone extract of the whole plant offered little protection from feeding by *Spilosoma obliqua* larvae.[194]
Hedychium spicatum **Hamilton:** This plant was reported to protect clothes from insect attack in India.[191]
Kaempferia galanga **L.**, galanga: Treatment of produce with a powdered formulation at 0.25 g/100 g was more effective in repelling *Sitophilus oryzae, Trogoderma granarium, Tribolium castaneum, Rhyzopertha dominica*, and

Callosobruchus chinensis than were ether extracts at 0.125 g/100 g.[715,1212]

Zingiber officinale Roscoe, ginger: The ingestion of ginger in East Africa is said to give the body an odor repellent to mosquitoes,[116] but extracts of the plant were nontoxic to *Aedes* larvae.[182] A 5% kerosene spray of the petroleum ether extract of the rhizomes tested on *Musca domestica* did not cause knockdown in 10 min and killed only 27% in 24 h.[1213] Ginger oil repelled adult *Blattella germanica* for only 24 h.[61]

FAMILY ZYGOPHYLLACEAE

***Balanites aegyptiaca* L., Del.,** desert date: An ether extract of the twigs and a pentane extract of the seeds tested at 0.1 and 0.5% did not deter feeding by adult *Acalymma vittatum*.[89]

***Balanites roxburghii* Planch:** A steroidal saponin isolated from the stem bark pre-

vented feeding by *Diacresia obliqua* at a concentration of 500 ppm.[1214]

***Guaiacum officinale* L.,** lignum-vitae: Guiacol derived from this tree was considered to be one of the best repellents for *Cochliomyia hominivorax*.[155] The heartwood is reported to be "practically immune" to *Cryptotermes brevis* termites in the West Indies.[370] Filter papers impregnated with an ethanol extract of the wood killed 100% of exposed *Reticulitermes flavipes*.[368]

***Guaiacum sanctum* L.,** lignum-vitae, holywood lignum-vitae: The heartwood is reported to be immune from attack by *Cryptotermes brevis* termites.[370] It was completely repellent to *Reticulitermes flavipes* termites.[462]

***Peganum harmala* L.,** African rue: Wheat containing 1% of the powdered seeds was only slightly effective in reducing the adult emergence of an infestation of *Corcyra cephalonica* larvae.[220]

Arthropods Deterred from Feeding by Neem

Insect order	Target insect	Stage	Test substance	Ref.
ACARI	Panonychus citri	Adult	Seed oil	719,725
	Tetranychus cinnabarinus	Adult	Kernel pentane, acetone extracts	726
COLEOPTERA	Tetrarychus urticae	Adult	Kernel aq. & ethane exts.	727
	Acalymma vittatum	Adult	Ethanol ext., azadirachtin, salannin	51,52 728,729 729
	Anthrenus flavipes	Adult	Seed oil	730
	Attagenus megatoma	Larva	Ethanol ext.	725
	Aulacophora foveicollis	Adult	Seed oil, Thionimone[a]	731-733 734
	Callosobruchus chinensis	Adult	Seed oil	132,291,730
	Callosobruchus maculatus	Adult	Powdered seed	735,736
			Powdered seed,	732,733, 735-736
			seed oil,	440,739
	Carpophilus hemipterus	Larva, adult	powdered leaf	738
	Chilo partellus	Adult	Seed, ether ext.	725
		Adult	Salannin	52
	Conotrachelus nenuphar	Adult	Various compds.	140
	Diabrotica undecimpunctata	Adult	Kernel, hexane ext.	725
		Adult	Seed, ethanol ext.	52,725, 729,740
	Diaprepes abbreviatus	Adult	Seed, methanol ext.	741
	Epilachna varivestis	Adult	Leaf, methanol ext.	742
			seed, methanol ext.,	743

TABLE 1 (continued)
Arthropods Deterred from Feeding by Neem

Insect order	Target insect	Stage	Test substance	Ref.
	Lasioderma serricorne	Adult	various compds.	744-747
		Adult	Powdered seed	748
	Latheticus oryzae	Adult	Powdered seed	748
	Leptinotarsa decemlineata	Larva	Leaf, methanol ext.	742
		Adult	Seed, methanol ext.	741
		Larva	Fruit, ethanol ext.	749
		Adult	Kernel, aq. ext.	750
	Ootheca bennigseni	Adult	Leaf, seed, aq. ext.	751
	Phyllostreta striolata	Adult	Kernel, ethanol ext., azadirachtin	752
				752
	Podagrica uniforma	Adult	Fruit, aq. ext.	753
	Popillia japonica	Adult	Seed, ether ext.	754-757
			Seed, ethanol ext.	754-757
	Rhyzopertha dominica	Adult	Powdered leaf	606,733
		Adult	Powdered seed	748,758—760
	Sitophilus granarius	Adult	Powdered leaf	606
	Sitophilus oryzae	Adult	Powdered seed	732,733,748, 758,761
	Sitophilus zeamais	Adult	Powdered leaf	762
			Kernel, flower	762
			seed oil	763
	Tribolium castaneum	Adult	Powdered seed	512,732, 748
	Tribolium confusum	Adult	Powdered seed, oil	763
	Trogoderma granarium	Larva	Salannin	725
		Adult	Powdered seed	333,764

Order	Species	Stage	Treatment	Ref.
	…roderema modacella	Adult	Seed,leaf, aq. ext., seed oil	767
	Atherigona soccata	Larva	Seed, ethanol ext.	767
	Liriomyza sativae	Larva, adult	Seed, ethanol ext.	768
	Liriomyza trifolii	Larva	Seed, aq. ext.	741,769
			Seed, ethanol ext.	773
			Azadirachtin	773
		Larva	Leaf distillate	772
		Adult	Seed, ethanol ext.	774
				741,769—
				772
HETEROPTERA	Musca domestica	Adult	Salannin	775
	Oncopeltus fasciatus	Larva	Azadirachtin	43
HOMOPTERA	Rhodnius prolixus	Nymph	Azadirachtin	778
	Aleurothrixus floccosus	Adult	Seed, hexane ext.	49,725
	Aonidiella aurantii	Adult	Seed, hexane ext.	49,725,779
	Aonidiella citrina	Adult	Seed, hexane ext.	49,725,730
	Aphis citricola	Adult	Seed, ethanol ext.	780
			Seed oil,	780
			azadirachtin	780
	Aphis gossypii	Adult	Seed oil	730,781
	Bemisia tabaci	Adult	Seed, aq. ext.	781
	Gnaphalocrocis medinalis	Adult	Seed oil	782
	Jacobiasca lybica	Adult	Seed, aq. ext.	781
	Nephotettix virescens	Adult	Seed oil	54,134-137, 139,783
	Nilaparvata lugens	Adult	Seed oil	53,131,785—
				789
	Planococcus citri	Adult	Seed, hexane ext.	49,725
	Sogatella furcifera	Adult	Seed oil	53,782,784
HYMENOPTERA	Apis florea	Adult	Seed oil	475
	Apis mellifera	Adult	Seed, ethanol ext.	790
LEPIDOPTERA	Achaea janata	Larva	Powdered seed	791
	Agrotis ipsilon	Larva	Powdered seed	791
			Seed, aq. ext.	781

TABLE 1 (continued)
Arthropds Deterred from Feeding by Neem

Insect order	Target insect	Stage	Test substance	Ref.
	Amsacta moorei	Larva	Leaf, aq. ext.	781
			Seed, aq. ext.	791,792
			Seed oil	372
	Amyelois transitella	Larva	Seed, ethanol ext.	49
			Azadirachtin	725
	Argyrotaenia veluinana	Larva	Azadirachtin	725
	Boarmia selenaria	Larva	Seed, ethanol ext.	793
	Chilo partellus	Larva	Vepaol	740
	Corcyra cephalonica	Larva	Powdered seed	748,794,795
	Crocidolomia binotalis	Larva	Seed, ethanol ext.,	796
			seed oil,	797
			azadirachtin	798
	Cydia pomonella	Larva	Azadirachtin	741
	Diacrisia obliqua	Larva	Powdered seed	791
	Diaphania hyalinata	Larva	Seed, aq. ext.	741
	Earias insulana	Larva	Seed, aq. ext.,	799
			azadirachtin,	800
			salannin	800
	Ephestia cautella	Larva	Powdered seed	748
	Euproctis fraterna	Larva	Seed, aq. ext.	768
	Euproctis lunata	Larva	Leaf, ethanol ext.	801
	Eupterote mollifera	Larva	Powdered seed	791
	Galleria mellonella	Larva	Azadirachtin	804
	Heliothis armigera	Larva	Seed oil,	805
			azadirachtin	806
	Heliothis virescens	Larva	Kernel, ethanol ext.,	807,808
			kernel, aq. ext.	809

Species	Stage	Material	Ref.
Heliothis zea	Larva	Azadirachtin	810
Hellula undalis	Larva	Seed oil	811
Homoeosoma electellum	Larva	Seed, ethanol ext.	812
Hypsipyla grandella	Larva	Powdered seed	60
Laspeyresia pomonella	Larva	Seed, hexane ext.	49
Lerodea eufala	Larva	Seed oil	439
Mamestra brassicae	Larva	Kernel, methanol ext.	813
Manduca sexta	Larva	Kernel, ethanol ext.	741
Mythimna separata	Larva	Seed oil	719,782
		vepaol[b]	740
Nephantis serinopa	Larva	Seed, aq. ext.	768
Ostrinia nubilalis	Larva	Azadirachtin	814
Phthorimaea operculella	Larva	Neemrich[c]	815
Pieris brassicae	Larva	Kernel, methanol ext.	813
Pieris rapae	Larva	Seed oil	816
Plutella xylostella	Larva	Kernel, ethanol ext.	60,808, 817,818
		kernel, methanol ext.	47
		kernel, aq. ext.	47,809, 818,819
		powdered kernel	811
		seed oil	811
		azadirachtin	804
Spodoptera exempta	Larva	Azadirachtin	806
Spodoptera frugiperda	Larva	Seed, ethanol ext.	725,820-822
Spodoptera littoralis	Larva	Seed, ethanol ext.	825
		Seed oil	44
Spodoptera litura	Larva	Seed, aq. ext.	826-829
		Seed, methanol ext.	830
		seed oil	221
		azadirachtin	800
		nimbidin	140
		isovepaol[b]	740
Tineola bisselliella	Larva	Azadirachtin	831

TABLE 1 (continued)
Arthropods Deterred from Feeding by Neem

Insect order	Target insect	Stage	Test substance	Ref.
ORTHOPTERA	*Utetheisa pulchella*	Larva	Seed, aq. ext.	733,791
	Acheta domesticus	Adult	Azadirachtin	831
	Acrida exaltana	Adult	Seed oil	732,733
	Blatta orientalis	Nymph	Margosan O[d]	832
	Blattella germanica	Nymph	Margosan O[d]	832
	Byrsotria fumigata	Nymph	Margosan O[d]	832
	Chrotogonus trachypterus	Adult	Seed oil	730
	Diapheromera femorat	Adult	Margosan O[d]	59
	Dissosteira carolina	Adult	Margosan O[d]	59
	Gryllus pennsylvanicus	Adult	Margosan O[d]	59
	Gromphadorhin portentosa	Nymph	Margosan O[d]	832
	Kraussaria angulifera	Nymph	Leaf, fruit	833
		Adult	Leaf, fruit	833
	Locusta migratoria	Adult	Seed oil	731-734
	Oedaleus senegalensis	Nymph	Leaf, fruit	832
	Schistocerca gregaria	Adult	Seed, ethanol ext.	385-388,731, 748,834,835
			leaf, ethanol ext.	387,836-838
		Nymph	Azadirachtin	300
			Meliantriol	839
	Supella longipalpa	Nymph	Margosan O[d]	832
	Zonocerus variegatus	Adult	Seed, aq. ext.	840

a Fractionated neem extract.

b A neem seed component.

c An enriched neem seed formulation.

d A commercial formulation of neem seed ethanol extract.

IV. Scientific and Common Names of Referenced Insects

SCIENTIFIC AND COMMON NAMES OF REFERENCED INSECTS

Acalymma vittatum (Fabricius), striped cucumber beetle

Acanthoscelides obtectus (Say), bean weevil

Achaea janata (Linnaeus), croton caterpillar, castor semi-looper

Acheta domesticus (Linnaeus), house cricket

Acrae eponina Cramer

Acrida exaltata (Walker)

Acromyrmex octospinosus (Reich), leafcutting ant

Adelges cooleyi (Gillette), Cooley spruce gall adelgid

Aedes aegypti (Linnaeus), yellowfever mosquito, northwest coast mosquito

Aedes atropalpus (Coquillett), mosquito

Aedes catator (Coquillett), brown saltmarsh mosquito

Aedes sticticus (Meigen), floodwater mosquito

Aedes triseriatus (Say), mosquito

Aedes vexans (Meigen), vexans mosquito

Aemida gahani (Banks), timber borer

Agrotis ipsilon (Hufnagel), black cutworm

Agrotis orthogonia Morrison, pale western cutworm

Ahasverus advena (Waltl.), foreign grain beetle

Aleurocanthus soiniferus (Quaintance), orange spiny whitefly

Aleurothrixus floccosus (Maskell), wooly whitefly

Altica ambiens (LeConte), alder flea beetle

Amblyomma americanum (Linnaeus), lone star tick

Amphicerus cornutus (Pallas), powderpost beetle

Amrasca devastans (Distant), leafhopper

Amsacta moorei Butler, red hairy caterpillar

Amyelois transitella (Walker), navel orangeworm

Anacanthotermes ahngerianus (Jacobsen), termite

Anacanthotermes ochraceus (Burmeister), termite

Anastrepha ludens Loew. Caribbean fruit fly

Anopheles claviger (Say), mosquito

Anopheles funastus Meigen, mosquito

Anopheles gambiae Giles, mosquito

Anopheles quadrimaculatus Say, malaria mosquito

Anopheles stephensi (Say), mosquito

Anthonomus grandis grandis Boheman, boll weevil

Anthrenus flavipes LeConte, furniture carpet beetle

Anthrenus vorax Waterhouse

Aonidiella aurantii (Maskell), California red scale

Aonidiella citrina (Coquillett), yellow scale

Aphis citricola Vander Groot, spirea aphid

Aphis crassivora Koch, cowpea aphid

Aphis fabae Scopoli, bean aphid

Aphis gossypii Glover, cotton aphid, melon aphid

Apis florea Fabricius, bee

Apis mellifera Linnaeus, honey bee

Aproaerema modicella (D.), groundnut leafminer

Archips cerasivorana (Fitch), uglynest caterpillar

Argyrotaenia velutinana (Walker), redbanded leafroller

Armigeres obturbans (DeGeer)

Athalia proxima Klug, mustard sawfly

Atherigona soccata Rondani, sorghum shoot fly

Atta cephalotes Linnaeus, leafcutting ant

Atta versicolor Pergande, leafcutting ant

Attagenus megatoma (Fabricius), black carpet beetle

Attagenus piceus Olivier, black carpet beetle

Atteva fabriciella Swederus

Aulacophora foveicollis Lucas, red pumpkin beetle

Austrotortrix postvittana (Walker), lightbrown apple moth

Bagrada cruciferarum Kirk

Bemisia tabaci (Gennadius), sweetpotato whitefly

Bithynus gibbosus (De Geer), carrot beetle

Blastophaga piniperda Linnaeus, bark beetle

Blatta orientalis Linnaeus, oriental cockroach

Blattella germanica (Linnaeus), German cockroach

Boarmia selenaria (Denis & Schiffermüller), giant looper

Bombyx mori (Linnaeus), silkworm

Boophilus annulatus australis Fuller, Australian cattle tick

Boophilus microplus (Canestrini), southern cattle tick

Brachypterus glaber Hërbst

Bruchophagus roddi (Gussakovsky), alfalfa seed chalcid

Byrsotria fumigata (Guérin-Méneville), cockroach

Cadra cautella (Walker), almond moth

Calandra granaria Linnaeus, granary weevil

Calandra oryzae Linnaeus, rice weevil

Calliphora vicina (Linnaeus), blue blowfly

Callosobruchus chinensis Lucas, cowpea weevil, adzuki bean weevil

Callosobruchus maculatus (Fabricius), stored cowpea beetle

Camnula pellucida (Scudder), clearwinged grasshopper

Camponotus pennsylvanicus De Geer, black carpet beetle

Carpophilus hemipterus (Herbst), dried-fruit beetle

Caryedes brasiliensis Thunberg

Caryedon serratus (Oliver)

Ceratitis capitata (Wiedemann), Mediterranean fruit fly

Ceratomia catalpae (Boisduval), catalpa sphinx

Chilo suppressalis (Walker), Asiatic rice borer, rice stem borer

Choristoneura fumiferana (Clemens), eastern spruce budworm

Choristoneura rosaceana (Harris), obliquebanded leafroller

Chrysopa carnea Stephens, common green lacewing

Cimex lectularius Linnaeus, bed bug

Cnaphalocrocis medinalis (Guenée), rice leaffolder

Coccinella septempunctata Linnaeus

Cochliomyia hominivorax (Coquerel), screwworm

Conotrachelus nenuphar (Herbst), plum curculio

Coptotermes acinaciformis Froggatt, termite

Coptotermes exitiosus Hill, termite

Coptotermes formosanus Shiraki, Formosan subterranean termite

Coptotermes lacteus (Froggatt), termite

Corcyra cephalonica (Stainton), rice moth

Crocidolomia binotalis Zeller, cabbage webworm, mustard webworm

Cryptotermes brevis (Walker), West Indian drywood termite

Ctenocephalides canis (Curtis), dog flea

Ctenocephalides felis (Bouché), cat flea

Cuclotogaster heterographus (Nitzsch), chickenhead louse

Culex fatigans Wiedemann, mosquito

Culex pipiens Linnaeus, northern house mosquito

Cydia pomonella (Linnaeus), codling moth

Dacus cucurbitae Coquillet, melon fly

Dacus dorsalis Hendel, oriental fruit fly

Danaus plexippus (Linnaeus), monarch butterfly

Delia antiqua (Meigen), onion maggot

Delia radicum (Linnaeus), cabbage maggot

Dendroctonus frontalis Zimmermann, southern pine beetle

Dendroctonus jeffreyi Hopkins, Jeffrey pine beetle

Dendroctonus ponderosae Hopkins, mountain pine beetle

Dendroctonus rufipennis (Kirby), spruce beetle

Dermacentor marginatus Say, tick

Dermacentor variabilis (Say), American dog tick

Diabrotica undecimpunctata howardii, Barber, spotted cucumber beetle, southern corn rootworm

Diacresia obliqua (Walker), Bihar hairy caterpillar

Diaphania hyalinata (Linnaeus), melonworm

Diaphania nitidalis (Stoll), pickleworm

Diaphorina citri (Kuway), Asiatic citrus psyllid

Diaprepes abbreviatus (Linnaeus), sugarcane rootstalk borer weevil

Dicladispa armigera Olivier, rice hispa

Dissosteira carolina (Linnaeus), Carolina grasshopper

Dolichovespula aretaria (Fabricius), aerial yellowjacket

Drosophila melanogaster Meigen, vinegar fly

Dysdercus cingulatus Fabricius, red cotton bug

Dysdercus similis Stal.

Dysdercus suturellus (Herrich-Schäffer), cotton stainer

Earias chlorana Linnaeus

Earias fabia Stoll

Earias insulana (Boisduval), spiny bollworm

Earias vitella(Fabricius), spotted bollworm

Echidnophaga gallinacea (Westwood), sticktight flea

Elatobium abietinum (Walker), spruce aphid

Eldana saccharina Walker

Eligma narcissus Cramer

Empoasca abrupta Delong, eastern potato leafhopper

Empoasca fabae (Harris), potato leafhopper

Empoasca kerri Pruthi

Empoasca kraemeri (Ross & Moore)

Entomoscelis americana Brown, red turnip beetle

Ephestia cautella Walker, fig moth, warehouse moth, almond moth

Epilachna borealis (Fabricius), squash beetle

Epilachna tredecimnotata Fabricius

Epilachna varivestis Mulsant, Mexican bean beetle

Epilachna vigintioctomaculata Fabricius

Epilachna vigintioctopunctata Fabricius

Ergates spiculatus LeConte, pine sawyer

Eriosoma lanigerum (Hausmann), wooly apple aphid

Eucosma gloriola Heinrich, eastern pineshoot borer

Euproctis fraterna (Moore)

Euproctis lunata Hampson, castor hairy caterpillar

Euproctis terminalis Linnaeus

Eupterote geminata Walker

Eupterote mollifera Walker

Eurema hecabe mandarina De L' orza, yellow butterfly

Euxoa messoria (Harris), darksided cutworm

Evergestis rimosalis (Guenée), cross-striped cabbageworm

Fenusa pucilla (Lepeletier), birch leafminer

Galleria mellonella (Linnaeus), greater wax moth

Gilpinia abieticola (Linnaeus)

Gilpinia hercynaie (Hartig), European spruce sawfly

Gilpinia polyoma Hartig

Gromphadorhina portentosa (Schaum), cockroach

Gryllus pennsylvanicus Burmeister, field cricket

Haemaphysalis punctata (Packard), tick

Haematobia irritans (Linnaeus), horn fly

Haemonchus arator (Fabricius)

Heliothis armigera Hübner, African armyworm, bollworm, gram pod borer

Heliothis virescens (Fabricius), tobacco budworm

Heliothis zea (Boddie), corn earworm, cotton bollworm

Hellula undalis Fabricius, imported cabbage webworm

Hemileuca oliviae Cockerell, range caterpillar

Herpetogramma bipunctalis (Fabricius), southern beet webworm

Heteronychus arator (Fabricius), black beetle

Homoeosoma electellum (Hulst), sunflower moth

Hylobius abietis Boheman

Hypera brunneipennis (Boehman), Egyptian alfalfa weevil

Hypera postica (Gyllenhal), alfalfa weevil

Hypderma lineata De Villiers, warble fly

Hyposoter exigua (Fiereck)

Hypsipyla grandella Zeller, shootborer

Incisitermes minor (Hagen), western drywood termite

Ixodes redikurzevi Banks, tick

Jacobiasca lybica (Jac.), leafhopper

Kraussaria angulifera (Krauss), grasshopper

Lasioderma serricorne (Fabricius), cigarette beetle

Lasius alienus (Foerster), cornfield ant

Latheticus oryzae Waterhouse, longheaded flour beetle

Lepisma saccharina Linnaeus, silverfish

Leptinotarsa decemlineata Say, Colorado potato beetle

Leptocarsia chinensis (Fabricius), rice bug

Leptoglossus corculus (Say), leaffooted pine seed bug

Lerodea eutala (Edwards), rice leaffolder

Lipaphis erysimi Kalt., mustard aphid

Liriomyza sativa Blanchard, vegetable leafminer

Liriomyza trifolii (Burgess), serpentine leafminer

Lochmaeae capreae cribrata Bates

Locusta migratoria (Linnaeus), locust

Locusta migratoria migratorioides R. & F., African migratory locust

Lucilia cuprina (Wiedemann), sheep blow fly

Lygus lineolaris (Palisot de Beauvois), tarnished plant bug

Lymantria dispar (Linnaeus), gypsy moth

Macrosiphum euphorbiae (Thomas), potato aphid

Macrosiphum sp., aphid

Malacosoma disstria Hübner, forest tent caterpillar

Malacosoma pluvialis (Dyer), coast tent caterpillar

Mamestra brassicae (Linnaeus), armyworm

Manduca sexta (Linnaeus), tobacco hornworm

Maruca testulalis Geyer, bean pod borer

Masonaphis masoni (Knowlton), aphid

Mastotermes darwiniensis Froggatt, termite

Mayetiola destructor (Say), Hessian fly

Megachile rotundata (Fabricius), alfalfa leafcutting bee

Megoura viciae Bust., aphid

Melanoplus differentialis (Thomas), differential grasshopper

Melanoplus femurrubrum (DeGeer), redlegged grasshopper

Melanoplus mexicanus mexicanus (Scudder)

Melanoplus sanguinipes (Fabricius), migratory grasshopper

Melanotus communis Gyllenhal, corn wireworm

Menacanthus stramineus (Nitzsch), chicken body louse

Messor structor Letrelle

Microcerotermes beesoni Snyder, termite

Microcerotermes crassus Snyder, termite

Monochamus alternatus Hope, sawyer beetle

Mosomorphus villiger Brulle, tobacco beetle

Musca autumnalis De Geer, face fly

Musca domestica Linnaeus, house fly

Musca nebulo Fabricius

Musca vetustisima De Geer, bush fly

Mythimna separata Walker, rice earcutting caterpillar

Mythimna unipuncta Schiffermuller

Myzus persicae (Sulzer), green peach aphid

Nasutitermes costalis (Holmgren), termite

Nasutitermes exitiosus (Hill), termite

Nasutitermes nigriceps (Haldemann), termite

Neodiprion rugifrons Middleton, redheaded jackpine sawfly

Neodiprion swainei Middleton, Swaine jackpine sawfly

Nephantis serinopa Meyrick, blackheaded coconut caterpillar

Nephotettix nigropictus (Stäl.), rice leafhopper

Nephotettix virescens (Distant), green leafhopper

Nilaparvata lugens (Stäl.), brown planthopper

Oedaleus senegalensis Krauss, African grasshopper

Oncopeltus fasciatus (Dallas), large milkweed bug

Ootheca bennigseni Weise, foliar beetle

Ootheca mutabilis Latrelle, cowpea flea beetle

Opatroides frater Brulle, tobacco beetle

Oryzaephilus surinamensis (Linnaeus), sawtoothed grain beetle

Oscinella frit (Linnaeus), frit fly

Ostrinia nubilalis (Hübner), European corn borer

Oulema melanopus (Linnaeus), cereal leaf beetle

Paleacrita vernata (Peck), spring cankerworm

Panonychus citri (McGregor), citrus red mite

Papilio ajia Linnaeus

Papilio polyxenes astericus Stoll, black swallowtail

Pectinophora gossypiella (Saunders), pink bollworm

Pediculus humanus capitis De Geer, head louse

Pediculus humanus humanus Linnaeus, body louse

Pericalia ricini (Forskal)

Peridroma saucia (Hübner), variegated cutworm

Periplaneta americana (Linnaeus), American cockroach

Perthida glyphopa Common, jarrah leafminer
Phaedon cochliariae Fabricius
Pheidole dentata Mayr, ant
Philosamia ricini Hutton, eri silkworm
Phoracantha semipunctata Fabricius, eucalypt borer
Phormia regina (Meigen), black blow fly
Phormia terrae-novae Desvoidy, blowfly
Phthorimaea operculella (Zeller), potato tuberworm
Phyllobius argentatus Germar
Phyllobius pyri Germar
Phyllotreta cruciferae (LeConte), cabbage flea beetle
Phyllotreta memorum Linnaeus, leaf beetle
Phyllotreta striolata (Fabricius), striped flea beetle
Phyllotreta tetrastigma Corn
Pieris brassicae (Linnaeus), large cabbage white
Pieris rapae crucivora (Linnaeus), cabbage butterfly
Piesma quadratum Fieber, beet leaf bug
Pissodes notatus Fabricius
Pissodes strobi (Peck), white pine weevil
Plagiodera versicolora Laicharting, imported willow leaf beetle
Planococcus eitri (Risso), citrus mealybug
Plathypena scabra Fabricius, green cloverworm
Plodia interpunctella (Hübner), Indian meal moth
Plusia gamma (Speyer), looper
Plutella xylostella (Linnaeus), diamondback moth
Podagrica uniforma Latrelle, flea beetle
Popillia japonica Newman, Japanese beetle
Pristiphora erichsonii (Hartig), larch sawfly
Psammotermes assuarensis (Sjöstedt), termite
Psammotermes fuscofemoralis (Sjöstedt), termite
Psila rosae (Fabricius), soybean looper
Psylla pyricola Foerster, pear psylla
Pyrausta machaeralis Walker
Pyrrhocoris apterus Linnaeus
Reticulitermes flavipes (Kollar), eastern subterranean termite
Reticulitermes lucifugus (Rossi), lucifugous termite

Reticulitermes speratus (Kolbe), termite
Reticulitermes virginicus Banks, termite
Rhagoletis pomonella (Walsh), apple maggot
Rhipicephalus rossius Banks, tick
Rhodnius prolixus (Stål.)
Rhopalosiphum fitchii (Sanderson), apple grain aphid
Rhopalosiphum maidis (Fitch), corn aphid, apple grain aphid
Rhyacionia buoliana (Denis & Schiffermüller), European pine shoot moth
Rhyzopertha dominica (Fabricius), lesser grain borer
Riptortus densipes (Fabricius), pod-sucking bug
Schistocerca canellata (Scudder)
Schistocerca gregaria Forskal, desert locust
Schistocerca vaga (Scudder), vagrant grasshopper
Schizaphis graminum (Rondani), greenbug
Sciopithes chacurus Horn, obscure root weevil
Scirpophaga incertula (Walker), rice yellow stemborer
Scolytus multistriatus (Marcham), smaller European elm bark beetle
Seleron latipes Brulle, tobacco beetle
Sipha flava (Forbes), yellow sugarcane aphid
Sirex noctilio (Cresson), wood wasp
Sitoma cylindricollis Fabricius, sweetclover weevil
Sitophilus granarius (Linnaeus), granary weevil
Sitophilus oryzae (Linnaeus), rice weevil
Sitophilus zeamais Motschulsky, maize weevil
Sitotroga cerealella Olivier, Anjoumois grain moth
Sogatella furcifera (Horvath), whitebacked planthopper
Spilosoma obliqua Walker, Bihar hairy caterpillar
Spilosoma virginica (Fabricius), yellow woolybear
Spodoptera abyssina Guenée
Spodoptera eridania (Cramer), southern armyworm
Spodoptera exempta` (Walker), African armyworm, nutgrass armyworm

Spodoptera exigua (Hübner), beet armyworm

Spodoptera frugiperda (J. E. Smith), fall armyworm

Spodoptera littoralis Boisduval, Egyptian cotton leafworm

Spodoptera litura (Fabricius), tobacco cutworm

Spodoptera maritia (Boisduval), paddy swarming caterpillar

Spodoptera ornithogalli (Guenée), yellow-striped armyworm

Spoladea recurvalis (Fabricius), Hawaiian beet webworm

Stegobium paniceum (Linnaeus), drugstore beetle

Stomoxys calcitrans (Linnaeus), stable fly

Supella longipalpa (Fabricius), brownbanded cockroach

Syrphus corollae Fabricius

Tenebrio molitor Linnaeus, yellow mealworm

Tetranychus cinnabarinus (Boisduval), carmine spider mite

Tetranychus teralius Linnaeus, carmine mite

Tetranychus urticae Koch, twospotted spider mite

Tetyra bipunctata (Herrick-Schäffer), shieldbacked seed bug

Therioaphis riehmi Hörner, sweetclover aphid

Thlaspida japonica (Cockerell)

Tinea granella Linnaeus, grain moth

Tineola bisselliella Hummel, webbing clothes moth

Tribolium castaneum (Herbst), red flour beetle

Tribolium confusum Jacquelin duVal, confused flour beetle

Trichoplusia ni Hübner, cabbage looper

Trimeresia miranda Butler

Trogoderma granarium Everts, khapra beetle

Udea rubigalis Guenée, celery leaf tier

Uraba lugens Walker, gum leaf skeletonizer

Urbanus proteus (Linnaeus), bean leafroller

Urentius echinus Distant

Utetheisa pulchella (Linnaeus), red spotted ermine moth

Xyleborus fornicatus Eichh.

Xyloryctes jamaicensis (Drury), rhinoceros beetle

Zabrotes subfasciatus (Boheman), Mexican bean weevil

Zonocerus variegatus Linnaeus, variegated grasshopper

Zygogramma exclamationis (Fabricius), sunflower beetle

References

REFERENCES

1. **Pathak, M. D. and Saxena, R. C.,** *Insect Resistance in Crop Plants,* Int. Rice Res. Inst., Los Banos, Laguna, Philippines, 1976, 1233.
2. **Sutherland, O. R. W.,** Invertebrate- plant relationships and breeding pest-resistant plants, in *Proc. 2nd Australasian Conf. on Grassland Invertebrate Ecology,* Crosby, T. K. and Pottinger, R. P., Eds., Government Printer, Wellington, New Zealand, 1980, 84.
3. **Wolff, A.,** How plants protect themselves, *Rockefeller Foundation Illustrated,* 9, June 1982.
4. **Waiss, A. C., Jr.,** unpublished data, 1979.
5. **Beck, S. D.,** Resistance of plants to insects, *Ann. Rev. Entomol.,* 10, 207, 1965.
6. **Reynolds, T.,** Aggressive chemicals in plants, *Chem. Ind.,* 603, 1975.
7. **Schalk, J. M. and Ratcliffe, R. H.,** Evaluation of the United States Department of Agriculture program on alternate methods of insect control: host plant resistance to insects, *FAO Plant Prot. Bull.,* 25, 9, 1977.
8. **Pathak, M. D.,** Defense of the rice crop against insect pests, *Ann. New York Acad. Sci.,* 287, 287, 1977.
9. **Meinwald, J., Prestwich, G. D., Nakanishi, K., and Kubo, I.,** Chemical ecology: studies from East Africa, *Science,* 199, 1167, 1978.
10. **Anderson, B. A., Lundgren, L., Och, G. N., and Stenhagen, G.,** The production of allelochemics in different plants as a criterium of selection in the search for genes of resistance against insects and other pests, *Entomol. Tidskr.,* 3, 130, 1979 (in Swedish).
11. **Wiseman, B. R. and Davis, F. M.,** Plant resistance to the fall armyworm, *Florida Entomol.,* 62, 123, 1979.
12. **Benedict, J. H. and George, D. M.,** A bibliography of host plant resistance literature for the boll weevil, *Anthonomus grandis, Bull. Entomol. Soc. Am.,* 25, 19, 1979.
13. **Smith, A. E. and Secoy, D. M.,** Plants used for agricultural pest control in western Europe before 1850, *Chem. Ind.,* 12, 1981.
14. **Waiss, A. C., Jr., Chan, B. G., Elliger, C. A., Dreyer, D. L., Binder, R. G., and Gueldner, R. C.,** Insect growth inhibitors in crop plants, *Bull. Entomol. Soc. Am.,* 27, 217, 1981.
15. **Maugh, T. H.,** Exploring plant resistance to insects, *Science,* 216, 722, 1982.
16. **Schneiderman, H. A.,** What entomology has in store for biotechnology, *Bull. Entomol. Soc. Am.,* 30, 55, 1983.
17. **Jermy, T.,** Evaluation of insect/host plant relationships, *Am. Naturalist,* 124, 609, 1984.
18. **Pickett, J. A.,** Production of behavior-controlling chemicals by crop plants, *Phil. Trans. R. Soc. London,* 310B, 235, 1985.
19. **Rosenthal, C. A.,** The chemical defenses of higher plants, *Sci. Am.,* 254, 94, 1986.
20. **Schoonhoven, L. M.,** Biological aspects of antifeedants, *Entomol. Exp. Appl.,* 31, 57, 1982.
21. **Painter, R. H.,** *Insect Resistance in Crop Plants,* Macmillan, New York, 1951, 121.
22. **Pathak, M. D.,** Utilization of insect-plant interactions in pest control, in *Insects, Science and Society,* Pimentel, D., Ed., Academic, New York, 1975, 121.
23. **Waiss, A. C., Chan, B. G., and Elliger, C. A.,** Host plant resistance to insects, in *Host Plant Resistance to Pests,* Hedin, P. A., Ed., Am. Chem. Soc. Symp. Ser. No. 62, Washington, D.C., 1977, 115.
24. **McIndoo, N. E.,** *Plants of Possible Insecticidal Value. A Review of the Literature up to 1941,* USDA Handbook No. E-661, U.S. Government Printing Office, Washington D.C., 1947, 286 pp.
25. **Heal, L. E., Rogers, E. F., Wallace, R. T., and Starnes, O.,** A survey of plants for insecticidal activity, *Lloydia,* 13, 89, 1950.
26. **Jacobson, M.,** *Insecticides From Plants. A Review of the Literature, 1941-1953,* USDA Handbook No. 154, U.S. Government Printing Office, Washington, D.C., 1958, 299 pp.
27. **Jacobson, M.,** *Insecticides From Plants. A Review of the Literature, 1954-1971,* USDA Handbook No. 461, U.S. Government Printing Office, Washington, D.C. 1975, 138 pp.
28. **Golob, P. and Webley, D. J.,** *The Use of Plants and Minerals as Traditional Protectants of Stored Products,* Tropical Products Inst. Pub. No. G138, London, 1980.

29. **Grainge, M., Ahmed, S., Mitchell, W. C., and Hylin, J. W.,** *Plant Species Reportedly Possessing Pest-Control Properties- a Database,* East-West Center, Honolulu, 1984.

30. **Schoonhoven, L. M.,** Perception of azadirachtin by some lepidopterous larvae, in *Natural Pesticides From the Neem Tree (Azadirachta indica A. Juss),* Schmutterer, H. and Ascher, K. R. S., Eds., GTZ Press, Eschborn, FRG, 1981, 105.

31. **Kubota, T. and Kubo, I.,** Bitterness and chemical structure, *Nature,* 223, 97, 1969.

32. **Feeny, P.,** Defensive ecology of the Cruciferae, *Ann. Missouri Bot. Garden,* 64, 221, 1977.

33. **Payne, T. L.,** Nature of insect and host tree interactions, *Z. Angew. Entomol.,* 96, 105, 1983.

34. **Metcalf, R. L., Metcalf, R. A., and Rhodes, A. M.,** Cucurbitacins as kairomones for diabroticite beetles, *Proc. Natl. Acad. Sci., U.S.A.,* 77, 3769, 1980.

35. **Carroll, C. R. and Hoffman, C. A.,** Chemical feeding deterrent mobilized in response to insect herbivory and counteradaptation by *Epilachna tredecimnotata, Science,* 209, 414, 1980.

36. **Metcalf, R. L.,** Plants, chemicals, and insects: some aspects of coevolution, *Bull. Entomol. Soc. Am.,* 25, 30, 1979.

37. **Dowd, R. F., Smith, C. M., and Sparks, T. C.,** Detoxification of plant toxins by insects, *Insect Biochem.,* 13, 453, 1983.

38. **Duffey, S. S.,** Sequestration of plant natural products by insects, *Ann. Rev. Entomol.,* 25, 447, 1980.

39. **Grisebach, H. and Ebel, J.,** Phytoalexins, chemical defense substances of higher plants, *Angew. Chem., Intl. Ed. English,* 17, 635, 1978.

40. **Rembold, H. and Winter, E.,** The chemist's role in host-plant resistance studies, in *Proc. Int. Workshop on Heliothis Management,* Nov. 15-20, 1981, Patancheru, India, 1982, 241.

41. **Picman, A. K.,** Biological activities of sesquiterpene lactones, *Biochem. System. Ecol.,* 14, 255, 1986.

42. **Smeljanez, W. P.,** A classification of resistance of plants to insect pests, *Anz. Schädlingsk., Pflanzenschutz, Umweltschutz,* 51, 34, 1978.

43. **Redfern, R. E., Warthen, J. D., Jr., Uebel, E. C., and Mills, G. D., Jr.,** The antifeedant and growth-disrupting effects of azadirachtin on *Spodoptera frugiperda and Oncopeltus fesciatus,* in *Natural Pesticides From the Neem Tree (Azadirachta indica A. Juss),* Schmutterer, H., Ascher, K. R. S., and Rembold, H., Eds., GTZ Press Eschborn, FRG, 1981, 129

44. **Meisner, J., Ascher, K. R. S., and Aly, R.,** The residual effect of some products of neem seeds on larvae of *Spodoptera littoralis* in laboratory and field trials, in *Natural Pesticides From the Neem Tree (Azadirachta indica A. Juss),* Schmutterer, H., Ascher, K. R. S., and Rembold, H., Eds., GTZ Press, Eschborn, FRG, 1981, 157.

45. **Ley, S. V., Santafianos, D., Blaney, W. M., and Simmonds, M. S. J.,** Synthesis of a hydroxy dihydrofuran acetal related to azadirachtin: a potent insect antifeedant, *Tetrahedron Lett.,* 28, 221, 1987.

46. **Freedman, B., Nowak, L. J., Kwolek, W. F., Berry, E. C., and Guthrie, W. D.,** A bioassay for plant-derived pest control agents using the European corn borer, *J. Econ. Entomol.,* 72, 541, 1979.

47. **Adhikary, S.,** Results of field trials to control the diamondback moth, *Plutella xylostella* L., by application of crude methanolic extracts and aqueous suspensions of seed kernels and leaves of neem, *Azadirachta indica* A. Juss in Togo, *Z. Angew. Entomol.,* 100, 27, 1985.

48. **Higgins, R. A. and Pedigo, L. P.,** A laboratory antifeedant simulation bioassay for phytophagous insects, *J. Econ. Entomol.,* 72, 238, 1979.

49. **Jacobson, M., Reed, D. K., Crystal, M. M., Moreno, D. S., and Soderstrom, E. L.,** Chemistry and biological activity of insect feeding deterrents from certain weed and crop plants, *Entomol. Exp. Appl.,* 24, 448, 1978.

50. **Reed, D. K., and Jacobson, M.,** Evaluation of aromatic tetrahydropyranyl ethers as feeding deterrents for the striped cucumber beetle, *Acalymma vittatum* (F.), *Experientia,* 39, 378, 1983.

51. **Reed, D. K.,** unpublished data, 1983.

52. **Reed, D. K., Jacobson, M., Warthen, J. D., Jr., Uebel, E. C., Tromley, N. J., Jurd, L., and Freedman, B.,** Cucumber beetle antifeedants: laboratory screening of natural products, *USDA Tech. Bull.* No. 1641, 1981, 13 pp.

53. **Von der Heyde, J., Saxena, R. C., and Schmutterer, H.,** Neem oil and neem extracts as potential insecticides for control of hemipterous rice pests, in *Natural Pesticides From the Neem Tree and Other Tropical Plants,* Schmutterer, H., and Ascher, K. R. S., Eds., GTZ Press, Eschborn, FRG, 1984, 377.

54. **Saxena, R. C. and Khan, Z. R.,** Electronically recorded disturbances in the feeding behavior of green leafhopper (GLH) on neem oil-treated rice plants, *Int. Rice Res. Newslett.,* 9(5), 17, 1984.

55. **Saxena, R. C. and Khan, Z. R.,** Electronically recorded disturbances in feeding behavior of *Nephotettix virescens (Homoptera: Cicadellidae)* on neem oil-treated rice plants, *J. Econ. Entomol.,* 78, 222, 1985.

56. **Antonious, A. G., Saito, T., and Nakamura, K.,** Electrophysiological response of the tobacco cutworm, *Spodoptera litura* (F.), to antifeeding compounds, *J. Pesticide Sci.,* 9, 143, 1984.

57. **Dethier, V. G.,** Evaluation of receptor sensitivity to secondary plant substances with special reference to deterrents, *Am. Naturalist,* 115, 45, 1980.

58. **Norris, D. M.,** Chemical deterrence of feeding by insect pests, *Soap Chem. Specialties,* 50(10), 48, 1970.

59. **Adler, V. E. and Uebel, E. C.,** Antifeedant bioassays of neem extract against the Carolina grasshopper, walkingstick, and field cricket, *J. Environ. Sci. Health,* A19, 393, 1984.

60. **Inazuka, S.,** New methods of evaluation for cockroach repellents and repellancy of essential oils against German cockroach (*Blattella germanica* L.), *J. Pesticide Sci.,* 7, 133, 1982.

61. **Schoonhoven, L. M. and Jermy, T.,** A behavioral and electrophysiological analysis of insect feeding deterrents, in *Crop Protection Agents, Their Biological Evaluation,* McFarland, N. R., Ed., Academic, New York, 1977, 133.

62. **Goodhue, L. D., and Tissel, C. L.,** Determining the repellent action of chemicals to the American Cockroach, *J. Econ. Entomol.,* 45, 133, 1952.

63. **Bodenstein, O. F. and Fales, J. H.,** Laboratory evaluations of compounds as repellents to cockroaches, 1953-1974, *USDA Production Research Report* No. 164, 1976, 28 pp.

64. **Anon.** *Materials Evaluated as Insecticides, Repellents, and Chemosterilants at Orlando and Gainesville, Florida, 1952-1964,* unpublished data, 1965.

65. **Salek, M. A., Abdel-Moein, N. M., and Ibrahim, N. A.,** Insect antifeeding azulene derivative from the brown alga, *Dictyota dichotoma, J. Agric Food Chem.,* 52, 1432, 1984.

66. **Kubo, I., Matsumoto, T., and Ochikawa, N.,** Absolute configuration of crinitol, an acyclic diterpene insect growth inhibitor from the brown alga *Sargassum tortile, Chem. Lett. (Tokyo),* 249, 1985.

67. **Slansky, F., Jr.,** Effect of the lichen chemicals atranorin and vulpinic acid upon feeding and growth of larvae of the yellow-striped armyworm, *Spodoptera ornithogalli, Environ. Entomol.,* 8, 865, 1979.

68. **Nawrot, J., Bloszyk, E., Harmatha, J., Novotny, L., and Drozdz, B.,** Action of antifeedants of plant origin on beetles infesting stored products, *Acta Entomol. Bohemoslov.,* 83, 327, 1986.

69. **Asakawa, Y., Toyota, M., and Takemoto, T.,** Three *ent*-secoaromadendrane-type sesquiterpene hemiacetals and a bicyclogermacrene from *Plagiochila ovalifolia* and *Plagiochila yokogurensis, Phytochemistry,* 19, 2141, 1980.

70. **Asakawa, Y., Toyota, M., Takemoto, T., Kubo, I., and Nakanishi, K.,** Insect antifeedant secoaromadendrane-type sesquiterpenes from *Plagiochila* species, *Phytochemistry,* 19, 2147, 1980.

71. **Jones, C. G. and Firn, R. D.,** Resistance of *Pteridium aquilinum* to attack by non-adapted phytophagous insects, *Biochem. System. Ecol.,* 7, 187, 1979.

72. **Jones, C. G. and Firn, ΣR. D.,** Some allelochemicals of *Pteridium aquilinum* and their involvement in resistance to *Pieris brassicae, Biochem. System. Ecol.,* 7, 187, 1979.

73. **Pandey, N. D., Singh, S. R., and Tewari, G. C.,** Use of some plant powders, oils and extracts as protectants against pulse beetle, *Callosobruchus chinensis* Linn., *Indian J. Entomol.,* 38, 110, 1976.

74. **Srivastava, A. S., Saxena, H. P., and Singh, D. R.,** *Adhatoda vasica,* a promising insecticide against pests of storage, *Labdev,* 3(2), 138, 1965.

75. **Abraham, C. C., Thomas, B., Karumakaran, K., and Gopalakrishbuan, R.,** Relative efficiency of some plant products in controlling infestations by the Angoumois grain moth (*Sitotroga cerealella* Oliver) (Gelechidae): Lepidoptera) infesting stored paddy in Kerala, *Bull. Grain Technol.,* 10(4), 263, 1972.

76. **Challapya, K. and Chelliah, S.,** Studies on the efficiency of malathion and certain plant products in the control of *Sitotroga cerealella* and *Rhyzopertha dominica* infesting rice grains, *Madras Agr. J.,* 63, 190, 1976.

77. **Saxena, B. P., Tikku, K., Ayal, C. K., and Koul, O.,** Insect antifertility and antifeedant allelochemics in *Adhatoda vasica, Insect Sci. Appl.,* 7, 489, 1986.

78. **Srivastava, A. S. and Awasthi, G. P.,** An insecticide from the extract of a plant, *Adhatoda vasica* Nees. harmless to man, in *Proc. 10th Int. Congr. of Entomol.,* 1956, 2, 245.

79. **Abrol, B. K. and Chopra, I. C.,** Development of indigenous vegetable insecticides and insect repellents, *Bull. Jammu Regional Res. Lab.,* 1, 156, 1963.

80. **Drury, H.,** *The Useful Plants of India,* Madras, 1858, 233.

81. **Puttarudriah, M. and Subramaniam, T. V.,** Work Done in Mysore on the Insecticidal Value of Fish-Poisons and Other Forest Products, Part I, Entomological Investigations, Mysore Agr. Dept. Rept. Jan. 1935-Mar. 1936, 1936.

82. **Murakoshi, S., Chang, C. F., Kamikado, T., and Tamura, S.,** Effects of the extracts from leaves of various plants on the growth of silkworm larvae, *Bombyx mori* L., *Jpn. J. Appl. Entomol. Zool.,* 19, 208, 1975.

83. **Murakoshi, S., Kamikado, T., Chang, C. F., Sakurai, A., and Tamura, S.,** Effects of several components from the leaves of four species of plants on the growth of silkworm larvae, *Jpn. J. Appl. Entomol. Zool.,* 20, 26, 1976.

84. **Doskotch, R. W., Odell, T. M., and Godwin, P. A.,** Feeding responses of gypsy moth larvae, *Lymantria dispar,* to extracts of plant leaves, *Environ. Entomol.,* 6, 563, 1977.

85. **Carter, F. L., Amburgey, T. L., and Manwillar, F. G.,** Resistance of 22 southern hardwoods to wood-decay fungi and subterranean termites, *Wood Sci.,* 8, 223, 1976.

86. **Wolcott, G. N.,** Inherent natural resistance of woods to the attack of the West Indian drywood termite, *Cryptotermes brevis* Walker, *Puerto Rico Univ. J. Agric.,* 41, 259, 1957.

87. **Mannesmann, R.,** Comparison of twenty-one commercial wood species from North America in relation to feeding rates of the Formosan termite, *Coptotermes formosanus* Shiraki, *Mater. Org.,* 8, 107, 1973.

88. **Oliver, A. D.,** Control studies of the forest tent caterpillar, *Malacosoma disstria,* in Louisiana, *J. Econ. Entomol.,* 57, 157, 1964.

89. **Reed, D. K. and Jacobson, M.,** unpublished data, 1986.

90. **Numata, A., Katsuno, T., Yamamoto, K., Nishida, T., Takemura, T., and Seto, K.,** Plant constituents biologically active to insects. IV. Antifeedants for the larvae of the yellow butterfly, *Eurema hecabe mandarina,* in *Arachniodes standishii, Chem. Pharm. Bull.,* 32, 325, 1984.

91. **Wada, K., Matsui, K., Enomoto, Y., Ogiso, O., and Munakata, A.,** Insect feeding inhibitors in plants. I. Isolation of three new sesquiterpenoids in *Parabenzoin trilobus* Kakai, *Agric. Biol. Chem.,* 34, 941, 1970.

92. **Muckensturm, B., Duplay, D., Mohamnadi, F., and Moradi, A.,** The role of natural phenylpropanoids as insect feeding deterrents, Proc. Chem. Colloq., Versailles, France, Nov. 16-20, 1981, 131 (in French).

93. **Chopra, R. N. and Badhwar, R. L.,** Poisonous plants in India, *Indian J. Agr. Sci.,* 10, 1, 1940.

94. **Quisumbing, E.,** Vegetable poisons of the Philippines, *Philippine J. Forestry,* 5, 146, 1947.

95. **Anon.,** Three methods for deterring flies, *Rev. Hort. Belg. Etrange,* 25, 170, 1899.

96. **Ellis, P. R. and Eckenrode, C. J.,** Factors influencing resistance in *Allium* sp. to onion maggot, *Bull. Entomol. Soc. Am.,* 25, 151, 1979.

97. **Fletcher, T. B. and Ghosh, C. C.,** Stored grain pests, in *Rept. Proc. Third Entomol. Mtg.,* Pusa, India, Vol. 2, 1919, 712.

98. **Nassem, M. O.,** The effect of crude extracts of *Allium sativum* L. on feeding activity and metamorphosis of *Epilachna varivestis* Muls. (Col., Coccinellidae), *Z. Angew. Entomol.,* 92, 464, 1981.

99. **Singh, R. P.,** Search for antifeedants in some botanicals for desert locust, *Schistocerca gregaria* Forskal, *Z. Angew. Entomol.,* 96, 316, 1983.

100. **Celli, A. and Casagrandi, O.,** Destruction of mosquitoes, *Atti Stud. Malaria,* 1, 73, 1899.

101. **Greenstock, D. L. and Larrea, Q.,** *Garlic as an Insecticide,* Doubleday Research Association, Braintree, Engl., 1972, 12 pp.

102. **Corey, F.,** Garlic: the new DDT?, The Washington (D.C.) Post, Dec. 28, 1971.

103. **Anon.,** Garlic: a killer of insect pests, Egyptian Gazette, Dec. 29, 1971.

104. **Greenstock, D. L.,** Garlic as pesticide, *New Scientist,* 54, 790, 1972.

105. **Novak, D.,** Soyabean extracts of garlic as mosquito larvicide, *Acta Univ. Carolinae (Biol.),* 1977, 367, 1980.

106. **Deb Kirtaniya, S., Ghosh, M. R., Adityachaudhury, N., and Chatterjee, A.,** Extracts of garlic as possible source of insecticides, *Indian J. Agr. Sci.,* 50, 507, 1980.

107. **Nassem, M. O.,** Effect of garlic extract on *Syrphus corollae* F., *Chrysopa carnea* Steph. and *Coccinella septempunctata* L., *Z. Angew. Entomol.,* 94, 123, 1982 (in German).

108. **Rajendran, B. and Gopalan, M.,** Note on the insecticidal properties of certain plant extracts, *Indian J. Agr. Sci.,* 49, 290, 1979.

109. **Rajendran, B. and Gopalan, M.,** Note on juvenomimetic activity of some plants, *Indian J. Agr. Sci.,* 48, 306, 1978.

110. **Amonkar, S. V. and Banerji, A.,** Isolation and characterization of larvicidal principle of garlic, *Science,* 174, 1343, 1971.

111. **Singh, R. P. and Pant, N. C.,** Investigation on the antifeedant property of subfamily Amaryllidoideae (Amaryllidaceae)against desert locust, *Schistocerca gregaria* Forsk., *Indian J. Entomol.,* 42, 465, 1980.

112. **Singh, R. P. and Pant, N. C.,** *Hymenocallis littoralis* Salisb. as antifeedant to desert locust *Schistocerca gregaria* Forsk., *Indian J. Entomol.,* 42, 460, 1980.

113. **Singh, R. P. and Pant, N. C.,** Lycorine—a resistance factor in the plants of subfamily Amaryllidoideae (Amaryllidaceae) against desert locust, *Schistocerca gregaria* F., *Experientia,* 36, 552, 1980.

114. **Marini Bettolo, G. B., Patamia, M., Nicoletti, M., Galeffi, C., and Messana, I.** Research on African medicinal plants. II. Hypoxoside, a new glycoside of uncommon structure from *Hypoxis obtusa* Busch, *Tetrahedron,* 38, 1683, 1982.

115. **Wolcott, G. N.,** The resistance to dry-wood termite attack of some Central American woods, *Caribbean Forester,* 9, 53, 1948.

116. **Watt, J. M. and Breyer-Brandwijk, Breyer-Brandwijk, M. G.,** *The Medicinal and Poisonous Plants of Southern and Eastern Africa,* 2nd ed. Edinburgh, 1962.

117. **Streets, R. J.,** *Exotic Forest Trees in the British Commonwealth,* Champion, H., Ed., Oxford, 1962.

118. **Wolcott, G. N.,** Factors in the natural resistance of woods to termite attack, *Caribbean Forester,* 7, 121, 1946.

119. **Lepage, E. S. and De Lelis, A. T.,** Protecting wood against dry-wood termites with cashewnut shell oil, *Forest Prod. J.,* 30, 35, 1980.

120. **Roonwald, M., Chatterjee, P. N., and Thapa, R. S.,** Prophylactic efficacy of various insecticides in the protection of freshly felled and converted timber (planks) against insect borers, *Indian Forest Bull. Entomol.,* (N.S.), No. 215, 1959.

121. **Irvine, F. R.,** West African insecticides, *Colonial Plant Animal Prod.,* 5(1), 34, 1955.

122. **Wats, R. C. and Bharucha, K. H.,** Larvicides for antimosquito work, with special reference to cashewnut shell oil, *Malaria Inst. India J.,* 1, 217, 1938.

123. **Gopakumar, B., Ambika, B., and Prabhu, K. K.,** Juvenomimetic activity in some South Indian plants and the probable cause of this activity in *Morus alba, Entomon,* 2, 259, 1977.

124. **Wolcott, G. N.,** What to do about polilla, *Puerto Rico Univ. Agr. Expt. Bull.,* 68, 1946, 29 pp.

125. **Kirtikar, K. R. and Basu, B. D.,** *Indian Medicinal Plants,* Vol. 1, Bahadurganj, India, 1918, 375.

126. **Metzger, F. W. and Grant, D. H.,** Repellency to the Japanese beetle of extracts made from plants immune to attack, USDA Tech. Bull. 299, 1932, 21 pp.

127. **Tattersfield, F. and Potter, C.,** The insecticidal properties of certain species of *Annona* and of an Indian strain of *Mundulea sericea* ("supli"), *Ann. Appl. Biol.,* 27, 262, 1940.

128. **Bottger, G. T. and Jacobson, M.,** Preliminary tests of plant materials as insecticides, USDA Bureau Entomol. Plant Quarantine Pub. E-796, 1950, 35 pp.

129. **Pandey, G. P. and Varma, B. K.,** *Annona* (custard apple) seed powder as a protectant of mung against pulse beetle, *Callosobruchus maculatus* Fabr., *Bull. Grain Technol.,* 15, 100, 1977.

130. **Pandey, G. P. and Varma, B. K.,** Attapulgite dust for the control of pulse beetle, *Callosobruchus maculatus,* on black gram (*Phaseolus mungo*), *Bull. Grain Technol.,* 15, 188, 1977.

131. **Islam, R. N.,** Pesticidal action of neem and certain indigenous plants and weeds of Bangladesh, in *Natural Pesticides From the Neem Tree and Other Tropical Plants,* Schmutterer, H., and Ascher, K. R. S., Eds., GTZ Press, Eschborn, FRG, 1984, 263.

132. **Islam, R. N.,** Use of some extracts from Meliaceae and Annonaceae for control of rice hispa, *Dicladispa armigera,* and the pulse beetle, *Callosobruchus chinensis,* in *Natural Pesticides From the Neem Tree and other Tropical Plants,* Schmutterer, H. and Ascher, K. R. S., Eds., GTZ Press, Eschborn, FRG, 1987, 217.

133. **Mukerjea, T. D. and Govind, R.,** Studies on indigenous insecticidal plants. Part II. *Annona squamosa, J. Sci. Ind. Res.,* 170, 9, 1959.

134. **Mariappan, V. and Saxena, R. C.,** Effect of custard-apple oil and neem oil on survival of *Nephotettix virescens* (Homoptera: Cicadellidae) and on rice tungro virus transmission, *J. Econ. Entomol.,* 76, 573, 1983.

135. **Mariappan, V. and Saxena, R. C.,** Effect of mixtures of custard-apple oil and neem oil on survival of *Nephotettix virescens* (Homoptera: Cicadellidae) and on rice tungro virus transmission, *J. Econ. Entomol.,* 77, 519, 1984.

136. **Saxena, R. C. and Justo, H. D., Jr.,** Effect of custard apple oil, neem oil, and neem cake on green leafhopper (GLH) population and on tungro (RTV) infection, *Int. Rice Res. Newslett.,* 11(2), 25, 1986.

137. **Narasimhan, V. and Mariappan, V.,** Effect of plant derivatives on green leafhopper (GLH) and rice tungro (RTV) transmission, *Int. Rice Res. Newslett.,* 13(1), 28, 1988.

138. **Brady, N. C., Khush, G. S., and Heinrichs, E. A.,** unpublished report, 1978.

139. **Mariappan, V. and Saxena, R. C.,** Custard apple oil, and their mixtures: effect on survival of *Nephotettix virescens* and on rice tungro virus transmission, in *Natural Pesticides From the Neem Tree and Other Tropical Plants,* Schmutterer, H. and Ascher, K. R. S., Eds., GTZ Press, Eschborn, FRG, 1984, 413.

140. **Wyllie, S. G., Cook, D., Brophy, J. J., and Richter, K. M.,** Volatile flavor components of *Annona atemoya* (custard apple), *J. Agr. Fd. Chem.,* 35, 768, 1987.

141. **Kutschabsky, L., Sandoval, D., and Ripperger, H.,** Bullatantriol, a sequiterpene from *Annona bullata, Phytochemistry,* 24, 2724, 1985.

142. **McCloud, T. G., Smith, D. L., Chang, C.-J., and Cassady, J. M.,** Annonacin, a novel, biologically active polyketide from *Annona densicoma, Experientia,* 43, 947, 1987.

143. **Sandoval, D., Preiss, A., Schreiber, K., and Ripperger, H.,** Annonelliptine, an alkaloid from *Annona elliptica, Phytochemistry,* 24, 375, 1985.

144. **Wu, T.-S., Jong, T.-T., Tien, H.-J., Kuoh, C.-S., Furukawa, H., and Lee, K.-H.,** Annoquinone-A, an antimicrobial and cytotoxic principle from *Annona montana, Phytochemistry,* 26, 1623, 1987.

145. **Etse, J. T., Gray, A. I., and Watterman, P. C.,** Chemistry in the Annonaceae. XXIV. Kaurane and kaur-16-ene diterpenes from the stem bark of *Annona reticulata, J. Natural Prod.,* 50, 979, 1987.

146. **Achenbach, H. and Franke, D.,** Constituents of tropical medicinal plants. XXIV. Syntheses of annonidines A, C, and D, *Arch. Pharm. (Weinheim),* 320, 91, 1987.

147. **El-Zayat, A. A. E., Ferrigni, N. R., McCloud, T. G., McKengie, A. T., Byrn, S. R., Cassidy, J. M., Chang, C.-J., and McLaughlin, J. L.,** Goniothalenol, a novel, bioactive, tetrahydrofurano-2-pyrone from *Goniothalamus giganteus* (Annonaceae), *Tetrahedron Lett.,* 26, 955, 1985.

148. **Howard, L. O.,** Preventive and remedial work against mosquitoes, USDA Bureau Entomol. Bull. No. 88, 1910, 126.

149. **Gupta, M.,** Essential oils: a new source of bee repellents, *Chem. Ind.,* 162, 1987.
150. **Ranasinghe, M. A. S. K.,** Neem and Other Promising Botanical Pest Control Materials From Sri Lanka, Rept. Int. Workshop on the Use of Botanical Pesticides, Los Banos, Philippines, Aug. 6-10, 1984, 15 pp.
151. **Agbakwuru, E. O. P., Osisiogu, I. U. W., and Ugochukwu, E. N.,** Insecticides of Nigerian vegetable origin. II. Some nitroalkanes as protectants of stored cowpeas and maize against insect pests, *Niger. J. Sci.,* 12, 493, 1978.
152. **Olaifa, J. I., Erhun, W. O., and Akingbohungbe, A. E.,** Insecticidal activity of some Nigerian plants, *Insect Sci. Appl.,* 8, 221, 1987.
153. **Reznik, P. A. and Imbs, Y. G.,** Ixodid ticks and phytoncides, *Zool. Zhur.,* 44, 1861, 1965.
154. **Back, E. A.,** Clothes moths and their control, USDA Farmers Bull. No. 1353, 1923, 28 pp.
155. **Parman, D. C., Bishopp, F. C., Laake, E. W., Cook, F. C., and Roark, R. C.,** Chemotropic tests with the screw-worm fly, USDA Agr. Bull. No. 1472, 1927, 32 pp.
156. **Abbott, W. S.,** Results of experiments with miscellaneous substances against chicken lice and the dog flea, USDA Agr. Bull. No. 888, 7, 11, 1920.
157. **Scott, E. W., Abbott, W. I., and Dudley, J. E., Jr.,** Results of experiments with miscellaneous substances against bedbugs, cockroaches, clothes moths, and carpet beetles, USDA Agr. Bull. No. 7075, 13, 26, 1918.
158. **Muckensturm, B., Duplay, D., Robert, P. C., Simonis, M. T., and Kienlen, J. C.,** Antifeedant substances for phytophagous insects present in *Angelica silvestris* and *Heracleum sphondyllium, Biochem. System. Ecol.,* 9, 289, 1981 (in French).
159. **Nawrot, J., Bloszyk, E., Harmatha, J., and Novotny, L.,** The effect of bisabolangelone, helenalin and bakkenolide A on development and behavior of some stored product beetles, *Z. Angew. Entomol.,* 98, 394, 1984.
160. **Meisner, J., Fleischer, A., and Eizick, C.,** Phagodeterrency induced by (−) carvone in the larva of *Spodoptera littoralis* (Lepidoptera:Noctuidae), *J. Econ. Entomol.,* 75, 462, 1982.
161. **Moore, W. and Hirschfelder, A. D.,** An investigation of the louse problem, Minnesota Univ. Res. Publ. 8(4), 1919, 86 pp.
162. **Cole, A. C., Jr.,** The olfactory responses of the cockroach (Blatta orientalis) to the more important essential oils and a control pressure formulated from the results, *J. Econ. Entomol.,* 25, 902, 1932.
163. **Novak, D.,** Toxicity of plants to mosquito larvae (Diptera, Culicidae), *Polon. Entomol. Bull.,* 38, 617, 1968.
164. **Hartzell, A. and Wilcoxon, F.,** A survey of plant products for insecticidal properties, *Boyce Thompson Inst. Contribs.,* 13, 243, 1944.
165. **Hartzell, A.,** Further tests on plant products for insecticidal properties, *Boyce Thompson Inst. Contribs.,* 13, 243, 1944.
166. **Roark, R. C. and Busbey, R. L.,** A third index of patented mothproofing materials, USDA Bureau of Entomol. and Plant Quarantine, 1936, 104 pp.
167. **Hartzell, A.,** Additional tests of plant products for insecticidal properties and summary of results to date, *Boyce Thompson Inst. Contribs.,* 15, 21, 1947.
168. **Cook, F. C. and Hutchison, R. H.,** Experiments during 1915 in the destruction of fly larvae in horse manure, USDA Bull. No. 408, 1916, 20 pp.
169. **Fraenkel, G. S.,** The chemistry of host specificity of phytophagous insects, in *Proc. 4th Int. Congress of Entomology,* 1956, 1.
170. **Kayumov, S. R.,** Tests of new vegetable poisons from plants that produce essential oils, *Sozial. Nauk Tekn.,* 5(6), 40, 1937 (in Russian).
171. **Von Gizycki, F. and Herrmanns, H.,** Examination of carrot greens (*Daucus carota*), *Arch. Pharm.,* 284, 8, 1951.
172. **Guerin, P. M. and Ryan, M. F.,** Insecticidal effect of *trans*-2-nonenal, a constituent of carrot root, *Experientia,* 36, 1387, 1980.
173. **Kubeczka, K. H.,** Germacrene-D from *Falcaria vulgaris, Phytochemistry,* 18, 1066, 1979.
174. **Crosby, C. R. and Leonard, M. D.,** The tarnished plant-bug, *Lygus pratensis* Linnaeus, New York (Cornell) Agr. Expt. Sta. Bull. No. 346, 1914, 463.

175. **Forbes, S. A.,** Recent Illinois work on the corn root-aphis and the control of its injuries, Illinois Agr. Expt. Sta. Bull. No. 178, 1915, 465.
176. **Chopra, R. L.,** unpublished results, 1927.
177. **Nawrot, J., Smitalova, Z., and Holub, M.,** Deterrent activity of sesquiterpene lactones from the Umbelliferae against storage pests, *Biochem. System. Ecol.,* 11, 243, 1983.
178. **Lichtenstein, E.M.P.,** Insecticides occurring naturally in crops, *Am. Chem. Soc. Adv. Chem. Ser.* No. 53, 34, 1966.
179. **Lichtenstein, E. P., and Casida, J. E.,** Myristicin, an insecticide and synergist occurring naturally in the edible parts of parsnips, *J. Agr. Fd. Chem.,* 11, 410, 1963.
180. **Kerr, R. W.,** Adjuvants for pyrethrins in fly sprays, Bull. Austral. Commonwealth Sci. Ind. Res. Organ. 261, 1951, 63.
181. **Singh, R. N. and Krishna, S. S.,** Effect of some common oilseeds and spices serving as adult food on the reproductive potential of *Tribolium castaneum* (Hbst.) (Coleoptera:Tenebrionidae),*Entomon,* 5, 161, 1980.
182. **Novak, D.,** Several volatile oils toxicity to mosquito larvae, *Arch. Roum. Path. Exp. Microbiol.,* 27, 721, 1968.
183. **Marcus, C. and Lichtenstein, E. P.,** Biologically active components of anise: Toxicity and interactions with insecticides in insects, *J. Agr. Fd. Chem.,* 27, 1217, 1979.
184. **Frankel, S.,** Further reports on lice-killing agents, *Munich Med. Wochenschr.,* 62, 624, 1915 (in German).
185. **Galavski, D.,** Handling and prophylaxis of clothing lice, *Deut. Med. Wochenschr.,* 41, 285, 1915.
186. **Abbassy, M. A., El-Shagli, A., and El-Gayar, F.,** A new antifeedant to *Spodoptera littoralis* Boisd. (Lepid., Noctuidae) from *Acokanthera spectabilis* Hook (Apocynaceae), *Z. Angew. Entomol.,* 83, 917, 1977.
187. **Martinez, J. B.,** Investigations with termites and substances resistant to them, *Madrid Inst. Forestal Invs. Expt. Bol.,* No. 81, 1963, 119 pp.
188. **Wallis, N. K.,** *Australian Timber Handbook,* 2nd Ed, Sydney, Austral., 1963, 391 pp.
189. **Patterson, B. D., Wahba Kalil, S. K., Schermeister, L. J., and Quraishi, M. S.,** Plant-insect interactions. I. Biological and phytochemical evaluation of selected plants,*Lloydia,* 38, 391, 1975.
190. **Villani, M. and Gould, F.,** Screening of crude plant extracts as feeding deterrents of the wireworm, *Melanotus communis, Entomol. Exp. Appl.,* 37, 69, 1985.
191. **Watt, G.,** *A Dictionary of the Economic Products of India,* London, 1899.
192. **Meisner, J., Weissenberg, M., Palevitch, D., and Aharonson, N.,** Phagodeterrency induced by leaves and leaf extracts of *Catharanthus roseus* in the larva of *Spodoptera littoralis, J. Econ. Entomol.,* 74, 131, 1981.
193. **Sukumar, K. and Osmani, Z.,** Insect sterilants from *Catharanthus roseus, Curr. Sci.,* 50, 552, 1981.
194. **Tripathi, A. K., Rao, S. M., Singh, D., Chakravarty, R. B., and Bhakuni, D. S.,** Antifeedant activity of plant extracts against *Spilosoma obliqua* Walker, *Curr. Sci.,* 56, 607, 1987.
195. **Gupta, P. D. and Thorsteinson, A. J.,** Food plant relationships of the diamondback moth (Pl. mac. Curt.), *Entomol. Exp. Appl.,* 3, 241, 1960.
196. **Abo Elghar, G.E.S. and El-Sheikh, A. E.,** Effectiveness of some plant extracts as surface protectants of cowpea seeds against the pulse beetle, *Callosobruchus chinensis, Phytoparasitica,* 15, 109, 1987.
197. **McIndoo, N. E. and Sievers, A. F.,** Plants tested for or reported to possess insecticidal preperties, USDA Bull. No. 1201, 1924, 61 pp.
198. **Reed, D. K., Friedman, B., and Ladd, T. L., Jr.,** Insecticidal and antifeedant activity of neriifolin against codling moth, striped cucumber beetle, and Japanese beetle, *J. Econ. Entomol.,* 75, 1093, 1982.
199. **McLaughlin, J. L., Freedman, B., Powell, R. G., and Smith, C. R., Jr.,** Neriifolin and 2-acetylneriifolin, insecticidal and cytotoxic agents of *Thevetia thevetioides* seeds, *J. Econ. Entomol.,* 73, 398, 1980.
200. **Singh, R. P., Tomar, S. S., Attri, B. S., Parmar, B. S., Maheshwari, M. L., and Mukerjee, S. K.,** Search for new pyrethrum synergists in some botanicals, *Pyrethrum Post,* 13(3), 91, 1976.

201. **Gattefossé, J.,** New agricultural insecticide, *Thevetia neriifolia, Bull. Soc. Sci. Natur. (Morocco),* 25, 1949.

202. **Svoboda, G. H.,** Antitumoral effect of *Vinca rosea* alkaloids, in *Proc. 1st Symp. G.E.C.A., Paris,* Excerpta Medical Foundation, Amsterdam, 1965, 9.

203. **Meisner, J. and Skattula, U.,** Phagostimulation and phagodeterrency in the larva of the gypsy moth, *Porthetria dispar, Phytoparasitica,* 3, 19, 1975.

204. **Warthen, J. D., Jr., Redfern, R. E., Uebel, E. C., and Mills, G. D., Jr.,** Antifeedant screening of thirty-nine local plants with fall armyworm larvae, *J. Environ. Sci. Health,* A17, 885, 1982.

205. **Dalzell, N. A. and Gibson, A.,** *The Bombay Flora,* Bombay, India, 1861, 332, pp.

206. **Ridley, A.,** Odorous acorus as an insecticide, *Rev. Hort. (Paris),* 76, 536, 1904.

207. **Anon.,** Neem from around the world, *India Grain Storage Newslett.,* 2(2), 3, 1960.

208. **Roark, R. C.,** An index of patented mothproofing materials, U.S. Bureau Chem. Soils, 1931, 125 pp.

209. **Adler, V. E. and Jacobson, M.,** Evaluation of selected natural and synthetic products as house fly repellents, *J. Environ. Sci. Health,* A17, 667, 1982.

210. **Dixit, R. S., Perti, S. L., and Agarwal, P. N.,** New repellents, *Labdev (India),* 3, 273, 1965.

211. **Trehan, K. N. and Pingle, S. V.,** Moths infesting stored grain and their control, *Indian Fmg.,* 8, 404, 1947.

212. **Saxena, B. P. and Srivastava, J. B.,** Effects of *Acorus calamus* L. oil vapours on *Dysdercus koenigii* F., *Indian J. Expt. Biol.,* 10, 391, 1972.

213. **Sudhakar, T. R., Pandey, N. D., and Tewari, G. C.,** Antifeeding property of some indigenous plants against mustard sawfly, *Athalia proxima* Klug, (Hymenoptera:Tenthredinidae), *Indian J. Agr. Sci.,* 48, 16, 1978.

214. **Khare, B. P., Gupta, S. B., and Chandra, S.,** Biological efficacy of some plant materials against *Sitophilus oryzae* Linnaeous, *Indian J. Agr. Res.,* 8, 243, 1974.

215. **Subramaniam, T. V.,** *Acorus calamus,* the sweet flag—a new indigenous insecticide for the household, *Indian J. Entomol.,* 4, 238, 1942.

216. **Subramaniam, T. V.,** Sweet flag (*Acorus calamus*)—a potential source of valuable insecticide, *J. Bombay Natur. Hist. Soc.,* 48, 338, 1949.

217. **Dixit, R. S., Perti, S. L., and Ranganathan, S. K.,** Evaluation of *Acorus calamus* Linn., an insecticidal plant of India, *J. Sci. Ind. Res.,* 15C, 1622, 1956.

218. **Tikku, K., Saxena, B. P., and Koul, O.,** Oogenesis in *Callosobruchus chinensis* and induced sterility by *Acorus calamus* L. oil vapors, *Ann. Zool. Ecol. Anim.,* 10, 545, 1978.

219. **Koul, O., Tikku, K., and Saxena, B. P.,** Follicular regression in *Trogoderma granarium* due to sterilizing vapours of *Acorus calamus* oil, *Curr. Sci.,* 46, 724, 1977.

220. **Chander, H. and Ahmed, S. M.,** Effect of some plant materials on the development of rice moth, *Corcyra cephalonica* (Staint.), *Entomon,* 11, 273, 1986.

221. **Koul, O.,** Antifeedant and growth inhibitory effects of calamus oil and neem oil on *Spodoptera litura* under laboratory conditions, *Phytoparasitica,* 15, 169, 1987.

222. **Agarwal, D. C., Deshpande, D. C., and Tipnis, H. P.,** Insecticidal activity of *Acorus calamus* on stored grain insects, *Pesticides (India),* 7(4), 21, 1973.

223. **Chopra, R. N., Vachist, V. R., and Handa, K. L.,** Chromatographic estimation of asarones in Indian *Acorus calamus* L. oil (tetraploid variety), *J. Chromatog.,* 17, 195, 1965.

224. **Raquibuddowla, M. and Haq, A.,** Pilot plant extraction of oil from *Acorus calamus* Linn., *Bangladesh J. Sci. Ind. Res.,* 9, 161, 1974.

225. **Patra, A. and Mitra, A. K.,** Constituents of *Acorus calamus* Linn., *Indian J. Chem.,* 17B, 412, 1979.

226. **Jacobson, M., Keiser, I., Miyashita, D. H., and Harris, E. J.,** Indian calamus root oil: attractiveness of the constituents to oriental fruit flies, melon flies, and Mediterranean fruit flies, *Lloydia,* 39, 412, 1976.

227. **Roark, R. C.,** *Excerpts From Consular Correspondence Relating to Insecticidal and Fish-poison Plants,* U.S. Bureau of Chemistry and Soils, 1931, 39 pp.

228. **Villani, M. G., Meinke, M. G., and Gould, F.,** Laboratory bioassay of crude extracts as antifeedants for the southern corn rootworm (Coleoptera: Chrysomelidae), *Environ. Entomol.,* 14, 617, 1985.

229. **Tschesche, R., Wulff, G., Weber, A., and Schmidt, H.,** Feeding deterrency of saponins on termites (Isoptera, Reticulitermes), *Z. Naturforsch.,* 25B, 999, 1970 (in German).

230. **Kondo, T., Shiro, K., and Teshima, M.,** The termiticidal wood-extractive from *Kalopanax septemlobus* Koidz., *Jap. Wood Res. Soc. J.,* 9(4), 125, 1963.

231. **Bigg, D. C. H. and Purvis, S. R.,** Muscarinic agonists provide a new class of acaricide, *Nature,* 262, 220, 1976.

232. **Tattersfield, F., Gimingham, C. T., and Morris, H. M.,** Studies on contact insecticides. IV. A quantitative examination of the toxicity of certain plants and plant products to *Aphis rumicis* (L.) (the bean aphis), *Ann. Appl. Biol.,* 13, 424, 1926.

233. **Devi, D. A. and Mohandas, N.,** Relative efficacy of some antifeedants and deterrents against insect pests of stored paddy, *Entomon,* 7, 261, 1982.

234. **Ali, S. I., Singh, O. P., and Misra, U. S.,** Effectiveness of plant oils against pulse beetle, *Callosobruchus chinensis* Linn., *Indian J. Entomol.,* 45, 6, 1983.

235. **Pereira, J.,** The effectiveness of six vegetable oils as protectants of cowpeas and bambara groundnuts against infestation by *Callosobruchus maculatus* (F.) (Coleoptera:Bruchidae), *J. Stored Prod. Res.,* 19, 57, 1983.

236. **Schoonhoven, A. V.,** Use of vegetable oils to protect stored beans from bruchid attack, *J. Econ. Entomol.,* 71, 254, 1978.

237. **Mummigati, S. G., and Ragunathan, A. N.,** Inhibition of the multiplication of *Callosobruchus chinensis* by vegetable oils, *J. Fd. Sci. Technol.,* 14, 184, 1977.

238. **Mehrotra, K. N. and Rao, P. J.,** Phagostimulants for locusts: studies with edible oils, *Entomol. Exp. Appl.,* 15, 208, 1972.

239. **Hill, J. and Schoonhoven, A. V.,** Effectiveness of vegetable oil fractions in controlling the Mexican bean weevil on stored beans, *J. Econ. Entomol.,* 74, 478, 1981.

240. **Shirokawa, E. A., Segal, G. M., and Torgov, I. V.,** Synthesis of asarones and its analogues, *Bioorg. Khim.,* 11, 270, 1985 (in Russian).

241. **Anonymous,** *Asclepias curassavica* as an insectifuge, Kew Roy. Bot. Garden Bull. Miscell. Informat. No. 130, 1897, 338.

242. **Bergey, D. H.,** *The Principles of Hygiene,* 4th Ed., Philadelphia, 1912, 529 pp.

243. **Seiber, J. N., Tuskes, P. M., Brower, L. P., and Nelson, C. J.,** Pharmacodynamics of some individual milkweed cardenolides fed to larvae of the monarch butterfly *(Danaus plexippus* L.), *J. Chem. Ecol.,* 6, 321, 1980.

244. **Sandquist, R. E., and Mulkern, G. B.,** Survival and development of *Melanoplus femurrubrum* and *M. sanguinipes* (Orthoptera:Acrididae) on plant-extract concentrations and fractions, *Ann. Entomol. Soc. Am.,* 65, 1005, 1972.

245. **Supavarn, P., Knapp, F. W., and Sigafus, R.,** Biologically active plant extracts for control of mosquito larvae, *Mosquito News,* 34, 398, 1974.

246. **Tarnopol, J. H. and Ball, H. J.,** A survey of some common midwestern plants for juvenile hormone activity, *J. Econ. Entomol.,* 65, 980, 1972.

247. **Villani, M. and Gould, F.,** Butterfly milkweed extracts as a feeding deterrent of the wireworm, *Melanoplus communis, Entomol. Exp. Appl.,* 37, 95, 1985.

248. **Rao, P. J.,** Phagostimulants and antifeedants from *Calotropis gigantea* for *Schistocerca gregaria* Forskal, *Z. Angew. Entomol.,* 93, 141, 1982.

249. **Rao, P. J., and Mehrotra, K. N.,** Phagostimulants and antifeedants from *Calotropis gigantea* for *Schistocerca gregaria* F., *Indian J. Exp. Biol,* 15, 148, 1977.

250. **Sievers, A. F., Archer, W. A., Moore, R. H., and McGovran, E. R.,** Insecticidal tests of plants from tropical America, *J. Econ. Entomol.,* 42, 549, 1949.

251. **Rao, D. S.,** Insecticide potential of the corolla of flowers, *Indian J. Entomol.,* 17, 121, 1955.

252. **Rao, D. S.,** The insecticidal property of petals of several common plants of India, *Econ. Bot.,* 11, 274, 1957.

253. **Granich, M. S., Halpern, B. P., and Eisner, T.,** Gymnemic acids: secondary plant substances of dual defensive action?, *J. Insect Physiol.,* 20, 435, 1974.

254. **Verma, G. S., Ramakrishnan, V., Mulchandani, N. B., and Chadha, M. S.,** Insect feeding deterrents from the medicinal plant *Tylophora asthmatica, Entomol. Exp. Appl.,* 40, 99, 1986.

255. **Nawrot, J., Bloszyk, E., Grabarczyk, H., Drozdz, B., Daniewski, W. M., and Holub, M.,** Further evaluation of feeding deterrency of sesquiterpene lactones to storage pests, *Prace Nauk. Inst. Ochrony Roslin.,* 25, 91, 1983.

256. **Hansberry, R. and Clausen, R. T.,** Insecticidal properties of miscellaneous plants, *J. Econ. Entomol.,* 38, 305, 1945.

257. **Lalonde, R. T., Wong, C. F., Hofstead, S. J., Morris, C. D., and Gardner, L. C.,** N-(2-methylpropyl)-*E,E*-2,4-decadienamide, a mosquito larvicide from *Achillea millefolium* L., *J. Chem. Ecol.,* 6, 35, 1980.

258. **Hosozawa, S., Kato, N., Munakata, A., and Chen, Y. L.,** Antifeeding active substances for insects in plants, *Agric. Biol. Chem.,* 38, 1045, 1974.

259. **Nawrot, J., Harmatha, J., and Novotny, L.,** Insect feeding deterrent activity of bisabolangelone and of some sesquiterpenes of eremophilane type, *Biochem. System. Ecol.,* 12, 99, 1984.

260. **Lu, R.,** Study of insect antijuvenile hormones. Chemical composition of *Ageratum conyzoides* L. and its effects on insects, *Kunchong Zhishi,* 19(4), 22, 1982.

261. **Bowers, W. S., Ohta, T., Cleare, J. S., and Marcelle, P. A.,** *Science,* 195, 542, 1976.

262. **Deb, D. C. and Chakravorty, S.,** Effect of precocene II, applied independently or subsequent to hydroprene treatment, on the morphogenesis of female reproductive organs of *Corcyra cephalonica, J. Insect Physiol.,* 28, 703, 1982.

263. **Azambuga, P. D., Bowers, W. S., Ribeiro, J. M. C., and Garcia, E. P. S.,** Antifeedant activity of precocenes and analogs on *Rhodnius prolixus, Experientia,* 38, 1054, 1982.

264. **Hodkova, M. and Socha, R.,** Why is *Pyrrhocoris apterus* insensitive to precocene II?, *Experientia,* 38, 977, 1982.

265. **Brooks, G. T., Pratt, G. E., and Jennings, R. C.,** The action of precocenes on milkweed bugs (*Oncopeltus fasciatus*) and locusts (*Locusta migratoria*), *Nature,* 281, 570, 1979.

266. **Deese, W. H., Sonenshine, D. E., Breidling, E., Buford, N. P., and Khalil, G. M.,** Toxicity of precocene-2 for the American dog tick, *Dermacentor variabilis* (Acari:Ixodidae), *J. Med. Entomol.,* 19, 734, 1982.

267. **Jacobson, M.,** The structure of pellitorine, *J. Am. Chem. Soc.,* 71, 366, 1949.

268. **Crombie, L.,** The structure of an insecticidal isobutylamide from pellitory root, *Chem. Ind.,* 1034, 1952.

269. **Jente, R., Bonnet, P. H., and Bohlmann, F.,** The constituents of *Anacyclus pyrethrum* DC, *Chem. Ber.,* 105, 1694, 1972 (in German).

270. **Sharma, S. D., Aggarwal, R. C., Soni, B. R., and Sharma, M. L.,** Syntheses of N-isobutyldeca-2(*E*),4-(*E*)-dienamides and N-isobutyldodeca-2(*E*),4(*E*)-dienamides, *Indian J. Chem.,* 18B, 81, 1979.

271. **Tsuji, J., Nagashima, H., Takahashi, T., and Masaoka, K.,** A simple synthesis of pellitorine (*N*-isobutyl-*E,E*-2,4-decadienamide) from the butadiene telomer, *Tetrahedron Lett.,* 1917, 1977.

272. **Mandai, T., Gotoh, J., Otera, J., and Kawada, M.,** A new synthetic method for pellitorine, *Chem. Lett.,* 313, 1980.

273. **Bohlmann, F. and Inhoffen, E.,** Synthesis of anacyclin, *Chem. Ber.,* 89, 1276, 1956.

274. **Crombie, L. and Mansour-i-Khuda, M.,** Amides of vegetable origin. IX. Total synthesis of anacyclin and related compounds, *J. Chem. Soc.,* 2767, 1957.

275. **Mulkern, G. B. and Toczek, D.,** Effect of plant extracts on survival and development of *Melanoplus differentialis* and *M. sanguinipes* (Orthoptera:Acrididae), *Ann. Entomol. Soc. Am.,* 65, 662, 1972.

276. **Ciaravellini, D.,** An insecticide from a plant, *Le Plante Officinali,* 3(9), 9, 1948 (in Italian).

277. **Yano, K.,** Insect antifeeding phenylacetylenes from growing buds of *Artemisia capillaris, J. Agr. Fd. Chem.,* 31, 667, 1985.

278. **Yano, K.,** Minor components from growing buds of *Artemisia capillaris* that act as insect antifeedants, *J. Agr. Fd. Chem.,* 35, 889, 1987.

279. **Yano, K.,** Relationships between chemical structure of phenylalkynes and their antifeeding activity for larvae of a cabbage butterfly, *Insect Biochem.,* 16, 717, 1986.

280. **Kaminski, J., Leitch, L. C., Morand, P., and Lam, J.,** Phototoxicity of naturally occurring and synthetic thiophene and acetylene analogues to mosquito larvae, *Phytochemistry,* 25, 1609, 1986.

136 *Glossary of Plant-Derived Insect Deterrents*

281. **Jermy, T., Butt, B. A., McDonough, L., Dreyer, D. L., and Rose, A. F.,** Antifeedants for the Colorado potato beetle. I. Antifeeding constituents of some plants from the sagebrush community, *Insect Sci. Appl.,* 1, 237, 1981.
282. **Nawrot, J., Bloszyk, E., Grabarczyk, H., and Drozdz, B.,** Feeding deterrent activity of the Compositae plant extracts for the selected storage pests, *Prace Nauk Inst. Ochrony Roslin,* 24, 37, 1982.
283. **Scarone, F.,** Some American and Asiatic toxic plants with insecticidal properties, *Agron. Colon.,* 28, 174, 1939.
284. **Kupchan, S. M., Steelman, D. R., Jarvis, B. B., Dailey, R. G., Jr., and Sneden, A. T.,** Isolation of potent new antileukemic trichothecenes from *Baccharis megapotamica, J. Org. Chem.,* 42, 4221, 1977.
285. **Wagner, H., Seitz, R., Lotter, H., and Herz, W.,** New furanoid *ent*-clerodanes from *Baccharis tricuneata, J. Org. Chem.,* 43, 3339, 1978.
286. **McLachlan, D., Arnason, J. T., Philogene, B. J. R., and Champagne, D.,** Antifeedant activity of the polyacetylene, phenylheptatriyne (PHT), from the Asteraceae to *Euxoa mesoria* (Lepidoptera:Noctuidae), *Experientia,* 38, 1061, 1982.
287. **Irvine, F. R.,** *Plants of the Gold Coast,* London, 1930, 521 pp.
288. **Dongre, T. K. and Rahalkar, G. W.,** Effect of *Blumea eriantha* (Compositae) oil on reproduction in *Earias vitella* F., *Experientia,* 38, 98, 1982.
289. **Gupta, S. C., Khanolkar, U. M., Koul, O., and Saxena, B. P.,** Pyrethrin synergistic activity by the essential oils of a few *Blumea* species, *Curr. Sci.,* 46, 304, 1977.
290. **Arnason, J. T., Philogene, B. J. R., Berg, C., MacEachern, A., Kaminski, J., Leitch, L. C., Morand, P., and Lam, J.,** Phototoxicity of naturally occurring and synthetic thiophene and acetylene analogues to mosquito larvae, *Phytochemistry,* 25, 1609, 1986.
291. **Sangappa, H. K.,** Effectiveness of oils as surface protectants against the bruchid, *Callosobruchus chinensis* Linnaeus infestation on redgram, *Mysore J. Agr. Sci.,* 11, 391, 1977.
292. **Tada, M. and Chiba, K.,** Novel plant growth inhibitors and an insect antifeedant from *Chrysanthemum coronarium* (Japanese name: shungiku), *Agr. Biol. Chem.,* 48, 1367, 1984.
293. **Rees, S. B. and Harborne, J. B.,** The role of sesquiterpene lactones and phenolics in the chemical defense of the chicory plant, *Phytochemistry,* 24, 2225, 1985.
294. **Willson, M. F., Anderson, P. K., and Thomas, P. A.,** Bracteal exudates in two *Cirsium* species as possible deterrents to insect consumers of seeds, *Am. Midland Naturalist,* 110, 212, 1983.
295. **Wellman, F. L. and Van Severen, M. L.,** Some plants used to combat pests by the native people in El Salvador, Am. Embassy, San Salvador Rept. 191, 1946.
296. **Fernandes, A. A. and Nagendrappa, G.,** Chemical constituents of *Dolichos lablab* (field bean) pod exudate, *J. Agr. Fd. Chem.,* 27, 795, 1979.
297. **Prakash, A., Pasalu, I. C., and Mathur, K. C.,** Ovicidal activity of *Eclipta alba* Hassk (Compositae), *Curr. Sci.,* 48, 1090, 1979.
298. **Wisdom, C. S., Smiley, J. T., and Rodriguez, E.,** Toxicity and deterrency of sesquiterpene lactones and chromenes to the corn earworm, *J. Econ. Entomol.,* 76, 993, 1983.
299. **Isman, M. B. and Proksch, P.,** Deterrent and insecticidal chromenes and benzofurans from *Encelia* (Asteraceae), *Phytochemistry,* 24, 1949, 1985.
300. **Kubo, I. and Nakanishi, K.,** Some terpenoid insect antifeedants from tropical plants, in *Advances in Pesticide Science,* Geissbühler, H., Brooks, G. T., and Kearney, P. C., Eds. Pergamon, New York, pt. 2, 1979, p. 293.
301. **Bordoloi, M. J., Shukla, V. S., and Sharma, R. P.,** Absolute stereochemistry of the insect antifeedant cadinene from *Eupatorium arenophorum, Tetrahedron Lett.,* 26, 509, 1985.
302. **Harmatha, J. and Nawrot, J.,** Comparison of the feeding deterrent activity of some sesquiterpene lactones and a lignan lactone towards selected insect storage pests, *Biochem. System. Ecol.,* 12, 95, 1984.
303. **Porcher, F. P.,** *Resources of the Southern Fields and Forests,* revised ed., Charleston, W. Va., 1869, 733 pp.
304. **Nakajima, S. and Kawazu, K.,** Coumarin and euponin, two inhibitors for insect development from the leaves of *Eupatorium japonicum, Agr. Biol. Chem.,* 44, 2893, 1980.

305. **Delobal, A. and Malonga, P.,** Insecticidal properties of six plant materials against *Caryedon serratus* (01.) (Coleoptera:Bruchidae), *J. Stored Prod. Res.,* 23, 173, 1987.

306. **Anon.,** Desert plants may be source of insecticides, *Chem. Engin. News,* 58, (36), 42, 1980.

307. **Nawrot, J., Bloszyk, E., Grabascyk, H., and Drozdz, B.,** Deterrent properties of sesquiterpene lactones for the selected storage pests, *Prace Nauk Inst. Ochrony Roslin,* 24, 27, 1982.

308. **Arnason, J. T., Isman, M. B., Philogene, B. J. R., and Waddell, T. G.,** Mode of action of the sesquiterpene lactone, tenulin, from *Helenium amarum* against herbivorous insects, *J. Nat. Prod.,* 50, 690, 1987.

309. **Rogers, C. E.,** Resistance of sunflower species to the western potato leafhopper, *Environ. Entomol.,* 10, 697, 1981.

310. **Rogers, C. E. and Thompson, T. E.,** Resistance of wild *Helianthus* species to an aphid, *Masonaphis masoni, J. Econ. Entomol.,* 71, 221, 1978.

311. **Rogers, C. E. and Thompson, T. E.,** *Helianthus* resistance to the carrot beetle, *J. Econ. Entomol.,* 71, 760, 1978.

312. **Waiss, A. C., Jr., Chan, B. G., Elliger, C. A., Garrett, V. H., Carlson, E. C., and Beard, B.,** unpublished results.

313. **Waiss, A. C., Jr.,** unpublished results.

314. **Elliger, C. A., Zinkel, D. F., Chan, B. G., and Waiss, A. C., Jr.,** Diterpene acids as larval growth inhibitors, *Experientia,* 32, 1364, 1976.

315. **Waiss, A. C., Jr.,** Helping plants defend themselves, *Agr. Res. (USDA),* 30,(1), 4, 1982.

316. **Gershenzon, J., Ohno, N., and Mabry, T. J.,** The terpenoid chemistry of sunflower (*Helianthus*), *Rev. Latinoamer. Quim.,* 13, (2), 53, 1981.

317. **Mitscher, L. A., Rao, G. S. R., Veysoglu, T., Drake, S., and Haas, T.,** Isolation and identification of trachyloban-19-oic and (-)-kaur-16-en-19-oic acids as antimicrobial agents from the prairie sunflower, *Helianthus annuus, J. Nat. Prod.,* 46, 745, 1983.

318. **Stipanovic, R. D., O'Brien, D. H., Rogers, C. E., and Thompson, T. E.,** Diterpenoid acids, (−)-*cis*- and (−)-*trans*-ozic acids, in wild sunflower, *Helianthus occidentalis, J. Agr. Fd. Chem.,* 27, 458, 1979.

319. **Mulkern, G. G. and Toczek, D. R.,** Bioassays of plant extracts for growthpromoting substances for *Melanoplus femurrubrum* (Orthoptera:Acrididae), *Ann. Entomol. Soc. Am.,* 63, 272, 1970.

320. **Klocke, J. A., Darlington, M. V., and Balandrin, M. F.,** 1,8-Cineole (eucalyptol), a mosquito feeding and ovipositional repellent from volatile oil of *Hemizonia fitchii* (Asteraceae), *J. Chem. Ecol.,* 13, 2131, 1987.

321. **Klocke, J. A., Balandrin, M. F., Adams, R. R., and Kingsford, E.,** Insecticidal chromenes from the volatile oil of *Hemizonia fitchii, J. Chem. Ecol.,* 11, 701, 1985.

322. **Waterhouse, D. F.,** Insectary tests of repellents for the Australian sheep blowfly *Lucilia cuprina,* Austral. Commonwealth Sci. Ind. Res. Organ. Bull. No. 218, 1947, p. 19.

323. **Tutin, F.,** Examination of plants for insecticidal principles, Bristol Univ. Agr. Hort. Res. Sta Ann. Repts. 1929, p. 96; 1930, p. 71.

324. **Picman, A. K., Elliott, R. H., and Towers, G. H. N.,** Insect feeding deterrent property of alantolactone, *Biochem. System. Ecol.,* 6, 333, 1978.

325. **Streibl, M., Nawrot, J., and Herout, V.,** Feeding deterrent activity of enantiomeric isoalantones, *Biochem. System. Ecol.,* 11, 381, 1983.

326. **Zalkow, L. H., Gordon, M. M., and Lanir, N.,** Antifeedants from rayless goldenrod and oil of pennyroyal: toxic effects for the fall armyworm, *J. Econ., Entomol.,* 72, 812, 1979.

327. **Fraenkel, G. S.,** The raison d'etre of secondary plant substances, *Science,* 129, 1466, 1959.

328. **Wiemer, D. F. and Ales, D. C.,** Lasidiol angelate: ant repellent sesquiterpenoid from *Lssiantheae fruticosa, J. Org. Chem.,* 46, 5449, 1981.

329. **Gajendran, G. and Gopalan, M.,** Note on the antifeedant activity of *Parthenium hysterophorus* Linn. on *Spodoptera litura* Fabricius (Lepidoptera:Noctuidae), *Indian J. Agr. Sci.,* 52, 203, 1982.

330. **Nandakumar, N. V., Prameela Devi, Y., Swami, K. S., and Majumder, S. K.,** Insecticidal properties of *Parthenium hysterophorus* extracts, *Comp. Physiol. Ecol.,* 5, 296, 1980.

331. **Rajendran, B. and Gopalan, M.,** Juvenile hormone-like activity of certain plant extracts on *Dysdercus cingulatus* Fabricius (Heteroptera:Pyrrhocoridae), *Indian J. Agr. Sci.,* 50, 781, 1980.

332. **Von Mueller, F.,** *Select Extra-Tropical Plants,* 9th ed., Melbourne, 1895, 654 pp.

333. **Saramma, P. U. and Verma, A. N.,** Efficacy of some plant products and magnesium carbonate as protectants of wheat seed against attack by *Trogoderma granarium, Bull. Grain Technol.,* 9, 207, 1971.

334. **Kalsi, P. S., Chhabra, B. R.,** Chhabra, A., **and Wadia, M. S.,** Chemistry of costunolide, *Tetrahedron,* 35, 1993, 1979.

335. **Pettei, M. J., Miura, I., Kubo, I., and Nakanishi, K.,** Insect antifeedant sesquiterpene lactones from *Schkuhria pinnata:* the direct obtention of pure compounds using reverse-phase preparative liquid chromatography, *Heterocycles,* 11, 471, 1978.

336. **Vig, B., Dehiya, S., Kad, G. L., and Ram, B.,** Synthesis of *N*-isobutyl-*trans*-4, *trans*-6-decadienamide, *J. Indian Chem. Soc.,* 53, 303, 1976.

337. **Yasuda, I., Takeya, K., and Itokawa, H.,** The geometric structure of spilanthol, *Chem. Pharm. Bull.,* 28, 2251, 1980.

338. **Nanayakhara, N. P. D., Klocke, J. A., Compadre, C. M., Hussain, R. A., Pezzuto, J. M., and Kinghorn, A. D.,** Characterization and feeding deterrent effects on the aphid, *Schizaphis graminum,* of some derivatives of the sweet compounds, stevioside and rebaudioside A, *J. Nat. Prod.,* 50, 434, 1987.

339. **Moennig, H. O.,** A new fly repellent and a blowfly dressing, *Onderstepoort J. Vet. Sci. Anim. Ind.,* 7, 419, 1936.

340. **Okoth, J.,** *Tagetes minuta* L. as a repellent and insecticide against adult mosquitoes, *East Afr. Med. J.,* 50, 317, 1973.

341. **Maradufu, A., Lubega, R., and Dorn, F.,** Isolation of (5*E*)-ocimenone, a mosquito larvicide from *Tagetes minuta, Lloydia,* 41, 181, 1978.

342. **McCulloch, R. N. and Waterhouse, D. F.,** Laboratory and field tests of mosquito repellents, Austral. Commonwealth Sci. Ind. Res. Organ. Bull. No. 213, 1947, 28 pp.

343. **Saxena, B. P. and Srivastava, J. B.,** *Tagetes minuta* L. oil—a new source of juvenile hormone mimicking substance, *Indian J. Exp. Biol.,* 11, 56, 1973.

344. **Brewer, G. J. and Ball, H. J.,** A feeding deterrent effect of a water extract of tansy (*Tanacetum vulgare* L., Compositae) on three lepidopterous larvae, *J. Kansas Entomol. Soc.,* 54, 733, 1981.

345. **Panasiuk, O.,** Response of Colorado potato beetles, *Leptinotarsa decemlineata* (Say) to volatile components of tansy, *Tanacetum vulgare, J. Chem. Ecol.,* 10, 1325, 1984.

346. **Forbes, S. A.,** Recent Illinois Work on the Corn Root-Aphis and the Control of its Injuries, Illinois Agr. Expt. Sta. Bull. No. 178, 1915, 466.

347. **Srbova, S. and Palaveyeva, M.,** Study of the insecticidal effect of some plants, *J. Hyg. Epidemiol., Microbiol., Immunol. (Prague),* 6, 498, 1962.

348. **Dutta, P., Bhattacharyya, P. R., Rabha, L. C., Bordoloi, D. N., Barua, N. C., Choudhury, P. K., Sharma, R. P., and Barua, J. N.,** Feeding deterrents for *Philosamia ricini* (*Samia cynthia* subsp. *ricini*) from *Tithonia diversifolia, Phytoparasitica,* 14, 77, 1986.

349. **Sarma, D. N., Barua, N. C., and Sharma, R. P.,** Structure-activity relationship: synthesis of 10-deoxytagitinin C and 10-deoxycyclotagitinin C and their antifeedant properties, *Chem. Ind.,* 167, 1985.

350. **Bentley, M. D., Leonard, D. E., Stoddard, W. F., and Zalkow, L. H.,** Pyrrolizidine alkaloids as larval feeding deterrents for spruce budworm, *Choristoneura fumiferana* (Lepidoptera:Tortricidae), *Ann. Entomol. Soc. Am.,* 77, 393, 1984.

351. **Ganjian, I., Kubo, I., and Fludzinski, P.,** Insect antifeedant elemanolide lactones from *Vernonia amygdalina, Phytochemistry,* 22, 2525, 1983.

352. **Burnett, W. C., Jones, S. B., Mabry, T. J., Padolina, W. G.,** Sesquiterpene lactones—insect feeding deterrents in *Vernonia, Biochem. System Ecol.,* 2, 25, 1974.

353. **Jones, S. B., Jr., Birnett, W. C., Jr., Coile, N. C., Mabry, T. J., and Betkouski, M. F.,** Sesquiterpene lactones of *Vernonia*—influence of glaucolide-A on the growth rate and survival of lepidopterous larvae, *Oecologia,* 39, 71, 1979.

354. **Nakajima, S. and Kawazu, K.,** Tridec-1-ene-3,5,7,9,11-pentayne, an ovicidal substance from *Xanthium canadense, Agr. Biol. Chem.,* 41, 1801, 1977.

355. **Jacobson, M., Crystal, M. M., and Kleiman, R.,** Effectiveness of several polyunsaturated seed oils as boll weevil feeding deterrents, *J. Am. Oil Chem. Soc.,* 58, 982, 1981.

356. **Konecky, M. S. and Mitlin, N.,** Chemical impairment of development in house flies, *J. Econ. Entomol.,* 48, 219, 1955.

357. **Mitlin, N., Butt, B. A., and Shortino, T. J.,** Effect of mitotic poisons on house fly oviposition, *Physiol. Zool.,* 30, 133, 1957.

358. **Lewis, W. H., and Elvin-Lewis, M. P. F.,** *Medical Botany,* Wiley, New York, 1977, 124.

359. **Hartwell, J. L.,** *Plants Used Against Cancer,* Quarterman, Lawrence Mass., 1982, 60.

360. **De Wilde, J., Mills, K., Lambers-Suverkropp, R., and Vontol, A.,** Response to air flow and airborne plant odour in the Colorado beetle, *Neth. J. Plant Pathol.,* 75, 53, 1969.

361. **Rust, M. K. and Reierson, D. A.,** Using wood extracts to determine the feeding preferences of the western drywood termite, *Incisitermes minor* (Hagen), *J. Chem. Ecol.,* 3, 391, 1977.

362. **Fiori, B. J. and Dolan, D. D.,** Resistance of *Betula davurica* to the birch leafminer, *Fenusa pusillo* (Hymenoptera:Tenthredinidae), *Can. Entomol.,* 116, 1275, 1984.

363. **Scriber, J. M.,** The behavior and nutritional physiology of southern armyworm larvae as a function of plant species consumed in earlier instars, *Entomol. Appl.,* 31, 359, 1982.

364. **Abramushkina, E. A.,** On the termite resistance of untreated and antiseptic-treated wood, *Proc. 13th Int. Cong. of Entomology,* Nauka, Leningrad, vol. 3, 1972, 7.

365. **El-Naggar, S. F. and Doskotch, R. W.,** Specioside: a new iridoid glycoside from *Catalpa speciosa, J. Nat. Prod.,* 43, 524, 1980.

366. **Chang, C. C. and Nakanishi, K.,** Specionin, an iridoid insect antifeedant from *Catalpa speciosa, J. Chem. Soc. Chem. Commun.* 605, 1983.

367. **Stephenson, A. G.,** Iridoid glycosides in the nectar of *Catalpa speciosa* are unpalatable to nectar thieves, *J. Chem. Ecol.,* 8, 1025, 1982.

368. **Von Sandermann, W. and Dietrichs, H. H.,** Research on termite-resistant woods, *Holz Roh-u. Wirkstoff.,* 15, 281, 1957 (in German).

369. **Carter, F. L., Beal, R. H., and Bultman, J. D.,** Extraction of antitermitic substances from 23 tropical hardwoods, *Wood Sci.,* 8, 406, 1975.

370. **Anonymous,** *A Handbook of Hardwoods,* H. M. Stationary Office, London, 1956, 269 pp.

371. **Scheffer, T. C. and Duncan, C. G.,** The decay resistance of certain Central American and Ecuadorian woods, Yale Univ. School of Forestry (Tropical Woods) Bull. No. 92, 1947, 24 pp.

372. **Verma, S. K. and Singh, M. P.,** Antifeedant effects of some plant extracts on *Amsacta moorei* Butler, *Indian J. Agr. Sci.,* 55, 298, 1985.

373. **Brar, K. S. and Sandhu, C. S.,** Comparative resistance of different *Brassica* species to the mustard aphid (*Lipaphis erysimi* Kalt.) under natural and artificial conditions, *Indian J. Agr. Res.,* 12, 198, 1978.

374. **Wats, R. C. and Singh, J.,** An investigation into the mosquitocidal value of indigenous derris and other drugs, *Rec. Malaria Survey India,* 7, 109, 1937.

375. **Nault, L. R. and Styer, W. E.,** Effects of sinigrin on host selection by aphids, *Entomol. Exp. Appl.,* 15, 423, 1972.

376. **Erickson, J. M. and Feeny, P.,** Sinigrin: a chemical barrier to the black swallowtail butterfly, *Papilio polyxenes, Ecology,* 55, 103, 1974.

377. **Hicks, K. L.,** Mustard oil glucosides: feeding stimulants for adult cabbage flea beetles, *Phyllotreta crucifera* (Coleoptera:Chrysomelidae), *Ann. Entomol. Soc. Am.,* 67, 261, 1974.

378. **Saxena, K. N. and Basit, A.,** Inhibition of oviposition by volatiles of certain plants and chemicals in the leafhopper *Amrasca devastans* (Distant), *J. Chem. Ecol.* 8, 329, 1982.

379. **Ediz, S. H. and Davis, G. R. F.,** Repellency of rapeseed extracts to adults of *Tribolium castaneum* and *Tribolium confusum* (Coleoptera:Tenebrionidae), *Can. Entomol.,* 112, 971, 1980.

380. **Lichtenstein, E. P., Strong, F. M., and Morgan, D. G.,** Identification of 2-phenylethyl isothiocyanate as an insecticide occurring naturally in the edible part of the turnip, *J. Agr. Fd. Chem.,* 16, 30, 1962.

381. **Nielsen, J. K.,** Host plant discrimination within Cruciferae: feeding responses of four leaf beetles (Coleoptera:Chrysomelidae) to glucosinolates, cucurbitacins, and cardenolides, *Entomol Exp. Appl.,* 24, 41, 1978.

382. **Meisner, J. and Mitchell, B. K.,** Phagodeterrency induced by two cruciferous plants in adults of the flea beetle *Phyllotreta striolata* (Coleoptera:Chrysomelidae), *Can. Entomol.,* 115, 1209, 1983.

383. **Renwick, J. A. A.,** Plant constituents as oviposition deterrents to lepidopterous insects, unpublished results.

384. **Nair, K. S. S. and McEwen, F. L.,** Host selection by the adult cabbage maggot, *Hylemya brassicae* (Diptera:Anthomyiidae): effect of glucosinolates and common nutrients on oviposition, *Can. Entomol.,* 108, 1021, 1976.

385. **Ayyar, T. V. R.,** Some local practices prevalent in south India in the control of insect pests, *Agr. J. India,* 16, 40, 1921.

386. **Anon.,** personal communication.

387. **Kassab, A., Chaarau, A. M., Hassan, I. I., and Shahwan, A. M.,** The Termite Problem in Egypt With Special Reference to Control, U.A.R. Ministry of Agr., Cairo, Rept. 91W, 1960, 52 pp.

388. **Bultman, J. D., Beal, R. H., and Ampong, F. F. K.,** Natural resistance of some tropical African woods to *Coptotermes formosanus* Shiraki, *Forest Prod. J.,* 29, (6), 46, 1979.

389. **Chatterjee, P. B.,** unpublished results.

390. **El-Feraly, F. S., McPhail, A. T., and Onan, K. D.,** X-ray crystal structure of canellal, a novel antimicrobial sesquiterpene from *Canella winterana, J. Chem. Soc., Chem. Commun.,* 75, 1978.

391. **El-Feraly, F. S. and Hoffstetter, M. D.,** Isolation, characterization and synthesis of 3-methoxy-4, 5-methylenedioxycinnamaldehyde: a novel constituent from *Canella winterana, J. Nat. Prod.,* 43, 407, 1980.

392. **Kubo, I., Lee, Y.-W., Pettei, M., Pilkiewics, F., and Nakanishi, K.,** Potent army worm antifeedants, from the East African *Warburgia* plants, *J. Chem. Soc., Chem. Commun.,* 1013, 1976.

393. **Kubo, I., and Ganjian, I.,** Insect antifeedant terpenes, hot-tasting to humans, *Experientia,* 37, 1063, 1981.

394. **Caprioli, V., Cimino, G., Colle, R., Gavaghnin, M., Sodano, G., and Spinella, A.,** Insect antifeedant activity and hot taste for humans of selected natural and synthetic 1,4-dialdehydes, *J. Nat. Prod.,* 50, 146, 1987.

395. **Blaney, W. M., Simmonds, M. S. J., Ley, S. V., and Katz, R. B.,** An electrophysiological and behavioural study of insect antifeedant properties of natural and synthetic drimane-related compounds, *Physiol. Entomol.,* 12, 281, 1987.

396. **Taniguchi, M., Adachi, T., Oi, S., Kimura, A., Katsumura, S., Isoe, S., and Kubo, I.,** Structure-activity relationship of the warburgia sesquiterpene dialdehydes, *Agr. Biol. Chem.,* 48, 73, 1984.

397. **Nakanishi, K. and Kubo, I.,** Studies on warburganal, muzigadial, and related compounds, *Israel J. Chem.,* 16, 28, 1977.

398. **Lam, P. Y.-S. and Frazier, J. L.,** Model study on the mode of action of muzigadial antifeedant, *Tetrahedron Lett.,* 28, 5477, 1987.

399. **Kubo, I., Miura, I., Pettei, M. J., Lee, Y.-W., Pilkiewicz, F., and Nakanishi, K.,** Muzigadial and warburganal, potent antifungal, antiyeast, and African army worm antifeedant agents, *Tetrahedron Lett.,* 4553, 1977.

400. **Cimino, G., De Rosa, S., De Steffano, S., and Sodano, G.,** Observations on the toxicity and metabolic relationships of polygodial, the chemical defense of the nudibranch *Dendrodoris limbata, Experientia,* 41, 1335, 1985.

401. **Kubo, I., Matsumoto, T., Kakooko, A. B., and Mubiru, N. K.,** Structure of mukaadial, a molluscide from the *Warburgia* plants, *Chem. Lett.,* 979, 1983.

402. **Taniguchi, M., Adachi, T., Haraguchi, H., Oi, S., and Kubo, I.,** Physiological activity of warburganal and its reactivity with sulfhydryl groups, *J. Biochem.,* 94, 149, 1983.

403. **D'Ischia, M., Prota, G., and Sodano, G.,** Reaction of polygodial with primary amines. An alternative explanation to the antifeedant activity, *Tetrahedron Lett.,* 23, 3295, 1982.

404. **Nakata, T., Akita, H., Naito, T., and Oishi, T.,** A total synthesis of (±)-warburganal, *J. Am. Chem. Soc.,* 101, 4400, 1979.

405. **Nakata, T., Akita, H., Naito, T, and Oishi, T.,** Stereospecific total synthesis of (±)-warburganal and related compounds, *J. Am. Chem. Soc.,* 101, 4398, 1979.

406. **Ohsuka, A., and Matsukawa, A.,** Syntheses of (±)-warburganal and (±)-isotadeonal, *Chem. Lett.,* 635, 1979.

407. **Goldsmith, D. J. and Kezar, H. S.,** A stereospecific total synthesis of warburganal, *Tetrahedron Lett.,* 21, 3543, 1980.

408. **Kende, A. S. and Blacklock, T. J.,** Stereoselective total synthesis of (±)-warburganal and (±)-isotadeonal, *Tetrahedron Lett.,* 21, 3119, 1980.

409. **Nakata, T., Akita, H., Naito, T., and Oishi, T.,** A total synthesis of (±)-warburganal, *Chem. Pharm. Bull.,* 26, 2172, 1980.

410. **Ley, S. V. and Mahon, M.,** Synthesis of (±)-warburganal, *Tetrahedron Lett.,* 3909, 1981.

411. **Hollinshead, D. M., Howell, S. C., Ley, S. V., Mahon, M., and Ratcliffe, N. M.,** The Diels-Alder route to drimane related sesquiterpenes: synthesis of cinnamolide, polygodial, isodrimeninol, drimenin and warburganal, *J. Chem. Soc., Perkin Trans. I,* 1579, 1983.

412. **Anon.,** Warburganal intermediates, Jap. Patent 58,131,990, Aug. 6, 1983.

413. **Peterse, A. J. G. M., Roskam, J. H., and DeGroot, A.,** A model synthesis of ring B of warburganal and muzigadial, *Rec. Trav. Chim. Pays-Bas,* 97, 249, 1978.

414. **Kojima, Y., and Kato, N.,** Synthesis of perhydrofuro [2,3-b] furan compounds and the structure-activity relationships of the antifeeding active compounds, *Agr. Biol. Chem.,* 44, 855, 1980.

415. **Oyarzun, M. L., Cortes, M., and Sierra, J.,** Synthesis of (−)-drim-7-ene-9α,11,12-triol, the direct precursor of (−)-warburganal, *Synth. Commun.,* 12, 951, 1982.

416. **Razmilic, I., Sierra, J., Lopez, J., and Cortes, M.,** A novel partial synthesis of (−)-warburganal, *Chem. Lett.,* 1113, 1985.

417. **Okawara, H., Nakai, H., and Ohno, M.,** Synthesis of (−)-warburganal and 4α-methoxycarbonyl congener from *1*-abietic acid, *Tetrahedron Lett.,* 23, 1087, 1982.

418. **Ototani, N., Kato, T., and Kitahara, Y.,** A synthesis of 2-formyl-5,5,9-trimethyl-trans-Δ^2-1-octalone, *Bull. Chem. Soc. Japan,* 40, 1730, 1967.

419. **Guillerm, D., Delarue, M., Jalali-Naini, M., Lemaitre, P., and Lallemand, J.-Y.,** Synthesis of all possible stereoisomers of polygodial, *Tetrahedron Lett.,* 25, 1043, 1984.

420. **Mori, K. and Watanabe, H.,** Synthesis of both the enantiomers of polygodial, an insect antifeedant sesquiterpene, *Tetrahedron,* 42, 273, 1986.

421. **Gruenwald, O.,** The Oleaginous Fruits of a Venezuelan Capparidacea, Venezuel. Ministry Agr., El Valle, Bull. No. 3, 1946, 33 pp.

422. **Adlung, K. G.,** Attraction of gypsy moth, *Z. Angew. Zool.,* 44, 61, 1957.

423. **Ohigashi, H., and Koshimizu, K.,** Chavicol, as a larva-growth inhibitor, from *Viburnum japonicum* Spreng, *Agr. Biol. Chem.,* 40, 2283, 1976.

424. **Chiu, S.-F.,** Effectiveness of Chinese insecticidal plants with reference to the comparative toxicity of botanical and synthetic insecticides, *J. Sci. Fd. Agr.,* 1, 276, 1950.

425. **Yamada, K., Shizuri, Y., and Hirata, Y.,** Isolation and structures of a new alkaloid alatamine and an insecticidal alkaloid wilfordine from *Euonymus alatus* forma *striatum* (Thunb.) Makino, *Tetrahedron,* 34, 1915, 1978.

426. **Sen-Sarma, P. K., Mishra, S. C., and Gupta, B. K.,** Studies on the natural resistance of timber to termite attack. VI. Laboratory evaluation of the resistance of "mundani," "anjan," and "Banati" to *Microcerotermes beesoni* Snyder (Insecta:Isoptera:Amitermitinae), *Indian Forester,* 96, 75, 1970.

427. **Delle Monache, F., Marini Bettolo, G. B., and Bernays, E. A.,** Isolation of insect antifeedant alkaloids from *Maytenus rigida* (Celastraceae), *Z. Angew. Entomol.,* 97, 406, 1984.

428. **Lee, C. S. and Hansberry, R.,** Toxicity studies of some Chinese plants. *J. Econ. Entomol.,* 36, 915, 1943.

429. **Cheng, T.-H.,** Field tests of the thundergod vine against melon leaf beetle, *J. Econ. Entomol.,* 38, 491, 1945.

430. **Tattersfield, F., Potter, C., Lord, K. A., Gillingham, E. M., Way, M. J., and Stoker, R. I.,** Insecticides derived from plants, *Kew Bull.,* 3, 329, 1948.

431. **Acree, F., Jr., and Haller, H. L., Wilfordine,** an insecticidal alkaloid from *Tripterygium wilfordii* Hook, *J. Am. Chem. Soc.,* 72, 1608, 1950.

432. **Beroza, M.,** Alkaloids from *Tripterygium wilfordii* Hook—wilforine and wilfordine, *J. Am. Chem. Soc.,* 73, 3656, 1951.

433. **Beroza, M.,** Alkaloids from *Tripterygium wilfordii*—wilforgine and wilfortrine, *J. Am. Chem. Soc.,* 74, 1585, 1952.
434. **Beroza, M.,** Alkaloids from *Tripterygium wilfordii* Hook. The isolation and structure of wilforzine, *J. Am. Chem. Soc.,* 75, 2136, 1953.
435. **Beroza, M.,** Alkaloids from *Tripterygium wilfordii* Hook. The chemical structure of wilfordic and hydroxywilfordic acids, *J. Org. Chem.,* 28, 3562, 1963.
436. **Nabokov, V. A.,** Anabasine sulfate: a protective agent against bites of malarial mosquitoes, *Am. Rev. Soviet Med.,* 2, 449, 1945.
437. **Font Quer, P.,** *Medicinal Plants,* Barcelona, 1962, 1,033 pp. (in Spanish).
438. **Klocke, J. A., Van Wagenen, B., and Balandrin, M. F.,** The ellagitannin geraniin and its hydrolysis products isolated as insect growth inhibitors from semi-arid land plants, *Phytochemistry,* 25, 85, 1986.
439. **Rajasekaran, B., Rajendran, R., Velusamy, R., and Sundara Babu, P. C.,** Effect of vegetable oil on rice leaffolder (LF) feeding behavior, *Int. Rice Res. Newslett.,* 12, (2), 34, 1987.
440. **Ketkar, C. M.,** Use of tree-derivated non-edible oils as surface protectants for stored legumes against *Callosobruchus maculatus* and *C. chinensis,* in *Natural Pesticides from the Neem Tree and Other Tropical Plants,* Schmutterer, H. and Ascher, K. R. S., Eds., GTZ Press, Eschborn, FRG, 1987, 535.
441. **Coudreau, J., Fougerousse, M., Bressy, O., and Lucas, S.,** Research to develop a new method for determining resistance of timber to destruction by termites (*Reticulitermes lucifugus* Rossi), *Holzforsch.,* 14, 50, 1960.
442. **Plank, H. K.,** Experiments with mamey for pests of man and animals, *Trop. Agr.,* 27, 38, 1950.
443. **Jones, M. A. and Plank, H. K.,** Chemical nature of the insecticidal principle in mamey seed, *J. Am. Chem. Soc.,* 67, 2266, 1945.
444. **Djerassi, C., Eisenbraun, E. J., Finnegan, R. A., and Gilbert, B.,** Naturally occurring oxygen heterocyclics. II. Characterization of an insecticidal principle from *Mammea americana* L., *J. Am. Chem. Soc.,* 80, 3686, 1958.
445. **Djerassi, C., Eisenbraun, E. J., Finnegan, R. A., and Gilbert, B.,** Naturally occurring heterocyclics. V. Mammein, *Tetrahedron Lett.,* 1, 10, 1959.
446. **Djerassi, C., Eisenbraun, E. J., Finnegan, R. A., and Gilbert, B.,** Naturally occurring oxygen heterocyclics. VII. The structure of mammein, *J. Org. Chem.,* 25, 2164, 1960.
447. **Finnegan, R. A. and Mueller, W. H.,** Constituents of *Mammea americana* L. IV. The structure of mammein, *J. Org. Chem.,* 30, 2342, 1965.
448. **Finnegan, R. A. and Djerassi, C.,** Naturally occurring oxygen heterocyclics. 4-Phenyl-5,7-dihydroxy-6-isovaleryl-8-isopentenylcoumarin, *Tetrahedron Lett.,* 13, 11, 1959.
449. **Finnegan, R. A., Morris, M. P., and Djerassi, C.,** Naturally occurring oxygen heterocyclics. X. 4-Phenyl-5,7-dihydroxy-6-isovaleryl-8-isopentenylcoumarin, *J. Org. Chem.,* 26, 1180, 1961.
450. **Pendse, C. S., Ghokale, V. G., Phalnikar, N. L., and Bhide, B. V.,** Investigation of new plant larvicides with special reference to *Spilanthes acmella, J. Univ. Bombay (Sci. Sect.),* A20, 26, 1946.
451. **Lesne, A.,** Small enemies of fruit trees, *J. Agr. Prat.,* 50, 507, 1886.
452. **Ferguson, J. E., Metcalf, E. R., Metcalf, R. L., and Rhodes, A. M.,** Influence of cucurbitacin content in cotyledons of Cucurbitaceae cultivars upon feeding behavior of Diabroticina beetles (Coleoptera:Chrysomelidae), *J. Econ. Entomol.,* 76, 47, 1983.
453. **Chambliss, O. L. and Jones, C. M.,** Chemical and genetic basis for insect resistance in cucurbits, *Proc. Am. Soc. Hort. Sci.,* 89, 397, 1966.
454. **Knipping, P. A., Patterson, C. G., Knavel, D. E., and Rodriguez, J. C.,** Resistance of cucurbits to twospotted spider mite, *Environ. Entomol.,* 4, 507, 1975.
455. **Anon.,** Ladybug, ladybug, *The Sciences (N.Y. Acad. Sci.),* 20(9), 5, 1980.
456. **Jackson, L. E. and Wassell, H. E.,** Mothproofing fabrics and furs, *Ind. Engin. Chem.,* 19, 1175, 1927.
457. **Elsey, K. D.,** Pickleworm: survival, development, and oviposition in selected hosts, *Ann. Entomol. Soc. Am.,* 74, 96, 1981.
458. **Giles, P. H.,** The storage of cereals by farmers in Northern Nigeria, *Trop. Agr. (Trinidad),* 41 (3), 197, 1964.

459. **Harayama, T., Cho, H., and Inubushi, Y.,** Total synthesis of *dl*-chamaecynone, a termiticidal norsesquiterpene, *Tetrahedron Lett.,* 37, 3273, 1977.

460. **Harayama, T., Cho, H., and Inubushi, Y.,** Total synthesis of *dl*-chamaecynone, a termiticidal norsesquiterpene, *Chem. Pharm. Bull.,* 26, 1201, 1978.

461. **Carter, F. L. and Smythe, R. V.,** Feeding and survival responses of *Reticulitermes flavipes* (Kollar) to extractives of wood from 11 coniferous genera, *Holzforsch.,* 28, 41, 1974.

462. **Esenther, G. R.,** Nutritive supplement method to evaluate resistance of natural or preservative-treated wood to subterranean termites, *J. Econ. Entomol.,* 70, 341, 1977.

463. **Jacobson, M.,** Chemistry of natural products with juvenile hormone activity, *Mitt. Schweiz. Entomol. Ges.,* 44, 73, 1971.

464. **Saeki, I., Sumimoto, M., and Kondo, T.,** The termiticidal substances from the wood of *Chamaecyparis pisifera* D. Don, *Holzforsch.,* 27, 93, 1973.

465. **Weissmann, G. and Dietrichs, H. H.,** The activity of *l*-citronellic acid and related compounds against *Reticulitermes* species, *Holzforsch.,* 29, 68, 1975.

466. **Novak, D. and Potocek, V.,** Juniper oil as a mosquito repellent, *Acta Univ. Carolinae (Biol.),* 12, 369, 1977 (Pub. 1980) (in German).

467. **Harrow, K. M.,** A comparative study of the antitermitic value of boric acid, zinc chloride, and tanalith H, Austral. Commonwealth Sci. Ind. Res. Organ. Tech. Paper 4, 14 pp.

468. **Oda, J., Ando, N., Nakajima, Y., and Inouye, Y.,** Studies on insecticidal constituents of *Juniperus recurva* Buch., *Agr. Biol. Chem.,* 41, 201, 1977.

469. **Hale, T.,** *A Compleat Body of Husbandry,* London, 2nd ed., Vol. 1, 1758.

470. **Back, E. A. and Rabak, P.,** Red cedar chests as protectors against moth damage, USDA Bull. No. 1051, 1922, 14 pp.

471. **Bishopp, F. C.,** Fleas, USDA Bull. No. 248, 1915, 31 pp.

472. **Anon.,** The Work of the Department of Insecticides and Fungicides, Rothamsted Expt. Sta. Rept. 1950, 104.

473. **Huddle, H. B., and Mills, A. P.,** Toxicity of cedar oil vapor to clothes moths, *J. Econ. Entomol.,* 45, 40, 1952.

474. **Sweetman, H. L., Benson, D. A., and Kellery, R. W., Jr.,** Efficacy of aroma of cedar in control of fabric pests, *J. Econ. Entomol.,* 46, 29, 1953.

475. **Gupta, M.,** Olfactory response of *Apis florea* F. to some essential oils, *J. Apicult. Res.,* 26, 3, 1987.

476. **Russell, G. B. Singh, P., and Fenemore, P. G.,** Insect control chemicals from plants. III. Toxic lignans from *Libocedrus bidwilli, Austral. J. Biol. Sci.,* 29, 99, 1976.

477. **Alfaro, R. I., Pierce, H. D., Jr., Borden, J. H., and Oehlschlager, A. C.,** Insect feeding and oviposition deterrents from western red cedar foliage, *J. Chem. Ecol.,* 7, 39, 1981.

478. **Verma, G. S., Pandey, U. K., and Pandey, M.,** Note on insecticidal properties of some plants against *Bagrada cruciferarum* Kirk. (Hemiptera:Pentatomidae), *Indian J. Agr. Sci.,* 52, 263, 1982.

479. **Kisskalt, K.,** Protection of man from mosquitoes and other insects in the sixteenth century, *Arch. Schiffe Trop. Hyg.,* 17, 85, 1913.

480. **Anon.,** *A Handbook of Softwoods,* H. M. Stationary Office, London, 1957, 73 pp.

481. **Becker, G.,** Testing and evaluation of the natural resistance of wood against termites, *Holz Roh Werkstoffe,* 19, 278, 1961 (in German).

482. **Pandey, G. P., Doharey, R. B., and Varma, B. K.,** Efficacy of some vegetable oils for protecting greengram against the attack of *Callosobruchus maculatus* (Fabr.), *Indian J. Agr. Sci.,* 51, 910, 1980.

483. **Sharma, H. L., Vimal, O. P., and Atri, B. S.,** Effect of processing oilcakes on plant growth and use of extracts as insecticide, *Indian J. Agr. Sci.,* 51, 896, 1981.

484. **Becker, C. and Puckett, D.,** A basic study on laboratory tests with two species of Reticulitermes, *Holzforsch. Holzverwert.,* 13, 110, 1961.

485. **Anonymous,** *Common Sarawak Timbers,* 2nd ed., Borneo Literature Bureau, Malaya, 1964, 53 pp.

486. **Carter, F. L., Carlo, A. M., and Stanley, J. B.,** Termiticidal components of wood extracts: 7-methyljuglone from *Diospyros virginiana, J. Agr. Fd. Chem.,* 26, 869, 1978.

487. **Matteson, J. W., Taft, H. M., and Rainwater, C. F.,** Chemically induced resistance in the cotton plant to attack by the boll weevil, *J. Econ. Entomol.,* 56, 189, 1963.

488. **El-Naggar, S. F., Doskotch, R. W., Odell, T. M., and Girard, L.,** Antifeedant diterpenes for the gypsy moth larvae from *Kalmia latifolia:* isolation and characterization of ten grayanoids, *J. Nat. Prod.,* 43, 617, 1980.

489. **El-Naggar, S. F., El-Feraly, F. S., Foos, J. S., and Doskotch, R. W.,** Flavanoids from the leaves of *Kalmia latifolia, J. Nat. Prod.,* 43, 739, 1980.

490. **Sakata, K., Hattori, M., Sakurai, A., and Hosotsuji, T.,** Isolation and identification of biologically active constituents of *Leucothoe catesbai* A. Gray, *J. Pesticide Sci.,* 2, 453, 1977.

491. **Bell, H. T. and Clarke, R. G.,** Resistance among rhododendron species to obscure root weevil feeding, *J. Econ. Entomol.,* 71, 869, 1978.

492. **Doss, R. P.,** Investigation of the bases of resistance of selected *Rhododendron* species to foliar feeding by the obscure root weevil (*Sciopithes obscurus*), *Environ. Entomol.,* 9, 549, 1980.

493. **Doss, R. P., Luthi, R., and Hrutfiord, B. F.,** Germacrone, a sesquiterpene repellent to obscure root weevil from *Rhododendron edgeworthii, Phytochemistry,* 19, 2379, 1980.

494. **Hardee, D. D. and Davich, T. B.,** A feeding deterrent for the boll weevil, *Anthonomus grandis,* from tung meal, *J. Econ. Entomol.,* 59, 1267, 1966.

495. **Jacobson, M., Crystal, M. M., and Warthen, J. D., Jr.,** Boll weevil feeding deterrents from tung oil, *J. Agr. Fd. Chem.,* 29, 591, 1981.

496. **Jacobson, M. and Crystal, M. M.,** Effectiveness of several polyunsaturated seed oils as boll weevil feeding deterrents, *J. Am. Oil Chem. Soc.,* 58, 982, 1981.

497. **Ratovelomanana, V. and Linstrumelle, G.,** Stereoselective preparation of 1-chloro-(*E,E*)-1,3-dienes. Application to the synthesis of methyl eleostearate, *Tetrahedron Lett.,* 25, 6001, 1984.

498. **Hooper, D.,** The bark of *Cleistanthus collinus* as a fish poison, *Pharm. J. Pharm. (London),* 61, 74, 1898.

499. **Sesseler, M. and Spoon, W.,** Usefulness of wild plants in the Windward Islands, *West Indies Gids.,* 33, 49, 1952 (in Flemish).

500. **McIndoo, N. E.,** The castor bean as a source of insecticides. A review of the literature, USDA Bureau Entomol. Plant Quarantine Pub. E-572, 1945, 15 pp.

501. **Krishna, B. S. and Vishwapremi, K. K. C.,** Effect of treating okra fruits with extracts of castor leaves on oviposition, *Natl. Acad. Sci. Lett. (India),* 1, 117, 1978.

502. **Haller, H. L. and McIndoo, N. E.,** The castor-bean plant as a source of insecticides, *J. Econ. Entomol.,* 36, 638, 1943.

503. **Siegler, E. H., Schechter, M. S., and Haller, H. L.,** Toxicity of ricin, ricinine, and some related compounds to codling moth larvae, *J. Econ. Entomol.,* 37, 416, 1944.

504. **Van Over, M. D. L.,** Insecticide, U.S. Patent 2,333,061, 1943.

505. **Freedman, B., Reed, D. K., Powell, R. G., Madrigal, R. V., and Smith, C. R., Jr.,** Biological activities of *Trewia nudiflora* extracts against certain economically important insect pests, *J. Chem. Ecol.,* 8, 409, 1982.

506. **Reed, D. K., Kwolek, W. F., and Smith, C. R., Jr.,** Investigation of antifeedant and other insecticidal activities of trewiasine towards the striped cucumber beetle and codling moth,*J. Econ. Entomol.,* 76, 641, 1983.

507. **Jacobson, M., Reed, D. K., Redfern, R. E., and Crystal, M. M.,** Insect feeding deterrents from plants of the Meliacidae and Euphorbiaceae, unpublished results.

508. **Meyer, B. N., Ferrigni, N. R., Jacobsen, L. B., Nichols, D. E., and McLaughlin, J. L.,** Brine shrimp: a convenient general bioassay for active plant constituents, *Planta Med.,* 45, 31, 1982.

509. **Ferrigni, N. R., McLaughlin, J. L., Powell, R., G., and Smith, C. R., Jr.,** Use of potato disc and brine shrimp bioassays to detect activity and isolate piceatannol as the antileukemic principle from the seeds of *Euphorbia lagascae, J. Nat. Prod.,* 47, 347, 1984.

510. **Ferrigni, N. R., Putnam, J. E., Anderson, B., Jacobsen, L. B., Nichols, D. E., Moore, D. S., and McLaughlin, J. L.,** Modification and evaluation of the potato disc assay and antitumor screening of Euphorbiaceae seeds, *J. Nat. Prod.,* 45, 679, 1982.

511. **Cross, J. R.,** Insecticidal repellent, U.S. Patent 2,159,550, May 23, 1939.

512. **Qadri, S. S. H.,** Some new indigenous plant repellents for storage pests,*Pesticides (India),* 1, (12), 18, 1973.

513. **Chellappa, K. and Chelliah, S.,** Studies on the efficiency of malathion and certain plant products to the control of *Sitotroga cerealella and Rhizopertha dominica* infesting rice grains, *Madras Agr. J.,* 63, 190, 1976.

514. **Rehr, S. S., Bell, E. A., Janzen, D. H., and Feeny, P. P.,** Insecticidal amino acids in legume seeds, *Biochem. System.,* 1, 63, 1973.

515. **Gombos, M. A. and Gasko, K.,** Extraction of natural antifeedants from the fruits of *Amorpha fruticosa* L., *Acta Phytopath. Acad. Sci. Hung.,* 12, 349, 1977.

516. **Gombos, M., Szendsei, K., Feuer, L., Toth, C., and Kecskee, M.,** Environmental aspects in the evaluation of the antifeedants extracted from *Amorpha fruticosa, Proc. Hung. Ann. Mtg. Biochem.,* 18, 23, 1978.

517. **Reed, W., Seshu Reddy, K. V., Lateef, S. S., Amin, P. W., and Davies, J. C.,** Contribution of ICRISAT to studies on plant resistance to insect attack, in *Proc. on the Use of Naturally Occurring Plant Products in Pest and Disease Control,* Nairobi, Kenya, May 12 to 15, 1980, 14 pp.

518. **Singh, S. R., Luse, R. A., Leuschner, K., and Nangju, D.,** Groundnut oil treatment for the control of *Callosobruchus maculatus* (F,) during cowpea storage, *J. Stored Prod. Res.,* 14, 77, 1978.

519. **Oi, Y.-T. and Burkholder, W. E.,** Protection of stored wheat from the granary weevil by vegetable oils, *J. Econ. Entomol.,* 74, 502, 1981.

520. **LeCato, C. L.,** Yield, development, and weight of *Cadra cautella* (Walker) and *Plodia interpunctella* (Hübner) on twenty-one diets derived from natural products, *J. Stored Prod. Res.,* 12, 43, 1976.

521. **Reddy, D. B.,** Scope for using vegetable oils as insecticides, Proc. Int. Cong. Entomol., 1956, 3, 353, 1958.

522. **Williams, S. W.,** On the indigenous medical botany of Massachusetts, *Trans. Am. Med. Assoc.,* 2, 863, 1849.

523. **Rosenthal, G. A.,** The protective action of a higher plant toxic product, *Bioscience,* 38, 104, 1988.

524. **Bell, E. A., Lackey, J. A., and Polhill, R. M.,** Systematic significance of canavanine in the Papilionoideae (Faboideae), *Biochem. System. Ecol.,* 6, 201, 1978.

525. **Isogai, A., Murakoshi, S., Suzuki, A., and Tamura, S.,** Isolation from "Astragali Radix" of L-canavanine as an inhibitory substance to metamorphosis of silkworm, *Bombyx mori* L., *J. Agr. Chem. Soc. Japan,* 47, 449, 1973.

526. **Boyar, A. and Marsh, R. E.,** L-Canavanine, a paradigm for the structures of substituted guanidines, *J. Am. Chem. Soc.,* 104, 1995, 1982.

527. **Natelson, S. and Bratton, C. R.,** Canavanine assay of some alfalfa varieties (*Medicago sativa*) by fluorescence: practical procedure for canavanine preparation, *Microchem. J.,* 29, 26, 1984.

528. **Natelson, S.,** Canavanine in alfalfa (*Medicago sativa*), *Experientia,* 41, 257, 1985.

529. **Natelson, S.,** Canavanine to arginine ratio in alfalfa (*Medicago sativa*), clover (*Trifolium*), and the jack bean (*Canavalia ensiformis*), *J. Agr. Fd. Chem.,* 33, 413, 1985.

530. **Koul, O.,** Foliage spray tests with *l*-canavanine for control of *Spodoptera litura, Phytoparasitica,* 13, 167, 1985.

531. **Mullenax, C. H.,** The use of jackbean (*Canavalia ensiformis*) as a biological control for leaf-cutting ants (*Atta* sp;.), *Biotropica,* 11, 313, 1979.

532. **Dahlmab, D. L., Herald, F., and Knapp, F. W.,** L-Canavanine effects on growth and development of four species of Muscidae, *J. Econ. Entomol.,* 72, 678, 1979.

533. **Horie, Y., Watanabe, K., Isogai, A., Suzuki, A., and Tamura, S.,** Inhibitory effect of *l*-canavanine on spinning and pupation of the silkworm, *Bombyx mori, Nippon Sahshigaku Zasshi,* 51, 407, 1982.

534. **Dahlman, D. L. and Rosenthal, C. A.,** Potentiation of *l*-canavanine-induced developmental anomalies in the tobacco hornworm, *Manduca sexta,* by some amino acids, *J. Insect Physiol.,* 28, 829, 1982.

535. **Dahlman, D. L.,** Field test of *l*-canavanine for control of tobacco hornworm, *J. Econ. Entomol.,* 73, 279, 1980.

536. **Janzen, D. H., Juster, H. B., and Bell, E. A.,** Toxicity of secondary compounds to the seed-eating larvae of the bruchid beetle *Callosobruchus maculatus, Phytochemistry,* 16, 223, 1977.

537. **Koul, O.,** L-Canavanine from *Canavalia ensiformis* seeds: effects on fertility of *Periplaneta americana* (L.) (Orthopt., Blattidae), *Z. Angew. Entomol.,* 96, 530, 1983.

538. **Koul, O.,** L-Canavanine, an antigonadal substance for *Dysdercus koenigii, Entomol. Exp. Appl.,* 34, 297, 1983.

539. **Rosenthal, C. A.,** Biochemical adaptations of the bruchid beetle, *Caryedes brasiliensis* to *l*-canavanine, a higher plant allelochemical, *J. Chem. Ecol.,* 9, 803, 1983.

540. **Jaipal, S., Singh, Z., and Chauhan, R.,** Juvenile-hormone-like activity in extracts of some common Indian plants, *Indian J. Agr. Sci.,* 53, 730, 1983.

541. **Lambert, J. D. H., Gale, J., Arnason, J. T., and Philogene, B. J. R.,** Bruchid control with traditionally used insecticidal plants *Hyptis spicigera* and *Cassia nigricans, Insect Sci. Appl.,* 6, 167, 1985.

542. **Byers, R. A., Gustine, D. L., and Meyer, B. G.,** Toxicity of β-nitropropionic acid to *Trichoplusia ni, Environ. Entomol.,* 6, 229, 1977.

543. **Dreyer, D. R., Jones, K. C., and Molyneux, R. J.,** Feeding deterrency of some pyrrolizidine, indolizidine, and quinolizidine alkaloids towards the pea aphid (*Acyrthosiphon pisum*) and evidence for phloem transport of indolizidine alkaloid swainsonine, *J. Chem. Ecol.,* 11, 1045, 1985.

544. **Bultman, J. D.,** An antiborer extractive from *Dalbergia retusa* (cocobolo), in *Proc. 5th Int. Symp. Controlled Release of Bioactive Materials,* Gaithersburg, Maryland, Aug. 14 to 16, 1978.

545. **Jurd, L. and Manners, G. D.,** Wood extractives as models for the development of new types of pest control agents, *J. Agr. Fd. Chem.,* 28, 163, 1980.

546. **Dietrichs, H. H. and Hausen, B. M.,** Dalbergiones—active compounds of the genus *Dalbergia, Holzforschung,* 25, 183, 1971.

547. **Saxena, S. C. and Yadav, R. S.,** A preliminary laboratory evaluation of an extract of leaves of *Delonix regia* Raf. as a disruptor of insect growth and development, *Trop. Pest Management,* 32, 58, 1986.

548. **Von Sandermann, W. and Lange, W.,** Tryptamine in the wood of angelique (*Dicorynia guianensis* Amsh.), *Naturwissenschaften,* 54, 249, 1967.

549. **Ito, T., Horie, Y., and Nakasone, S.,** Deterrent effect of soybean meal on feeding of the silkworm, *Bombyx mori, J. Insect Physiol.,* 21, 995, 1975.

550. **Tester, C. F.,** Constituents of soybean cultivars differing in insect resistance, *Phytochemistry,* 16, 1899, 1977.

551. **Waiss, A. C., Jr., Binder, R. G., Chan, B., Elliger, C. A., and Dreyer, D. L.,** Programs in research on chemical aspects of host-plant resistance to *Heliothis zea* in corn, soybean, and tomatoes, in *Proc. Int. Workshop Heliothis Management,* Patancheru, India, Nov. 15 to 20, 1981, 251.

552. **Su, H. C. F., Speirs, R. D., and Mahany, P. G.,** Toxic effects of soybean saponins and its calcium salt on the rice weevil, *J. Econ. Entomol.,* 65, 844, 1972.

553. **Grunwald, C. and Kogan, M.,** Sterols of soybeans differing in insect resistance and maturity group. *Phytochemistry,* 20, 765, 1981.

554. **Stubblebine, W. H. and Langenheim, J. H.,** Effects of *Hymenaea courbaril* leaf resin on the generalist herbivore *Spodoptera exigua* (beet armyworm), *J. Chem. Ecol.,* 3, 633, 1977.

555. **Hubbell, S. P., Wiesner, D. F., and Adejara, A.,** An antifungal terpenoid defends a neotropical tree (*Hymenaea*) against attack by fungus-growing ants (*Atta*), *Oecologia* 60, 321, 1983.

556. **Welter, A., Jadot, J., Dardenne, G., Marlier, M., and Casimir, J.,** 2,5-Dihydroxymethyl-3,4-dihydroxypyrrolidine in *Derris elliptica* leaves, *Phytochemistry,* 15, 747, 1976 (in French).

557. **Blaney, W. M., Simmonds, M. S. J., Evans, S. V., and Fellows, L. E.,** The role of the secondary plant compound 2,5-dihydroxymethyl 3,4-dihydroxypyrrolidine as a feeding inhibitor for insects, *Entomol. Exp. Appl.,* 36, 209, 1984.

558. **Scriber, J. M.,** Cyanogenic glycosides in *Lotus corniculatus.* The effect upon growth, energy budget, and nitrogen utilization of the southern armyworm, *Spodoptera eridania, Oecologia,* 34, 143, 1978.

559. **Sutherland, O. R. W., Mann, J., and Hillier, J. R.,** Feeding deterrents for the grass grub *Costelytra zealandica* (Coleoptera:Scarabaeidae) in the root of a resistant pasture plant, *Lotus pedunculatus, N. Z. J. Zool.,* 2, 509, 1975.

560. **Russell, C. B., Sutherland, O. R. W., Hutchins, R. F. N., and Christmas, P. E.,** Vestitol: a phytoalexin with insect feeding-deterrent activity, *J. Chem. Ecol.,* 4, 571, 1978.

561. **Sutherland, O. R. W., Russell, G. B., Biggs, D. R., and Lane, G. A.,** Insect feeding deterrent activity of phytoalexin isoflavonoids, *Biochem. System. Ecol.,* 8, 73, 1980.

562. **Lane, G. A., Sutherland, O. R. W., and Skipp, R. A.,** Isoflavonoids as insect feeding deterrents and antifungal components from root of *Lupinus angustifolius, J. Chem. Ecol.,* 13, 771, 1987.

563. **Sutherland, O. R. W. and Greenfield, W. J.,** Effect of root extracts of resistant pasture plants on the feeding and survival of black beetle larvae, *Heteronychus arator* (Scarabaeidae), *N. Z. J. Zool.,* 5, 173, 1978.

564. **Bentley, M. D., Leonard, D. E., Reynolds, E. K., Leach, S., Beck, A. B., and Murakoshi, I.,** Lupine alkaloids as larval feeding deterrents for spruce budworm, *Choristoneura fumiferana* (Lepidoptera:Tortricidae), *Ann. Entomol. Soc. Am.,* 77, 398, 1984.

565. **Vandenbosch, R.,** Observations on *Hypera brunneiperennis* (Coleoptera:Curculionidae) and certain of its natural enemies in the Near East, *J. Econ. Entomol.,* 57, 194, 1964.

566. **Manglitz, G. R. and Gorz, H. R.,** Host range studies with sweetclover weevil and sweetclover aphid, *J. Econ. Entomol.,* 57, 683, 1964.

567. **Kamm, J. A., and Fronk, W. D.,** Olfactory response of the alfalfa seed chalcid, *Bruchophagus roddi* Guss., to chemicals found in alfalfa, Wyoming Univ. Agr. Expt. Sta. Bull. No. 413, 1964, 36 pp.

568. **Johnson, K. J. R., Sorensen, E. L., and Horber, E. K.,** Resistance in glandularhaired annual *Medicago* species to feeding by adult alfalfa weevils (*Hypera postica*), *Environ. Entomol.,* 9, 133, 1980.

569. **Shade, R. E., Thompson, T. E., and Campbell, W. R.,** An alfalfa weevil larval resistance mechanism detected in *Medicago, J. Econ. Entomol.,* 68, 399, 1975.

570. **Thorp, R. W. and Briggs, D. L.,** Mortality in immature alfalfa leafcutter bees to alfalfa saponins, *Environ. Entomol.,* 1, 399, 1972.

571. **Nielson, M. W. and Schonhorst, M. H.,** Sources of alfalfa seed chalcid resistance in alfalfa, *J. Econ. Entomol.,* 60, 1506, 1968.

572. **Akeson, W. R., Haskins, F. A., Gorz, H. J., and Manglitz, C. R.,** Water-soluble factors in *Melilotus* leaves which influence feeding by the sweetclover weevil (*Sitona cylindricollis*), *Crop Sci.,* 8, 574, 1968.

573. **Ossipov, N.,** A remedy against the house-moth, family Tineidae, *The Horticulturist,* No. 12, 897, 1915 (in Russian).

574. **Subramaniam, T. V.,** Vegetable fish poisons as insecticides, *Mysore Agr. Calendar,* 41, 1934.

575. **Subramaniam, T. V.,** How to free stored grain from insect attack, *Mysore Agr. Calendar,* 21, 1935.

576. **Russell, C. B., Shaw, G. J., Christmas, P. E., Yates, M. B., and Sutherland, O. R. W.,** Two 2-arylbenzofurans as insect feeding deterrents from sainfoin (*Onobrychis viciifolia*), *Phytochemistry,* 23, 1417, 1984.

577. **Norton, L. B. and Hansberry, R.,** Constituents of the insecticidal resin of the yam bean (*Pachyrrhizus erosus*), *J. Am. Chem. Soc.,* 67, 1609, 1945.

578. **Kalra, A. J., Krishnamurti, M., and Nath, M.,** Chemical investigation of Indian yam bean (*Pachyrrhizus erosus*): isolation and structures of two new rotenoids and a new isoflavanone, erosenone, *Indian J. Chem.,* 15, 1084, 1977.

579. **Wilde, G. and Schoonhoven, A. V.,** Mechanism of resistance to *Empoasca kraemeri* on *Phaseolus vulgaris, Environ. Entomol.,* 5, 251, 1976.

580. **Wilde, G., Schoonhoven, A. V., and Gomez-Laverde, I.,** The biology of *Empoasca kraemeri* on *Phaseolus vulgaris, Ann. Entomol Soc. Am.,* 69, 442, 1976.

581. **Eskafe, F. M. and Schoonhoven, A. V.,** Comparison of greenhouse and field resistance of bean varieties to *Empoasca kraemeri* (Homoptera:Cicadellidae), *Can. Entomol.,* 110, 853, 1978.

582. **Appelbaum, S. W.,** Physiological aspects of host specificity in the Bruchidae. I. General considerations of developmental compatibility, *J. Insect Physiol.,* 10, 783, 1964.

583. **Buhr, H.,** Comparative studies on the resistance of some solanaceae against *Epilachna vigintiooctomaculata* Motsch. and *E. vigintioctopunctata* Fabr., *Proc. 4th Int. Cong. Crop Protect.,* 1, 707, 1957.

584. **Osborn, T. C., Alexander, D. C., Sun, S. S. M., Cardona, C., and Bliss, F. A.,** Insecticidal activity and lectin homology of arselin seed protein, *Science,* 240, 207, 1988.

585. **Janzen, D. H., Juster, H. B., and Liener, I. E.,** Insecticidal action of phytohaemagglutinin in black beans on a bruchid beetle, *Science,* 192, 795, 1976.

586. **Coombs, C. W., Billings, C. J., and Porter, J. E.,** The effect of yellow split peas (*Pisum sativum* L.) and other pulses on the productivity of certain strains of *Sitophilus oryzae* L. (Col. Curculionidae) and the ability of other strains to breed thereon, *J. Stored Prod. Res.,* 13, 53, 1977.

587. **Kon, R. T., Zabik, M. J., Webster, J. A., and Leavitt, R. A.,** Cereal leaf beetle response to biochemicals from barley and pea seedlings. I. Crude extract, hydrophobic and hydrophilic fractions, *J. Chem. Ecol.,* 4, 511, 1978.

588. **Osmani, Z. H. and Naidu, M. B.,** *Pongamia glabra* (karanja) as insecticide, *Sci. Culture,* 22, 235, 1956.

589. **Joshi, B. G. and Rao, R. S. N.,** *Pongamia* cake can control tobacco beetle, *Indian Fmg.,* 18, 33, 1968.

590. **Sahrawat, K. L. and Mukerjee, S. K.,** Nitrification inhibitors. I. Studies with karanjin, a furanolflavonoid from karanja (*Pongamia glabra*) seeds, *Plant Soil,* 47, 27, 1977.

591. **Parmar, B. S., Attre, S., Singh, R. P., and Mukherjee, S. K.,** Karanja oil as a synergist for chlorinated insecticides, *Pesticides,* 9(5), 29, 1975.

592. **Parmar, B. S.,** Karanja, *Pongamia glabra,* seed oil as a synergist for pyrethrins, *Pyrethrum Post,* 14(1), 22, 1977.

593. **Purushotham,Rao, A. and Niranjan, B.,** Juvenile-hormone-like activity of "karanjin" against larvae of red flour beetle *Tribolium castaneum* J., *Comp. Physiol. Ecol.,* 7, 234, 1982.

594. **Bhan, P., Soman, R., and Dev, S.,** Insect juvenile hormone mimics based on bakuchiol, *Agr. Biol. Chem.,* 44, 1483, 1980.

595. **Prabhu, V. K. K., John, M., and Ambikgamma, B.,** Juvenile hormone activity in some South Indian plants, *Curr. Sci.,* 42, 725, 1973.

596. **Dalziel, J. M.,** *The Useful Plants of West Africa,* London, 1937, 612 pp.

597. **Holland, T. H.,** *The Useful Plants of Nigeria,* Kew Roy. Bot. Gardens, London, 1908, 963 pp.

598. **Mitchell, B. K.,** Some aspects of gustation in the larval red turnip beetle, *Entomoscelis americana,* related to feeding and host-plant selection, *Entomol. Exp. Appl.,* 24, 340, 1978.

599. **Mitchell, B. K. and Sutcliffe, J. F.,** Sensory inhibition as a mechanism of feeding deterrence: effects of three alkaloids on leaf beetle feeding, *Physiol. Entomol.,* 9, 57, 1984.

600. **Borel, C. and Hostettmann, K.,** Molluscicidal saponins from *Swartzia madagascariensis* Desvaux, *Helv. Chim. Acta,* 70, 570, 1987.

601. **Beauquesne, L.,** Samagoura, an African kegume (*Swartzia madagascariensis* Desv.), *Ann. Pharm. France,* 5, 470, 1947.

602. **Thomson, R. H.,** *Naturally Occurring Quinones,* Academic, New York, 2nd ed., 1971, 200.

603. **Bentley, M. D., Hassanali, A., Lwande, W., Njoroge, P. E. W., Ole Sitayo, E. N., and Yatagai, M.,** Insect antifeedants from *Tephrosia elata* Deflers, *Insect Sci. Appl.,* 8, 85, 1987.

604. **Lwande, W., Hassanali, A., Njoroge, P. W., Bentley, M. D., Delle Monache, F., and Jondiko, J. I.,** A new 6a-hydroxypterocarpan with insect antifeedant and antifungal properties from the roots of *Tephrosia hildebrandtii* Vatke, *Insect Sci. Appl.,* 6, 537, 1985.

605. **Russell, G. B., Sutherland, O. R. W., Christmas, P. E., and Wright, H.,** Feeding deterrents for black beetle larvae, *Heteronychus arator* (Scarabaeidae), in *Trifolium repens, N. Z. J. Zool.,* 9, 145, 1982.

606. **Jilani, G., and Su, H. C. F.,** Laboratory studies on several plant materials as insect repellents for protection of cereal grains, *J. Econ. Entomol.,* 76, 154, 1983.

607. **Badawi, A., Faragalla, A. A., and Dabbour, A.,** The natural resistance of some imported wood species to subterranean termites in Saudi Arabia, *Z. Angew. Entomol.,* 98, 500, 1984.

608. **Pospisil, J.,** The response of *Leptinotarsa decemlineata* (Coleoptera) to tannin as an antifeedant, *Acta Entomol. Bohemoslov.,* 79, 429, 1982.

609. **Drummond, F. A. and Casagrande, R. A.,** Effect of white oak extracts on feeding by the Colorado potato beetle (Coleoptera:Chrysomelidae), *J. Econ. Entomol.,* 78, 1272, 1985.

610. **Von Schultze-Dewitz, G.,** Investigations of the difference in feeding by *Reticulitermes flavipes* Kollar, *Holzforsch. Holzverwert,* 13, 29, 1961 (in German).

611. **Anonymous,** 10th Ann. Rept. Indian Central Oilseed Committee, 1956-1957, Hyderabad, 1958, 166 pp.

612. **Sorenson, J. W., Kline, G. L., Redlinger, L. M., Davenport, G., and Alfred, W. H.,** Drying and storage of sorghum grain, Bull. Texas Agr. Expt. Sta. No. 885, 1957, 23 pp.

613. **La Rue, D. W., Clements, B. W., and Womack, H.,** In-store treatments for the protection of farmer's stock peanuts from insect attack. Exploratory tests, USDA Agr. Res. Service MRD No. 363, 1959.

614. **La Rue, D. W., Womack, H., and Clements, B. W.,** Treatments for the protection of stored southern-grown corn from rice weevil attack, USDA Agr. Res. Service MRD No. 272, 1958.

615. **Walkden, H. H. and Nelson, H. D.,** Evaluation of ryania for the protection of stored wheat and shelled corn from insect attack, USDA Agr. Res. Service MRS No. 245, 1958.

616. **Clark, P. H. and Laudani, H.,** Ryania compared to chlordane. A preliminary evaluation of Ryania dust against resistant and laboratory strains of roaches, *Pest Control Mag.,* No. 12, 1953.

617. **Rogers, E. F., Koniuszy, F. R., Shavel, J., Jr., and Folkers, K.,** Plant insecticides. I. Ryanodine, a new alkaloid from *Ryania speciosa* Vahl., *J. Am. Chem. Soc.,* 70, 3086, 1951.

618. **Wolcott, G. N.,** The most effective termite repellents, *J. Econ. Entomol.,* 42, 273, 1949.

619. **Wolcott, G. N.,** The termite resistance of pinosylvin and other new insecticides, *J. Econ. Entomol.,* 44, 263, 1951.

620. **Jacobson, M.,** *Insecticidal Plants,* USDA Bureau Entomol. Plant Quarantine, Div. Insecticide Invs. Special Rept. No. 26, 1953, 132 pp.

621. **Belanger, A., et al. (20 coauthors),** Total synthesis of ryanodol, *Can. J. Chem.,* 57, 3348, 1979.

622. **Osmani, Z., Anees, I., and Naidu, M. B.,** Effect of different temperatures on the repellency of certain essential oils against house flies and mosquitoes, *Pesticides (India),* 8, (9), 95, 1974.

623. **Osmani, Z., Anees, I., and Naidu, M. B.,** Insect repellent creams from essential oils, *Pesticides (India),* 6, (3), 19, 1972.

624. **Major, R. T.,** The ginkgo, the most ancient living tree, *Science,* 157, 1270, 1967.

625. **Matsumoto, T. and Sei, T.,** Antifeedant activities of *Ginkgo biloba* L. components against the larva of *Pieris rapae crucivora, Agr. Biol. Chem.,* 51, 249, 1987.

626. **Mullin, C. E.,** Moths and mothproofing, *Textile Colorist,* 47, 180, 1925.

627. **Carpenter, T. L., Neal, W. W., and Hedin, P. A.,** A review of host plant resistance of pecan, *Carya illinoensis,* to Insecta and Acarina, *Bull. Entomol. Soc. Am.,* 25, 251, 1979.

628. **Riley, C. V.,** Vegetable insecticides, U.S. Entomol. Committee Rept. No. 4, 1885, 164.

629. **Kubo, I.,** Pharmacies in the jungle, *World Book Science Annual,* 1982, 126.

630. **Hernandez, A., Pascual, C., Sanz, J., and Rodriguez, B.,** Diterpenoids from *Ajuga chamaepitys:* two neo-clerodane derivatives, *Phytochemistry,* 21, 2909, 1982.

631. **Camps, F., Coll, J., and Cortel, A.,** New clerodane diterpenoids from *Ajuga iva* (Labiatae), *Chem. Lett.,* 1053, 1982.

632. **Belles, X., Camps, F., Coll, J., and Dolors Piulacha, M.,** Insect antifeedant activity of clerodane diterpenoids against larvae of *Spodoptera littoralis* (Boisd.) (Lepidoptera), *J. Chem. Ecol.,* 11, 1439, 1985.

633. **Shimomura, H., Sashida, Y., Ogawa, K., and Iitaka, Y.,** Ajugamarin, a new bitter diterpene from *Ajuga nipponensis* Makino, *Tetrahedron Lett.,* 22, 1367, 1981.

634. **Shimomura, H., Sashida, Y., Ogawa, K., and Iitaka, Y.,** The chemical constituents of *Ajuga* plants. I. Neo-clerodanes from the leaves of *Ajuga nipponensis* Makino, *Chem. Pharm. Bull.,* 31, 2192, 1983.

635. **Camps, F., Coll, J., and Dargallo, O.,** Neo-clerodane diterpenoids from *Ajuga pseudoiva, Phytochemistry,* 23, 387, 1984.

636. **Kubo, I., Lee, Y.-W., Balogh-Nair, V., Nakanishi, K., and Chapya, A.,** Structure of ajugarins, *J. Chem. Soc. Chem. Commun.,* 949, 1976.

637. **Kubo, I., Klocke, J. A., and Asano, S.,** Insect ecdysis inhibitors from the East African medicinal plant *Ajuga remota* (Labiatae), *Agr. Biol. Chem.,* 45, 1925, 1981.

638. **Camps, F., Coll, J., and Cortel, A.,** Two new clerodane diterpenoids from *Ajuga reptans* (Labiatae), *Chem. Lett.,* 1093, 1981.

639. **Richter, K and Birkenbeil, H.,** The effect of extracts of *Ajuga reptans* on moult regulation in *Periplaneta americana, J. Insect Physiol.,* 33, 933, 1987.

640. **Schauer, M. and Schmutterer, H.,** The effect of squeezed fresh juices and crude extracts of *Ajuga remota* on the two-spotted spider mite, *Tetranychus urticae* Koch, *Z. Angew. Entomol.,* 91, 425, 1981 (in German).

641. **Schmutterer, H. and Tervooren, G.,** The effect of squeezed juices and crude extracts of *Ajuga* species on feeding activity and metamorphosis of *Epilachna varivestis* (Col., Coccinellidae), *Z. Angew. Entomol.,* 89, 470, 1980 (in German).

642. **Lineberger, R. D. and Wanstreet, A.,** Micropropagation of *Ajuga reptans* "Burgundy Glow," *Ohio Agr. Res. Circular,* 1983, 19.

643. **Kubo, I., Kido, M., and Fukuyama, Y.,** X-ray crystal structure of 12-bromoajugarin-1 and conclusion on the absolute configuration of ajugarins, *J. Chem. Soc. Chem. Commun.,* 897, 1980.

644. **Kubo, I., Klocke, J. A., Miura, I., and Fukuyama, Y.,** Structure of ajugarin-IV, *J. Chem. Soc. Chem. Commun.,* 618, 1982.

645. **Kubo, I., Fukuyama, Y., and Chapya, A.,** Structure of ajugarin-V, *Chem. Lett.,* 223, 1983.

646. **Luteijn, J. M. and De Groot, A.,** Stereospecific formation of a substituted *trans*-decaline as an intermediate for the synthesis of clerodane insect antifeedants, *Tetrahedron Lett.,* 22, 789, 1981.

647. **Kende, A. S. and Roth, B.,** Stereospecific total synthesis of ajugarin-IV, *Tetrahedron Lett.,* 23, 1751, 1982.

648. **Luteijn, J. M. and De Groot, A.,** Investigations into the total synthesis of insect antifeedant clerodanes. The total synthesis of (±)-4-epi-ajugarin, *Tetrahedron Lett.,* 23, 3421, 1982.

649. **Ley, S. V., Simpkins, N. S., and Whittle, A. J.,** The total synthesis of the clerodane diterpene insect antifeedant ajugarin I, *J. Chem. Soc. Chem. Commun.,* 503, 1983.

650. **Bowen, J. P.,** I: Synthetic Approaches to Ajugarin I and IV. II: Molecular Mechanics Studies on Ajugarin I and IV Intermediates and Related Model Compounds, Ph.D. thesis, Emory University, 1984, 189 pp (*Diss. Abstr. Int. B,* 45(9), 1985).

651. **Jones, P. S., Ley, S. V., Simpkins, N. S., and Whittle, A. J.,** Total synthesis of the insect antifeedant ajugarin I and degradation studies of related clerodane diterpenes, *Tetrahedron,* 42, 6519, 1986.

652. **Abivardi, C. and Renz, G.,** Tests with the extracts of 21 medicinal plants for antifeedant activity against larvae of *Pieris brassicae* L. (Lep., Pieridae), *Mitteil. Schweiz. Entomol. Ges.,* 57, 383, 1984.

653. **Brieskorn, C. H. and Hofmann, R.,** Bitter substances from Labiatae: a clerodane derivative from *Leonurus cardiaca* L., *Tetrahedron Lett.,* 2511, 1979.

654. **Aditychaudhury, N. and Chosh, D.,** Studies on insecticidal plants: chemical examination of *Leucas aspera, J. Indian Chem. Soc.,* 46, 95, 1969.

655. **Inaba, S.,** New methods of evaluation for cockroach repellents and repellency of essential oils against German cockroach (*Blattella germanica* L.), *Nippon Novaku Gakkaishi,* 7, 133, 1982.

656. **Kashyap, N. P., Gupta, V. K., and Kaushal, A. N.,** *Mentha spicata,* a promising protectant to stored wheat against *Sitophilus oryzae, Bull. Grain Technol.,* 12, 41, 1974.

657. **Meisner, J. and Kehat, M.,** The response of *Earias insulana* Boisd. larvae to phagodeterrent (-)-carvone incorporated in an artificial diet, *Z. Angew. Entomol.,* 90, 80, 1980.

658. **Inazuka, S.,** Cockroach repellents contained in Japanese mint and Scotch spearmint, *Nippon Novaku Gakkaishi,* 7, 145, 1982.

659. **Petrowitz, H. J.,** The effect of optical antipodes of menthol on termite species, *Naturwissenschaften,* 67, 43, 1980.

660. **Eisner, T.,** Catnip: its raison d'etre, *Science,* 146, 1318, 1964.

661. **Chavan, S. R. and Nikam, S. T.,** Mosquito larvicidal activity of *Ocimum basilicum* Linn., *Indian J. Med. Res.,* 75, 220, 1982.

662. **Bowers, W. S. and Nishida, R.,** Juvocimenes: potent juvenile hormone mimics from sweet basil, *Science,* 209, 1030, 1980.

663. **Nishida, R., Bowers, W. S., and Evans, P. H.,** Synthesis of highly active juvenile hormone analogs, juvocimene I and II, from the oil of sweet basil, *J. Chem. Ecol.,* 10, 1435, 1984.

664. **Rathore, H. S.,** Preliminary observations on the mosquito repellent efficacy of the leaf extract of *Ocimum sanctum, Pakistan J. Zool.,* 10, 303, 1978.

665. **Kubo, I., Matsumoto, T., Tori, M., and Asakawa, Y.,** Structure of plectrin, an aphid antifeedant diterpene from *Plectranthus barbatus, Chem. Lett.,* 1513, 1984.

666. **Kubo, I., Miura, I., Nakanishi, K., Kamikawa, T., Isobe, T., and Kubota, T.,** Structure of isodomedin, a novel *ent*-kaurenoid diterpene, *J. Chem. Soc. Chem. Commun.,* 555, 1977.

667. **Drummond, H. M.,** Patchouli oil, *Perfumery Essential Oil Rec.,* 51, 484, 1960.

668. **Macherras, I. M. and Mackerras, M. J.,** Sheep blowfly investigations. The attractiveness of sheep for *Lucilia cuprina,* Austral. Commonwealth Sci. Ind. Res. Organization Bull. No. 181, 44 pp.

669. **Hanson, J. R., Rivett, D. E. A., Ley, S. V., and Williams, D. J.,** The x-ray structure and absolute configuration of insect antifeedant clerodane diterpenoids from *Teucrium africanum, J. Chem. Soc. Perkin I,* 1005, 1982.

670. **Ono, M.,** Joint action in insecticides. I. Synergistic action of safrole with pyrethrins, *Botyu-Kagaku,* 15, 155, 1950.

671. **Verma, M., and Meloan, C. E.,** A natural cockroach repellent in bay leaves, *Am. Laboratory,* 13(10), 64, 1981.

672. **Matsui, K., Wada, K., and Munakata, K.,** Insect antifeeding substances in *Parabenzoin praecox* and *Piper futokadzura, Agr. Biol. Chem.,* 40, 1045, 1976.

673. **Morton, J. F.,** The mamey, *Proc. Florida State Hort. Soc.,* 75, 400, 1962.

674. **Chang, C.-F., Isogai, A., Kamikado, T., Nurakoshi, S., Sakurai, A., and Tamura, S.,** Isolation and structure elucidation of growth inhibitors for silkworm larvae from avocado leaves, *Ann. Biol. Chem.,* 39, 1167, 1975.

675. **Murakoshi, S., Chang, C.-F., Kamikado, T., Sakurai, A., and Tamura, S.,** Effects of two components from the avocado leaves (*Persea americana* Mill.) and the related compounds on the growth of silkworm larvae, *Bombyx mori* L., *Jpn. J. Appl. Entomol. Zool.,* 20, 87, 1976.

676. **Sneh, B. and Gross, S.,** Toxicity of avocado leaves (*Persea americana*) to young larvae of *Spodoptera littoralis* Boisd. (Lep., Noctuidae), *Z. Angew. Entomol.,* 92, 420, 1981.

677. **Aries, R. S.,** Extractives of northeastern woods, *Chemurgic Digest,* 4, 153, 155, 1947.

678. **Chesnut, V. K.,** Plants used by the Indians of Mendocino County, California, *U.S. Natl. Mus., Contrib. U.S. Natl. Herbarium,* 7, 295, 1902.

679. **McClintock, C. T., Hamilton, H. C., and Lowe, F. B.,** A further contribution to our knowledge of insecticides. Fumigants, *J. Am. Public Health Assoc.,* 1, 227, 1911.

680. **Martelli, G. M.,** Possible cultivation of squill in Tripoli and its industrial exploitation, *Rev. Bot. Appl. Agr. Colon.,* 17, 844, 1937 (in French).

681. **De Bussy, L. P.,** Tests with plant materials affecting *Prodenia litura* larvae, *Bijdr. Dierkunde,* 21, 1922 (in Flemish).

682. **Priselkov, A. M., Khatin, M. C., and Gmelina-Lure, M. Z.,** Early treatment of the warble-fly disease of cattle, *Veterinariya,* 25(1), 14, 1948.

683. **Nanobashvili, V. I.,** The effect of the decoction of the root and rhizome of hellebore on the quality of the wool of sheep, *Veterinariya,* 26(8), 39, 1948.

684. **Smith, D. S., Hanford, R. H., and Chfurka, W.,** Some effects of various food plants on *Melanoplus mexicanus mexicanus* (Saus.) (Orthoptera:Acrididae), *Can. Entomol.,* 84, 113, 1952.

685. **Suryakala, G., Rao, B. K., Thakur, S. S., and Rao, P. N.,** Insect growth regulators from some plants of Andhra Pradesh (India), *J. Reprod. Biol. Comp. Endocrinol.,* 3, (2), 33, 1983.

686. **Doskotch, R. W., El-Feraly, F. S., Fairchild, E. H., and Huang, C.-T.,** Isolation and characterization of peroxyferolide, a hydroperoxy sesquiterpene lactone from *Liriodendron tulipifera, J. Org. Chem.,* 42, 3614, 1977.

687. **Doskotch, R. W., Fairchild, E. H., Huang, C.-T., Wilton, J. H., Beno, M. A., and Christoph, G. C.,** Tulirinol, an antifeedant sesquiterpene lactone for the gypsy moth larvae from *Liriodendron tulipifera, J. Org. Chem.,* 45, 1441, 1980.

688. **Lukefahr, M. J.,** Insect resistant cotton, *Agr. Res. (USDA),* 27(5), 6, 1978.

689. **Maxwell, F. G.,** Plant resistance to cotton insects, *Bull. Entomol. Soc. Am.,* 23, 199, 1977.

690. **Stipanovic, R. D., Bell, A. A., and Lukefahr, M. J.,** Natural insecticides from cotton, unpublished results (review).

691. **Maxwell, F. G., Jenkins, J. N., and Keller, J. C.,** A boll weevil repellent from the volatile substance of cotton, *J. Econ. Entomol.,* 56, 894, 1963.

692. **Shaaya, E. and Ikan, R.,** Insect control using natural products, in *Advances in Pesticide Science,* Geissbuhler, H., Brooks, G. T., and Kearney, P. C., Eds., Pergamon, New York, pt. 2, 1979, 303.

693. **Meisner, J., Navon, A., Zur, M., and Ascher, K. R. S.,** The response of *Spodoptera littoralis* larvae to gossypol incorporated in an artificial diet, *Environ. Entomol.,* 6, 243, 1977.

694. **Meisner, J., Ascher, K. R. S., and Zur, M.,** Phagodeterrency induced by pure gossypol and leaf extracts of a cotton strain with high gossypol content in the larvae of *Spodoptera littoralis, J. Econ. Entomol.,* 70, 149, 1977.

695. **Zur, M., Meisner, J., Kabonci, E., and Ascher, K. R. S.,** Field evaluation of the response of *Spodoptera littoralis* and *Earias insulana* to cotton strains differing in gossypol content, *Phytoparasitica,* 8, 189, 1980.

696. **Meisner, J., Ascher, K. R. S., Zur, M., and Eizik, C.,** Synergistic and antagonistic interactions of gossypol with some OP-compounds demonstrated in *Spodoptera littoralis* larvae by topical application, *Z. Pflanzenkrankh. Pflanzenschutz,* 89, 571, 1982.

697. **Meisner, J., Kehat, M., Zur, M., and Ascher, K. R. S.,** The effect of gossypol on the larvae of the spiny bollworm, *Earias insulana, Entomol. Exp. Appl.,* 22, 301, 1977.

698. **Elliger, C. A., Chan, B. G., and Waiss, A. C., Jr.,** Relative toxicity of minor cotton terpenoids compared to gossypol, *J. Econ. Entomol.,* 71, 161, 1978.

699. **Lukefahr, M. J., Stipanovic, R. D., Bell, A. A., and Gray, J. R.,** Biological activity of new terpenoid compounds from *Gossypium hirsutum* against the tobacco budworm and pink bollworm, Proc. Beltwide Cotton Production Res. Conf., 1977.

700. **Bailey, J. C.,** Orange and yellow cotton pollens retard growth of tobacco budworm larvae, *J. Georgia Entomol. Soc.,* 18, 9, 1983.

701. **Gray, J. R., Mabry, T. J., Bell, A. A., Stipanovic, R. D., and Lukefahr, M. J.,** Para-hemigossypolone: a sesquiterpenoid aldehyde quinone from *Gossypium hirsutum J. Chem. Soc. Chem. Commun.,* 109, 1976.

702. **Stipanovic, R. D., Bell, A. A., O'Brien, D. H., and Lukefahr, M. J.,** Heliocide H_1. A new insecticidal C_{25} terpenoid from cotton (*Gossypium hirsutum*), *J. Agr. Fd. Chem.,* 26, 115, 1978.

703. **Stipanovic, R. D., Bell, A. A., O'Brien, D. H., and Lukefahr, M. J.,** Heliocide H_2: an insecticidal sesterterpenoid from cotton (*Gossypium*), *Tetrahedron Lett.,* 567, 1977.

704. **Stipanovic, R. D., Bell, A. A., O'Brien, D. H., and Lukefahr, M. J.,** Heliocide H_3: an insecticidal terpenoid from *Gossypium hirsutum, Phytochemistry,* 17, 151, 1978.

705. **Hedin, P. A., Jenkins, J. N., Collum, D. H., White, W. H., Parrott, W. I., and McGown, M. W.,** Cyanidin-3-α-glucoside, a newly recognized basis for resistance in cotton to the tobacco budworm *Heliothis virescens* (Fab.) (Lepidoptera:Noctuidae), *Experientia,* 39, 799, 1983.

706. **Klocke, J. A. and Chan, B. G.,** Effects of cotton condensed tannin on feeding and digestion in the cotton pest, *Heliothis zea, J. Insect Physiol.,* 28, 911, 1982.

707. **Reese, J. C., Chan, B. G., and Waiss, A. C., Jr.,** Effects of cotton condensed tannin, Maysin (corn), and pinitol (soybeans) on *Heliothis zea* growth and development, *J. Chem. Ecol.,* 8, 1429, 1982.

708. **Chan, B. G., Waiss, A. C., Jr., Lukefahr, M. J., and Grey, R.,** Condensed tannin, an antibiotic chemical from *Gossypium hirsutum* L., unpublished results.

709. **Lindroth, R. L. and Peterson, S. S.,** Effects of plant phenols on performance of southern armyworm larvae, *Oecologia,* 75, 185, 1988.

710. **Elliger, C. A., Chan, B. G., and Waiss, A. C., Jr.,** Flavonoids as larval growth inhibitors. Structural factors governing toxicity, *Naturwissenschaften,* 69, 358, 1980.

711. **Congthammakun, S., Sukhato, P., Pavasuthipaisit, K., and Puttipongse Varavudhi, M. R.,** Studies on the effects of gossypol in male cynomolgus monkey, I. Semen analysis, *J. Sci. Soc. Thailand,* 12, 213, 1986.

712. Everett, T. R., Feeding and oviposition reaction of boll weevils to cotton, althea, and okra flower buds, *J. Econ. Entomol.,* 57, 165, 1964.

713. Maxwell, F. G., Parrott, W. L., Jenkins, J. N., and LeFevre, H. N., A boll weevil feeding deterrent from the calyx of an alternate host, *Hibiscus syriacus, J. Econ. Entomol.,* 56. 985, 1965.

714. Bird, T. G., Hedin, P. A., and Burks, M. L., Feeding deterrent compounds to the boll weevil, *Anthonomus grandis* Boheman in Rose-of-Sharon, *Hibiscus syriacus* L., *J. Chem. Ecol.,* 13, 1087, 1987.

715. Qadri, S. S. H. and Majumder, S. K., Repellency of arrowroot (*Maranta arudinaceae* L.) to some important insect pests, *Pesticides,* 2(1), 25, 1968.

716. Bernays, E., Lupi, A., Marini Bettolo, R., Mastrofrancesco, C., and Tagliatesta, P., Antifeedant nature of the quinone primin and its quinol micondin from *Miconia* spp., *Experientia,* 40, 1010, 1984.

717. Mikolajczak, K. L. and Reed, D. K., Extractives of seeds of the Meliaceae: effects on *Spodoptera frugiperda* (J. E. Smith), *Acalymma vittatum* (F.), and *Artemia salina* Leach, *J. Chem. Ecol.,* 13, 99, 1987.

718. Purushothaman, K. K., Sarada, A., Connolly, J. D., and Akinniyi, J. A., The structure of roxburghilin, a bis-amide of 2-aminopyrrolidine from the leaves of *Aglaia roxburghiana* (Meliaceae), *J. Chem. Soc. Perkin Trans. I,* 3171, 1979.

719. Chiu, S.-F., Recent research findings on Meliaceae and other promising botanical insecticides in China, *Z. Pflanzenkrankh. Pflanzenschutz,* 92, 310, 1985.

720. Kundu, A. B., Ray, S., and Chatterjee, A., Abstr. Int. Conf. Natural Prods. as Regulators of Insect Reprod., Jammu, India, Sept. 29 to Oct. 1, 1984, 32.

721. Agarwala, S. B. D., Control of sugarcane termites (1946-1953), *J. Econ. Entomol.,* 48, 533, 1955.

722. Van Beek, T. A. and De Groot, A., Terpenoid antifeedants of natural origin, *Rec. Trav. Chim. Pays-Bas,* 105, 513, 1986.

723. Koul, O., Insect feeding deterrents in plants, *Indian Rev. Life Sci.,* 2, 97, 1982.

724. Das, M. F., Da Silva, G. F., Gottlieb, O. R., and Dreyer, D. L., Evolution of limonoids in the Meliaceae, *Biochem. System. Ecol.,* 12, 299, 1984.

725. Jacobson, M., Neem research in the U.S. Department of Agriculture: chemical, biological and cultural aspects, in *Natural Pesticides From the Neem Tree (Azadirachta indica A. Juss.),* Schmutterer, H., Ascher, K. R. S., and Rembold, H., Eds., GTZ Press, Eschborn, FRG, 1981, 33.

726. Mansour, F. A. and Ascher, K. R. S., Effects of neem (*Azadirachta indica*) seed kernel extracts from different solvents on the carmine spider mite, *Tetranychus cinnabarinus, Phytoparasitica,* 11, 177, 1983.

727. Schauer, M. and Schmutterer, H., Effects of neem kernel extracts on the two-spotted spider mite, *Tetranychus urticae,* in *Natural Pesticides From the Neem Tree (Azadirachta indica A. Juss.),* Schmutterer, H., Ascher, K. R. S., and Rembold, H., Eds., GTZ Press, Eschborn, FRG, 1981, 259.

728. Reed, D. K. and Reed, G. L., Natural products repel cucumber beetle, *Agr. Res. (USDA),* 30(2), 12, 1981.

729. Reed, D. K., Warthen, J. D., Jr., Uebel, E. C., and Reed, G. L., Effects of two triterpenoids from neem on feeding by cucumber beetles (Coleoptera:Chrysomelidae), *J. Econ. Entomol.,* 75, 1109, 1982.

730. Ketkar, C. M., *Utilization of Neem (Azadirachta indica Juss.) and Its Byproducts,* 1st Ed., Nana Dengle Press, Pooma, India, 1976, 234 pp.

731. Pradhan, S., Jotwani, M. G., and Rai, B. K., The repellent properties of some neem products, *Bull. Regional Res. Lab. Jammu (India),* 1, 149, 1963.

732. Pradhan, S. and Jotwani, M. G., Neem as an insect deterrent, *Chem. Age (India),* 19, 756, 1968.

733. Pradhan, S. and Jotwani, M. G., Neem kernel as antifeedant for locust, *Sneha-Sandesh,* 13, 1, 1971.

734. Chakravorty, D. P., Ghosh, G. C., and Dhua, S. P., Repellent properties of thionimone on red pumpkin beetle *Aplacophora foveicollis* L., *Technology (India),* 6, 48, 1969.

735. Yadav, T. D., Efficacy of neem (*Azadirachta indica* A. Juss.) kernel powder as seed treatment against pulse beetles, *Neem Newslett.,* 1, (1), 6, 1984.

736. **Yadav, T. D.,** Efficacy of neem (*Azadirachta indica* A. Juss.) kernel powder as seed treatment against pulse beetles, *Neem Newslett.,* 1, (2), 13, 1984.

737. **Jotwani, M. G. and Sircar, P.,** Neem seed as a protectant against bruchid *Callosobruchus maculatus* (Fabricius) infesting some leguminous seeds, *Indian J. Entomol.,* 29, 21, 1967.

738. **Tanzubil, P. B.,** The use of neem products in controlling the cowpea weevil, *Callosobruchus maculatus,* in *Natural Pesticides From the Neem Tree and Other Tropical Plants,* Schmutterer, H. and Ascher, K. R. S., Eds., GTZ Press, Eschborn, FRG, 1987, 517.

739. **Ketkar, C. M.,** Use of tree-derived non-edible oils as surface protectants for stored legumes against *Callosobruchus maculatus* and *C. chinensis,* in *Natural Pesticides From the Neem Tree and Other Tropical Plants,* Schmutterer, H. and Ascher, K. R. S., Eds., GTZ Press, Eschborn, FRG, 1987, 535.

740. **Sankaram, A. V. B., Marthanda Murthy, M. M., Bhaskaraiah, K., Subramanyam, M., Sultana, N., Sharma, H. C., Leuschner, K., Hamaprasad, G., Sitaramiaiah, S., Rukmini, C., and Rao, P. U.,** Chemistry, biological activity, and utilization of some promising neem extractives, in *Natural Pesticides From the Neem Tree and Other Tropical Plants,* Schmutterer, H. and Ascher, K. R. S., Eds., GTZ Press, Eschborn, FRG, 1987, 127.

741. **Jacobson, M., Stokes, J. B., Warthem, J. D., Jr., Redfern, R. E., Reed, D. K., Webb, R. E., and Telek, L.,** Neem research in the U.S. Department of Agriculture: an update, in *Natural Pesticides From the Neem Tree and Other Tropical Plants,* Schmutterer, H. and Ascher, K. R. S., Eds., GTZ Press, Eschborn, FRG, 1984, 31.

742. **Steets, R.,** The Effect of Constituents of Meliaceae and Anacardiaceae on Coleoptera and Lepidoptera, Ph.D. thesis, Justus Liebig University, Giessen, FRG, 1976, 173 pp.

743. **Schmutterer, H. and Rembold, H.,** Effects of some pure fractions of extracts from neem (*Azadirachta indica*) seeds on feeding activity and metamorphosis of *Epilachna varivestis* (Col., Coccinellidae), *Z. Angew. Entomol.,* 89, 179, 1980.

744. **Kraus, W. and Cramer, R.,** New tetranortriterpenoids with feeding deterrent activity for insects from neem oil, *Liebigs Ann. Chem.,* 181, 1981 (in German).

745. **Schwinger, M., Ehhammer, B., and Kraus, W.,** Methodology of the *Epilachna varivestis* bioassay of antifeedants demonstrated with some compounds from *Azadirachta indica* and *Melia azedarach,* in *Natural Pesticides From the Neem Tree and Other Tropical Plants,* Schmutterer, H. and Ascher, K. R. S., Eds., GTZ Press, Eschborn, FRG, 1984, 181.

746. **Lidert, Z., Taylor, D. O. H., and Thirugnanan, M.,** Insect antifeedant activity of four prieurianin-type limonoids, *J. Nat. Prod.,* 48, 843, 1985.

747. **Kraus, W., Klenk, A., Bokel, M., and Vogler, B.,** Tetranortriterpenoid lactams with insect antifeedant activity from *Azadirachta indica* A. Juss (Meliaceae), *Liebigs Ann. Chem.,* 337, 1987.

748. **Sarup, P. and Srivastava, U. S.,** Observations on the damage of neem (*Azadirachta indica* A. Juss.) seed kernel in storage by various pests and efficacy of the damaged kernel as an antifeedant against the desert locust, *Schistocerca gregaria* Forsk., *Indian J. Entomol.,* 33, 228, 1972.

749. **Steets, R.,** Activity of purified extracts of *Azadirachta indica* fruit on *Leptinotarsa decemlineata* Say (Coleoptera, Chrysomelidae), *Z. Angew. Entomol.,* 2, 169, 1976.

750. **Schmutterer, H.,** Fecundity-reducing and sterilizing effects of neem seed kernel extracts in the Colorado potato beetle, *Leptinotarsa decemlineata,* in *Natural Pesticides From the Neem Tree and Other Tropical Plants,* Schmutterer, H. and Ascher, K. R. S., Eds., GTZ Press, Eschborn, FRG, 1987, 351.

751. **Karel, A. K.,** Response of *Ootheca bennigseni* (Coleoptera:Chrysomelidae) to neem extracts, in *Natural Pesticides From the Neem Tree and Other Tropical Plants,* Schmutterer, H. and Ascher, K. R. S., Eds., GTZ Press, Eschborn, FRG, 1987, 393.

752. **Meisner, J. and Mitchell, B. K.,** Phagodeterrent effect of neem extracts and azadirachtin on flea beetles, *Phyllotreta striolata* (F.), *Z. Pflanzenkrankh. Pflanzenschutz,* 89, 463, 1982.

753. **Redknap, R. S.,** The use of crushed neem berries in the control of some insect pests in Gambia, in *Natural Pesticides From the Neem Tree (Azadirachta indica A. Juss.),* Schmutterer, H., Ascher, K. R. S., and Rembold, H., Eds., GTZ Press, Eschborn, FRG, 1981, 205.

754. **Ladd, T. L., Jr.,** Seed extracts repel Japanese beetles, *Agr. Res. (USDA),* 27(9), 8, 1979.

755. **Ladd, T. L., Jr.,** Neem seed extracts as feeding deterrents for the Japanese beetle, *Popillia japonica,* in *Natural Pesticides From the Neem Tree (Azadirachta indica A. Juss.),* Schmutterer, H., Ascher, K. R. S., and Rembold, H., Eds., GTZ Press, Eschborn, FRG, 1981, 149.
756. **Ladd, T. L., Jr., Jacobson, M., and Buriff, G. B.,** Japanese beetles: extracts from neem tree seeds as feeding deterrents, *J. Econ. Entomol.,* 71, 810, 1978.
757. **Hackett, K.,** Neem oil, *Rept. John Muir Inst. Environ. Studies,* 1981, 56, 1982.
758. **Jotwani, M. G. and Sircar, P.,** Neem seed as a protectant against stored grain pests infesting wheat seed, *Indian J. Entomon.,* 27, 160, 1965.
759. **Jilani, G. and Malik, M. M.,** Studies on neem plant as repellent against stored grain insects, *Pakistan J. Sci. Ind. Res.,* 16, 251, 1973.
760. **Girish, G. K. and Jain, S. K.,** Studies on the efficacy of neem seed kernel powder against stored grain pests, *Bull. Grain Technol.,* 12, 226, 1974.
761. **Rout, G.,** Comparative efficacy of neem seed powder and some common plant product admixtures against *Sitophilus oryzae* (Linn.), *Neem Newslett.,* 3, (2), 13, 1986.
762. **Karnavar, G. K. and Dlamini, S. V.,** Effect of neem products on survival and feeding of *Sitophilus zeamais* Motschulsky, a pest of stored maize, *Neem Newslett.,* 4, (3), 28, 1987.
763. **Akou-Edi, D.,** Effects of neem seed powder and oil on *Tribolium confusum* and *Sitophilus zeamais,* in *Natural Pesticides From the Neem Tree and Other Tropical Plants,* Schmutterer, H. and Ascher, K. R. S., Eds., GTZ Press, Eschborn, FRG, 1984, 445.
764. **Siddig, S. A.,** Efficacy and persistence of powdered neem seeds for treatment of stored wheat against *Trogoderma granarium,* in *Natural Pesticides From the Neem Tree (Azadirachta indica A. Juss.),* Schmutterer, H., Ascher, K. R. S., and Rembold, H., Eds., GTZ Press, Eschborn, FRG, 1981, 251.
765. **Weidhass, D. E.,** personal communication, 1978.
766. **Schiefer, B.,** personal communication, 1985.
767. **Mohan, N. C. and Sivasubramanian, P.,** Evaluation of neem products against leafminer, *Aproserema modicella, Neem Newslett.,* 4, (4), 44, 1987.
768. **Jotwani, M. G., Sadakathula, S., Venugopal, M. S., and Subramaniam, T. R.,** Efficacy of two organotin compounds and neem extract against the sorghum shoot fly, *Phytoparasitica,* 2, 127, 1974.
769. **Webb, R. E., Hinebaugh, M. A., Lindquist, R. K., and Jacobson, M.,** Evaluation of aqueous solution of neem seed extract against *Liriomyza sativae* and *L. trifolii* (Diptera:Agromyzidae), *J. Econ. Entomol.,* 76, 357, 1983.
770. **Lindquist, R. K.,** personal communication, 1983.
771. **Stein, U.,** The potential of neem for inclusion in a pest management program for control of *Liriomyza trifolii* on chrysanthemum, M. S. thesis, University of California, Riverside, 1984, 86 pp.
772. **Larew, H. G.,** Use of neem seed kernel extract in a developed country: *Liriomyza* leafminer as a model case, in *Natural Pesticides From the Neem Tree and Other Tropical Plants,* Schmutterer, H. and Ascher, K. R. S., Eds., GTZ Press, Eschborn, FRG, 1987, 375.
773. **Meisner, J. and Ascher, K. R. S.,** Extracts of neem (*Azadirachta indica*) and their effectiveness as pesticides for different insects, *Phytoparasitica,* 14, 171, 1986.
774. **Fagoonee, I. and Toory, V.,** Contribution to the study of the biology and ecology of the leaf-miner *Liriomyza trifolii* and its control by neem, *Insect Sci. Appl.,* 5, 23, 1984.
775. **Warthen, J. D., Jr., Uebel, E. C., Dutky, S. R., Lusby, W. R., and Finegold, H.,** Adult house fly feeding deterrent from neem seeds, *USDA Sci. Education Admin., Agr. Res. Results* ARR-NE-2, 1978.
776. **Wilps, H.,** Effects of azadirachtin on flight activity, reproduction rate, food intake, and energy metabolism in adult *Phormia terrae-novae,* in *Abstr. 3rd Int. Neem Conf., Nairobi, Kenya,* July 10 to 15, 1986, 35.
777. **Wilps, H.,** Growth and adult molting of larvae and pupae of the blowfly *Phormia terrae-novae* in relationship to azadirachtin concentrations, in *Natural Pesticides From the Neem Tree and Other Tropical Plants,* Schmutterer, H. and Ascher, K. R. S., Eds., GTZ Press, Eschborn, FRG, 1987, 299.

778. **Garcia, E. S., Azambuja, P. D., Forster, H., and Rembold, H.,** Feeding and molt inhibition by azadirachtins A, B, and 7-acetyl-azadirachtin A in *Rhodnius prolixus* nymphs, *Z. Naturforsch.,* 39C, 1155, 1984.

779. **Moreno, D. S. and Tanigoshi, L. K.,** personal communication, 1978.

780. **Chiu, S.-F. and Zeng, X.-N.,** Antifeeding activity of some materials from Meliaceae against the citrus aphid *Aphis citricola* Vander Groot, *Neem Newslett.,* 3, (4), 48, 1986.

781. **Siddig, S. A.,** A proposed pest management programme including neem treatments for combating potato pests in the Sudan, in *Natural Pesticides From the Neem Tree and other Tropical Plants,* Schmutterer, H. and Ascher, K. R. S., Eds., GTZ Press, Eschborn, FRG, 1987, 449.

782. **Saxena, R. C.,** Neem seed derivatives for management of rice insect pests—a review of recent studies, in *Natural Pesticides From the Neem Tree and Other Tropical Plants,* Schmutterer, H. and Ascher, K. R. S., Eds., GTZ Press, Eschborn, FRG, 1987, 81.

783. **Saxena, R. C. and Khan, Z. R.,** Neem oil disturbs *Nephotettix virescens* feeding, *Neem Newslett.,* 1, (3), 28, 1984.

784. **Saxena, R. C., Epino, P. B., Tu, C.-W., and Puma, B. C.,** Neem, chinaberry, and custard apple: antifeedant and insecticidal effects of seed oils on leafhopper and planthopper pests of rice, in *Natural Pesticides From the Neem Tree and Other Tropical Plants,* Schmutterer, H. and Ascher, K. R. S., Eds., GTZ Press, Eschborn, FRG, 1984, 403.

785. **Saxena, R. C., Liquido, N. J., and Justo, H. D.,** Neem seed oil, a potential antifeedant for the control of the rice brown planthopper, *Nilaparvata lugens,* in *Natural Pesticides From the Neem Tree (Azadirachta indica A. Juss.),* Schmutterer, H., Ascher, K. R. S., and Rembold, H., Eds., GTZ Press, Eschborn, FRG, 1981, 171.

786. **Saxena, R. C., Justo, H. D., Jr., and Epino, P. B.,** Evaluation and utilization of neem cake against the rice brown planthopper, *Nilaparvata lugens* (Homoptera:Delphacidae), *J. Econ, Entomol.,* 77, 502, 1984.

787. **Saxena, R. C. and Khan, Z. R.,** Survival of brown planthopper and its ability to transmit grassy stunt diseases on neem oil-treated rice seedlings, *Neem Newslett.,* 1, (3), 25, 1984.

788. **Saxena, R. C. and Khan, Z. R.,** Effect of neem oil on survival of *Nilaparvata lugens* (Homoptera:Delphacidae) and on grassy stunt and ragged stunt virus transmission, *J. Econ. Entomol.,* 78, 647, 1985.

789. **Saxena, R. C., Justo, H. D., Jr., and Epino, P. B.,** Evaluation and utilization of neem cake against the brown planthopper, *Nilaparvata lugens,* in *Natural Pesticides From the Neem Tree and Other Tropical Plants,* Schmutterer, H. and Ascher, K. R. S., Eds., GTZ Press, Eschborn, FRG, 1984, 391.

790. **Rieth, R. E.,** personal communication, 1982.

791. **Radwanski, S.,** Neem tree. 3: Further uses and potential uses, *World Crops Livestock,* 29, 167, 1977.

792. **Patel, H. K., Patel, V. C., Chari, M. S., Patel, J. C., and Patel, J. R.,** Neem seed paste suspension-a sure deterrent to hairy caterpillar (*Amsacta moorei* But.), *Madras Agr. J.,* 55, 509, 1968.

793. **Meisner, J., Wysoki, M., and Ascher, K. R. S.,** The residual effect of some products from neem (*Azadirachta indica* A. Juss.) seeds upon larvae of *Boarmia (Ascotis) selenaria* Schiff. in laboratory trials, *Phytoparasitica,* 4, 185, 1976.

794. **Saxena, R. C., Waldbauer, G. P., Liquido, N. J., and Puma, B. C.,** Effects of neem seed oil on the rice leaf folder, *Gnaphalocrocis medinalis,* in *Natural Pesticides From the Neem Tree (Azadirachta indica A. Juss.),* Schmutterer, H., Ascher, K. R. S., and Rembold, H., Eds., GTZ Press, Eschborn, FRG, 1981, 189.

795. **Krishnamurti, B. and Rao, D. S.,** Some important insect pests of stored grains and their control, Entomol. Ser. Bull. No. 14, Bangalore, India, 93 pp.

796. **Fagoonee, I.,** Behavioral response of *Crocidolomia binotalis* to neem, in *Natural Pesticides From the Neem Tree (Azadirachta indica A. Juss.),* Schmutterer, H., Ascher, K. R. S., and Rembold, H., Eds., GTZ Press, Eschborn, FRG, 1981, 109.

797. **Patnaik, N. C., Panda, N., Patro, E. R., and Mishra, B. K.,** Effect of neem (*Azadirachta indica* A. Juss.), *Neem Newslett.,* 4, (2), 15, 1987.

798. **Fagoonee, I.,** Effect of azadirachtin and of a neem extract on food utilization by *Crocidolomia binotalis,* in *Natural Pesticides From the Neem Tree and Other Tropical Plants,* Schmutterer, H. and Ascher, K. R. S., Eds., GTZ Press, Eschborn, FRG, 1984, 211.

799. **Meisner, J., Kehat, M., Zur, M., and Eizik, C.,** Response of *Earias insulana* Boisd. larvae to neem (*Azadirachta indica* A. Juss.) kernel extract, *Phytoparasitica,* 6, 85, 1978.

800. **Meisner, J., Ascher, K. R. S., Aly, R., and Warthen, J. D., Jr.,** Response of *Spodoptera littoralis* (Boisd.) and *Earias insulana* (Boisd.) larvae to azadirachtin and salannin, *Phytoparasitica,* 9, 27, 1981.

801. **Basak, S. P. and Chakraborty, D. P.,** Chemical investigation of *Azadirachta indica* leaf (*M. azadirachta*), *J. Indian Chem. Soc.,* 45, 466, 1968.

802. **Babu, T. H. and Beri, Y. P.,** Efficacy of neem (*Azadirachta indica*) seed extracts in different solvents as a deterrent to the larvae of *Euproctis lunata, Andhra Agr. J.,* 16, 107, 1969.

803. **Bernays, E. A.,** personal communication, 1978.

804. **Zanno, P. R., Miura, E., Nakanishi, K., and Elder, D. L.,** Structure of the insect phagorepellent azadirachtin. Applications of PRFT/CWD carbon-13 nuclear magnetic resonance, *J. Am. Chem. Soc.,* 97, 1975, 1975.

805. **Sinha, S. N. and Mehrotra, K. N.,** Diflubenzuron and neem (*Azadirachta indica*) oil in control of *Heliothis armigera* infesting chickpea (*Cicer aristimum*), *Indian J. Agr. Sci.,* 58, 238, 1988.

806. **Simmonds, M. S. J. and Blaney, W. M.,** Some neurophysiological effects of azadirachtin on lepidopterous larvae and their feeding response, in *Natural Pesticides From the Neem Tree and Other Tropical Plants,* Schmutterer, H. and Ascher K. R. S., Eds., GTZ Press, Eschborn, FRG, 1984, 163.

807. **Hedin, P. A.,** personal communication.

808. **Ruscoe, C. N. E.,** Growth disruption effects of an insect antifeedant, *Nature,* 236, 159, 1972.

809. **Kirsch, K.,** Studies of the efficacy of neem extracts in controlling major insect pests in tobacco and cabbage, in *Natural Pesticides From the Neem Tree and Other Tropical Plants,* Schmutterer, H. and Ascher, K. R. S., Eds., GTZ Press, Eschborn, FRG, 1987, 495.

810. **Kubo, I. and Klocke, J. A.,** Limonoids as insect control agents, Les Mediateurs Chimique, Versailles, Nov. 16 to 20, 1981, 1982, 117.

811. **Dreyer, D. M.,** Field and laboratory trials with simple neem products as protectants against pests of vegetable and field crops in Togo, in *Natural Pesticides From the Neem Tree and Other Tropical Plants,* Schmutterer, H. and Ascher, K. R. S., Eds., GTZ Press, Eschborn, FRG, 1987, 431.

812. **Rogers, C. E.,** personal communication, 1979.

813. **Feuerhake, K.,** Investigations of the Isolation and Formulation of the Seed Components of the Neem Tree (*Azadirachta indica* A. Juss.) in Relation to their Usefulness as Pesticides in Developing Countries, Ph.D. thesis, Justus Liebig University, Giessen, FRG, 1985, 132 pp. (in German).

814. **Arnason, J. T., Philogene, B. J. R., Donskov, N., Hudond, M., McDougall, C., Fortier, G., Morand, P., Gardner, D., Lambert, J., Morris, C., and Nozzolillo, C.,** Antifeedant and insecticidal properties of azadirachtin to the European corn borer, *Ostrinia nubilalis, Entomol. Exp. Appl.,* 38, 29, 1985.

815. **Sharma, R. N., Nagasampagi, B. A., Bhopale, B. A., Kulkarni, M. M., and Tungikar, V. B.,** "Neemrich": the concept of enriched fractions from neem for behavioral and physiological control of insects, in *Natural Pesticides From the Neem Tree and Other Tropical Plants,* Schmutterer, H. and Ascher K. R. S., Eds., GTZ Press, Eschborn, FRG, 1984, 115.

816. **Zhang, Y.-C. and Chiu, S.-F.,** Preliminary investigation of some meliaceous plants as feeding deterrents against the imported cabbage worm (*Pieris rapae*), *Neem Newslett.,* 2, (3), 30, 1985.

817. **Steets, R.,** The effect of crude extracts of the Meliaceae *Azadirachta indica* and *Melia azedarach* on various species of insects, *Z. Angew. Entomol.,* 77, 306, 1975.

818. **Tan, M. T. and Sudderuddin, K. I.,** Effects of neem tree (*Azadirachta indica*) extracts on diamond-back moth (*Plutella xylostella* L.), *Malayan Appl. Biol.,* 7, 1, 1978.

819. **Sombatsiri, K.,** Studies on the control of tobacco cutworm (*Spodoptera litura* L.) and diamond-back moth (*Plutella xylostella* L.) with an improved kernel extract of neem (*Azadirachta indica* var. *siamensis* Val.), Abstr. 3rd Int. Neem Conf., Nairobi, Kenya, July 10 to 15, 1986, 27.

820. **Chan, B. G.,** personal communication.

821. **Redfern, R. E., Warthen, J. D., Jr., Jacobson, M., and Stokes, J. B.,** Antifeeding potency of neem formulations, *J. Environ. Sci. Health,* A19, 477, 1984.

822. **Hellpap, G.,** Effects of neem kernel extracts on the fall armyworm, *Spodoptera frugiperda,* in *Natural Pesticides From the Neem Tree and Other Tropical Plants,* Schmutterer, H. and Ascher K. R. S., Eds., GTZ Press, Eschborn, FRG, 1984, 353.

823. **Warthen, J. D., Jr., Redfern, R. E., Uebel, E. C., and Mills, G. D., Jr.,** An antifeedant for fall armyworm larvae from neem seeds, USDA Sci. Educ. Admin., Agr. Res. Results ARR-NE-1, 1978.

824. **Klocke, J. A. and Kubo, I.,** Citrus limonoid byproducts as insect control agents, *Entomol. Exp. Appl.,* 32, 299, 1982.

825. **Ascher, K. R. S., Eliyahu, M., Nemny, N. E., and Meisner, J.,** Neem seeds kernel extract as an inhibitor of growth and fecundity in *Spodoptera littoralis,* in *Natural Pesticides From the Neem Tree and Other Tropical Plants,* Schmutterer, H. and Ascher, K. R. S., Eds., GTZ Press, Eschborn, FRG, 1984, 331.

826. **Joshi, B. G. and Ramaprasad, G.,** Neem kernel as an antifeedant against the tobacco caterpillar *(Spodoptera litura* F.), *Phytoparasitica,* 3, 59, 1975.

827. **Joshi, B. G., Ramaprasad, G., and Satyanarayana, S. V. V.,** Relative efficacy of neem kernel, fentin acetate and fentin hydroxide as antifeedants against tobacco caterpillar, *Spodoptera litura* Fabricius, in the nursery, *Indian J. Agr. Sci.,* 48, 19, 1978.

828. **Sitaramiah, S., Ramaprasad, G., and Joshi, B. G.,** Efficacy of diflubenzuron and neem seed kernel suspension against the tobacco caterpillar, *Spodoptera litura,* in tobacco nurseries, *Phytoparasitica,* 14, 265, 1986.

829. **Joshi, B. G.,** Use of neem products in tobacco, in *Natural Pesticides From the Neem Tree and Other Tropical Plants,* Schmutterer, H. and Ascher, K. R. S., Eds., GTZ Press, Eschborn, FRG, 1987, 479.

830. **Sombatsiri, K. and Tigvattanont, S.,** Effects of neem extracts on some insect pests of economic importance in Thailand, in *Natural Pesticides From the Neem Tree and Other Tropical Plants,* Schmutterer, H. and Ascher, K. R. S., Eds., GTZ Press, Eschborn, FRG, 1984, 95.

831. **Warthen, J. D., Jr. and Uebel, E. C.,** Effect of azadirachtin on house crickets *Acheta domesticus,* in *Natural Pesticides From the Neem Tree (Azadirachta indica a. Juss.),* Schmutterer, H. and Ascher K. R. S., Eds., GTZ Press, Eschborn, FRG, 1981, 137.

832. **Adler, V. E. and Uebel, E. C.,** Effects of Margosan O on six species of cockroaches (Orthoptera:Blaberidae, Blattidae) and Blatellidae, in *Natural Pesticides From the Neem Tree and Other Tropical Plants,* Schmutterer, H. and Ascher, K. R. S., Eds., GTZ Press, Eschborn, FRG, 1987, 387.

833. **Amatobi, C. I., Apeji, S. B., and Oyidi, O.,** Effects of farming practices on populations of two grasshopper pests *(Krausseria angulifera* Krauss and *Oedaleus senegalensis* Krauss) (Orthoptera:Acrididae) in Northern Nigeria, *Trop. Pest Management,* 34, 173, 1988.

834. **Sinha, N. P. and Gulati, K. C.,** Neem seed cake *(Azadirachta indica)* as a source of pest control chemicals, *Bull. Regional Res. Lab. Jammu (India),* 1, 176, 1963.

835. **Goyal, R. S., Gulati, K. C., Sarup, P., Kidwai, M. A., and Singh, D. S.,** Biological activity of various alcohol extractives and isolates of neem *(Azadirachta indica)* seed cake against *Rhopalosiphum nympheae* (Linn.) and *Schistocerca gregaria* Forsk., *Indian J. Entomol.,* 33, 67, 1971.

836. **Jain, H. H.,** Neem in insect control, *Indian Agr. Res. Inst. Res. Bull.,* 40, 19, 1983.

837. **Mishra, P. K.,** Studies on the biological efficacy of extracts of different parts of neem, *Azadirachta indica* A. Juss., MS thesis, *Indian Agricultural* Research Institute, New Delhi, 1983.

838. **Singh, R. P.,** Comparison of antifeedant efficacy and extract yields from different parts and ecotypes of neem *(Azadirachta indica* A. Juss.) trees, in *Natural Pesticides From the Neem Tree and Other Tropical Plants,* Schmutterer, H. and Ascher, K. R. S., Eds., GTZ Press, Eschborn, FRG, 1987, 185.

839. **Lavie, D., Jain, M. K., and Shpan-Gabrielith, S. R.,** A locust phagorepellent from two *Melia* species, *Chem. Commun.,* 910, 1967.

840. **Dreyer, M.,** Attempts to control insect plant pests with neem products in Togo, Ph.D. thesis, University of Giessen, FRG, 1985, (in German).

841. **Olaifa, J. I. and Akingbohungbe, A. E.,** Antifeedant and insecticidal effects of extracts of *Azadirachta indica, Petiveria alliacea,* and *Piper guineense* on the variegated grasshopper, *Zonocerus variegatus,* in *Natural Pesticides From the Neem Tree and Other Tropical Plants,* Schmutterer, H. and Ascher, K. R. S., Eds., GTZ Press, Eschborn, FRG, 1987, 405.

842. **Jones, P. S. Ley, S. V., Morgan, E. D., and Santafianos, D.,** The chemistry of the neem tree, in *Focus on Phytochemical Pesticides,* Vol. 1, Jacobson, M., Ed., *The Neem Tree,* CRC Press, Boca Raton, Florida, 1989, 19.

843. **Rice, M. J., Sexton, S., and Esmail, A. M.,** Antifeedant biochemical blocks oviposition by sheep blowfly, *J. Austral. Entomol. Soc.,* 24, 16, 1985.

844. **Joshi, B. G. and Sitaramiah, S.,** Neem kernels as an ovipositional repellent for *Spodoptera litura* (F.) moths, *Phytoparasitica,* 7, 199, 1979.

845. **Naqvi, S. N. H.,** Biological evaluation of fresh neem extracts and some new components, with reference to abnormalities and esterase activity, in *Natural Pesticides From the Neem Tree and Other Tropical Plants,* Schmutterer, H. and Ascher, K. R. S., Eds., GTZ Press, Eschborn, FRG, 1987, 315.

846. **Matemu, D. P. and Mosha, F. W.,** Toxic effects of neem (*Azadirachta indica*) berry extract on mosquitoes, *Neem Newslett.,* 3, (4), 44, 1986.

847. **Rembold, H., Sharma, G. K., Czoppelt, Ch., and Schmutterer, H.,** Evidence of growth disruption in insects without feeding inhibition by neem seed fractions, *Z. Pflanzenkrankh. Pflanzenschutz,* 87, 290, 1980.

848. **Patnaik, N. C., Panda, N., Bhuyan, K., and Mishra, B. K.,** Developmental aberrations and mortality of the mustard sawfly larvae, *Anthalia lugens proxima* (Klug) by neem oil, *Neem Newslett.,* 4, (2), 18, 1987.

849. **Bidmon, H. J., Kauser, G., Mobus, P., and Koolman, J.,** Effect of azadirachtin on blowfly larvae and pupae, in *Natural Pesticides From the Neem Tree and Other Tropical Plants,* Schmutterer, H. and Ascher, K. R. S., Eds., GTZ Press, Eschborn, FRG, 1987, 253.

850. **Fagoonee, I. and Lauge, G.,** Noxious effects of neem extract on *Crocidolomia binotalis, Phytoparasitica,* 9, 111, 1981.

851. **Rembold, H.,** Azadirachtin, a new class of insect growth regulators, in *Abstr. Int. Conf. on Natural Products as Regulators of Insect Reproduction,* Jammu, India, Sept. 29 to Oct. 1, 1984, 34.

852. **Ascher, K. R. S.,** Some physical (solubility) properties and biological (sterilant for *Epilachna varivestis* females) effects of a dried methanolic neem (*Azadirachta indica*) seed kernel extract, in *Natural Pesticides From the Neem Tree (Azadirachta indica A. Juss.),* Schmutterer, H., Ascher, K. R. S., and Rembold, H., Eds., GTZ Press, Eschborn, FRG, 1981, 63.

853. **Schulz, W. D.,** Pathological alterations in the ovaries of *Epilachna varivestis* induced by an extract from neem kernels, in *Natural Pesticides From the Neem Tree (Azadirachta indica A. Juss.),* Schmutterer, H., Ascher, K. R. S., and Rembold, H., Eds., GTZ Press, Eschborn, FRG, 1981, 81.

854. **Rembold, H., Forster, H., Czoppelt, Ch., Rao, P. J., and Sieber, K. P.,** The azadirachtins, a group of insect growth regulators from the neem tree, in *Natural Pesticides From the Neem Tree and Other Tropical Plants,* Schmutterer, H. and Ascher, K. R. S., Eds., GTZ Press, Eschborn, FRG, 1984, 153.

855. **Schlüter, U. and Schulz, W. D.,** Structural damages caused by neem in *Epilachna varivestis.* A summary of histological and ultrastructural data. I. Tissues affected in larvae, in *Natural Pesticides From the Neem Tree and Other Tropical Plants,* Schmutterer, H. and Ascher, K. R. S., Eds., GTZ Press, Eschborn, FRG, 1984, 227.

856. **Schlüter, U.,** Disturbance of epidermal and fat body tissue after feeding azadirachtin and its consequence on larval moulting in the Mexican bean beetle *Epilachna varivestis* (Coleoptera:Coccinellidae), *Entomol. Gener.,* 10, 97, 1984.

857. **Saxena, K. N. and Rembold, H.,** Orientation and ovipositional responses of *Heliothis armigera* to certain neem constituents, in *Natural Pesticides From the Neem Tree and Other Tropical Plants,* Schmutterer, H. and Ascher, K. R. S., Eds., GTZ Press, Eschborn, FRG, 1984, 199.

858. **Barnby, M. A. and Klocke, J. A.,** Effects of azadirachtin on the nutrition and development of the tobacco budworm, *Heliothis virescens* (Fabr.) (Noctuidae), *J. Insect Physiol.,* 33, 69, 1987.

859. **Kubo, I. and Klocke, J. A.,** Azadirachtin, insect ecdysis inhibitor, *Agr. Biol. Chem.,* 46, 1951, 1982.
860. **Rembold, H. and Sieber, K. P.,** Effect of azadirachtin on oocyte development in *Locusta migratoria migratorioides,* in *Natural Pesticides From the Neem Tree (Azadirachta indica A. Juss.),* Schmutterer, H., Ascher, K. R. S., and Rembold, H., Eds., GTZ Press, Eschborn, FRG, 1981, 75.
861. **Rembold, H., Uhl, M., and Mueller, T.,** Effect of azadirachtin A on hormone titers during the gonadotrophic cycle of *Locusta migratoria,* in *Natural Pesticides From the Neem Tree and Other Tropical Plants,* Schmutterer, H. and Ascher, K. R. S., Eds., GTZ Press, Eschborn, FRG, 1987, 289.
862. **Schlüter, U., Bidmon, H. J., and Grewe, S.,** Azadirachtin affects growth and endocrine events in larvae of the tobacco hornworm, *Manduca sexta, J. Insect Physiol.,* 31, 773, 1985.
863. **Haasler, C.,** Effects of neem seed extract on the post-embryonic development of the tobacco hornworm, *Manduca sexta,* in *Natural Pesticides From the Neem Tree and Other Tropical Plants,* Schmutterer, H. and Ascher, K. R. S., Eds., GTZ Press, Eschborn, FRG, 1984, 321.
864. **Sharma, H. C., Leuschner, K., Sankaram, A. V. B., Gunasekhar, D., Marthandamurthi, M., Bhaskariah, K., Subramanyam, M., and Sultana, N.,** Insect antifeedants and growth inhibitors from *Azadirachta indica* and *Plumbago zeylanica,* in *Natural Pesticides From the Neem Tree and Other Tropical Plants,* Schmutterer, H. and Ascher, K. R. S., Eds., GTZ Press, Eschborn, FRG, 1984, 291.
865. **Dorn, A., Rademacher, J. M., and Sehn, E.,** Effects of azadirachtin on reproductive organs and fertility in the large milkweed bug, *Oncopeltus fasciatus,* in *Natural Pesticides From the Neem Tree and Other Tropical Plants,* Schmutterer, H. and Ascher, K. R. S., Eds., GTZ Press, Eschborn, FRG, 1987, 273.
866. **Meisner, J., Melamed-Madjak, V., Yathom, S., and Ascher, K. R. S.,** The influence of neem on the European corn borer (*Ostrinia nubilalis*) and the serpentine leaf-miner (*Liriomyza trifolii*), in *Natural Pesticides From the Neem Tree and Other Tropical Plants,* Schmutterer, H. and Ascher, K. R. S., Eds., GTZ Press, Eschborn, FRG, 1987, 461.
867. **Gujar, G. T. and Mehrotra, K. N.,** Juvenilizing effect of azadirachtin on a noctuid moth, *Spodoptera litura* Fabr., *Indian J. Exptl. Biol.,* 21, 292, 1983.
868. **Koul, O.,** Azadirachtin interaction with development of *Spodoptera litura* Fab., *Indian J. Exptl. Biol.,* 23, 160, 1985.
869. **Uebel, E. C., Warthen, J. D., Jr., and Jacobson, M.,** Preparative reversed-phase liquid chromatographic isolation of azadirachtin from neem kernels, *J. Liquid Chromatog.,* 2, 875, 1979.
870. **Kraus, W. and Cramer, R.,** 17-Epiazadiradione and 17β-hydroxyazadiradione, two new components from *Azadirachta indica* A. Juss, *Tetrahedron Lett.,* 2395, 1978 (in German).
871. **Garg, H. S. and Bhakuni, D. S.,** Salannolide, a meliacin from *Azadirachta indica, Phytochemistry,* 23, 2383, 1984.
872. **Siddiqui, S., Faizi, S., Mahmood, T., and Siddiqui, B. S.,** Margosinolide and isomargosinolide, two new tetranortriterpenoids from *Azadirachta indica* A. Juss (Meliaceae), *Tetrahedron,* 42, 4849, 1986.
873. **Lavie, D., Levy, E. C., and Jain, M. K.,** Limonoids of biogenetic interest from *Melia azadirachta* L., *Tetrahedron,* 27, 3927, 1971.
874. **Forster, H.,** Isolation of azadirachtins form neem (*Azadirachta indica*) and radioactive labeling of azadirachtin A, Ph.D. thesis, Ludwig Maximilian University, Munich, FRG, 1983.
875. **Siddiqui, S., Mahmood, T., Siddiqui, B. S., and Faizi, S.,** Isolation of a triterpenoid from *Azadirachta indica, Phytochemistry,* 25, 2183, 1986.
876. **Siddiqui, S., Siddiqui, B. S., Faizi, S., and Mahmood, T.,** Isolation of a tetranortriterpenoid from *Azadirachta indica, Phytochemistry,* 23, 2899, 1984.
877. **Kraus, W. and Cramer, R.,** Pentanortriterpenoids from *Azadirachta indica* A. Juss (Meliaceae), *Chem. Ber.,* 114, 2375, 1981.
878. **Kraus, W., Cramer, R., and Sawitzki, G.,** Tetranortriterpenoids from the seeds of *Azadirachta indica, Phytochemistry,* 20, 117, 1981.
879. **Siddiqui, S., Mahmood, T., Siddiqui, B. S., and Faizi, S.,** Two new tetranortriterpenoids from *Azadirachta indica, J. Nat. Prod.,* 49, 1068, 1986.

880. **Garg, H. S. and Bhakuni, D. S.,** An isoprenylated flavanone from leaves of *Azadirachta indica, Phytochemistry,* 23, 2115, 1984.
881. **Narasimha Rao, K. and Parmar, B. S.,** *Neem Newslett.,* 1, (4), 39, 1984.
882. **Kraus, W., Bokel, M., Klenk, A., and Poehnl, H.,** The structure of azadirachtin and 22,23-dihydro-23β-methoxyazadirachtin, *Tetrahedron Lett.,* 26, 6435, 1985.
883. **Klenk, A., Bokel, M., and Kraus, W.,** 3-Tigloylazadirachtol (tigloyl=2-methylcrotonoyl), an insect growth regulating constituent of *Azadirachta indica, J. Chem. Soc. Chem. Commun.,* 523, 1986.
884. **Kraus, W., Bokel, M., Bruhn, A., Cramer, R., Klaiber, I., Klenk, A., Poehnl, H., Sadlo, H., and Vogler, B.,** Structure determination by NMR of azadirachtin and related compounds from *Azadirachta indica* A. Juss (Meliaceae), *Tetrahedron,* 43, 2817, 1987.
885. **Taylor, D. A. H.,** Azadirachtin, a study in the methodgy of structure determination, *Tetrahedron,* 43, 2779, 1987.
886. **Banerji, R., Misra, G., and Nigam, S. K.,** Identification of 24-methylenelophenol from heartwood of *Azadirachta indica, Phytochemistry,* 26, 2644, 1987.
887. **Majumder, P. L., Maiti, D. C., Kraus, W., and Bokel, M.,** Nimbidiol, a modified diterpenoid of the rootbark of *Azadirachta indica, Phytochemistry,* 26, 3021, 1987.
888. **Rembold, H., Forster, H., and Sonnenbichler, J.,** Structure of azadirachtin B, *Z. Naturforsch.,* 42C, 4, 1987.
889. **Siddiqui, S., Faizi, S., and Siddiqui, B. S.,** Studies on the chemical constituents of *Azadirachta indica* A. Juss (Meliaceae), part VII, *Z. Naturforsch.,* 41B, 922, 1986.
890. **Siddiqui, S., Siddiqui, B. S., and Faizi, S.,** Studies on the chemical constituents of *Azadirachta indica.* II. Isolation and structure of the new triterpenoid azadirachtol, *Planta Med.,* 478, 1985.
891. **Madhusudanan, K. P., Chaturvedi, R., Garg, H. S., and Bhakuni, D. S.,** Negative ion mass spectra of tetranortriterpenoids isolated from neem (*Azadirachta indica* A. Juss), *Indian J. Chem.,* 23B, 1082, 1984.
892. **Kubo, I., Matsumoto, T., Matsumoto, A., and Shoolery, J. N.,** Structure of deacetylazadirachtinol. Application of 2D^H-^{1}H and ^{1}H-^{13}C shift correlation spectroscopy, *Tetrahedron Lett.,* 25, 4729, 1984.
893. **Bilton, J. N., Broughton, H. B., Ley, S. V., Lidert, Z., Morgan, E. D., Rzapa, H. S., and Sheppard, R. N.,** Structural reappraisal of the limonoid insect antifeedant azadirachtin, *J. Chem. Soc. Chem. Commun.,* 968, 1985.
894. **Broughton, H. B., Ley, S. V., Slawin, A. M., Williams, D. J., and Morgan, E. D.,** X-ray crystallographic structure determination of detigloyldihydroazadirachtin and reassignment of the structure of the limonoid insect antifeedant azadirachtin, *J. Chem. Soc. Chem. Commun.,* 46, 1986.
895. **Yamasaki, R. B. and Klocke, J. A.,** Structure-bioactivity relationships of azadirachtin, a potential insect control agent, *J. Agr. Fd. Chem.,* 35, 467, 1987.
896. **Schroeder, D. R., and Nakanishi, K.,** A simplified isolation procedure for azadirachtin, *J. Nat. Prod.,* 50, 241, 1987.
897. **Fon Kimbu, S., Foyere Ayafor, J., Sondengam, B. L., Tsamo, E., Connolly, J. D., and Rycroft, D. S.,** Carapolides D, E and F, novel tetranortriterpenoids from the seeds of *Carapa grandiflora* (Meliaceae), *Tetrahedron Lett.,* 25, 1617, 1984.
898. **Sondengam, B. L., Kanga, C. S., and Connolly, J. D.,** Evodulone, a new tetranortriterpenoid from *Carapa procera, Tetrahedron Lett.,* 1357, 1979.
899. **Mikolajczak, K. L. and Powell, R. C.,** Some plant materials with activity in insects, Abstr. Am. Chem. Soc., Agrochem. Div., 194th Natl. Mtg., New Orleans, LA, Aug. 30 to Sept. 4, 1987, No. 106.
900. **Koul, O.,** Feeding deterrence induced by plant limonoids in the larvae of *Spodoptera litura* (F.) (Lepidoptera, Noctuidae), *Z. Angew. Entomol.,* 95, 166, 1983.
901. **Akinuiyi, J. A., Connolly, J. D., Rycroft, D. S., Sondengam, B. L., and Ifeadika, N. P.,** Tetranortriterpenoids and related compounds. Part 25. Two 3,4-secotirucallane derivatives and 2"-hydroxyhitukin from the bark of *Guarea cedrata* (Meliaceae),*Can. J. Chem.,* 58, 1865, 1980.
902. **Grijpma, P.,** Immunity of *Toona ciliata* M. Roem. var *australis* (F. v. M.) CDC and *Khaya ivorensis* A. Chev. to attacks of *Hypsipyla grandella* Zeller in Turrialba, Costa Rica, *Turrialba,* 20, 85, 1970.

903. **Chiu, S.-F.,** Experiments on the practical application of chinaberry, *Melia azedarach,* and other naturally occurring insecticides in China, in *Natural Pesticides From the Neem Tree and Other Tropical Plants,* Schmutterer, H. and Ascher, K. R. S., Eds., GTZ Press, Eschborn, FRG, 1987, 661.

904. **Husain, M. A.,** Ann. Rept. Punjab Dept. Agr. 1927-28, 1, 55, 1929.

905. **McMillian, W. W., Bowman, M. C., Burton, R. L., et al.,** Extract of chinaberry leaf as a feeding deterrent and growth retardant for larvae of the corn earworm and fall armyworm, *J. Econ. Entomol.,* 62, 708, 1969.

906. **LePage, H. S., Gianotti, O., and Orlando, A.,** Protection of cultures against grasshoppers by means of extracts of *Melia azedarach, Biologico (Sao Paulo),* 12, (1), 265, 1946.

907. **Pruthi, H. S. and Singh, M.,** Stored grain pests and their control, Imperial Council Agr. Res. Muscell. Bull. No. 57, 1944.

908. **Chiu, S.-F. and Chang, Y.-C.,** Effects of some plant materials of Meliaceae on fifth instar larvae of *Spodoptera litura* as feeding inhibitors, *Neem Newslett.,* 1, (3), 23, 1984.

909. **Chiu, S.-F.,** The active principles and insecticidal properties of some Chinese plants, with special reference to Meliaceae, in *Natural Pesticides From the Neem Tree and Other Tropical Plants,* Schmutterer, H. and Ascher K. R. S., Eds., GTZ Press, Eschborn, FRG, 1984, 255.

910. **Chiu, S.-F.,** Experiments on insecticidal plants as a source of insect feeding inhibitors and growth regulators with special reference to Meliaceae, South China Agr. College Special Rept., 1982.

911. **Ochi, M., Kotsuki, H., Hirotsu, K., and Tokoroyama, T.,** Sendanin, a new limonoid from *Melia azedarach* var. *Japonica* Makino, *Tetrahedron Lett.,* 2877, 1976.

912. **Schulte, K. E., Ruecker, G., and Materu, H. U.,** Some constituents of the fruits and root of *Melia azedarach* L., *Planta Med.,* 35, 76, 1979 (in German).

913. **Kraus, W. and Bokel, M.,** New tetranortriterpenoids from *Melia azedarach* Linn., *Chem. Ber.,* 114, 267, 1981 (in German).

914. **Srivastava, S.,** Limonoids from the seeds of *Melia azedarach, J. Nat. Prod.,* 49, 56, 1980.

915. **Nakatani, M.,** Azedarachol, a steroid ester antifeedant from *Melia azedarach* var. *japonica, Phytochemistry,* 24, 1945, 1985.

916. **Lee, S. M., Klocke, J. A., and Balandrin, M. F.,** The structure of 1-cinnamoyl-melianolone, a new insecticidal tetranortriterpenoid from *Melia azedarach* L. (Meliaceae), *Tetrahedron Lett.,* 28, 1543, 1987.

917. **De Silva, L. B., Stoecklin, W., and Geissman, T. A.,** The isolation of salannin from *Melia dubia, Phytochemistry,* 8, 1817, 1969.

918. **Purushothaman, K. K., Duraiswamy, K., and Connolly, J. D.,** Tetranortriterpenoids from *Melia dubia, Phytochemistry,* 23, 135, 1984.

919. **Inada, A., Kobayashi, M., and Nakanishi, T.,** Phytochemical studies on Meliaceaous plants. III. Structures of two new pregnane steroids, toosendansterols A and B, from leaves of *Melia toosendan* Sieb. et Zucc., *Chem. Pharm. Bull. (Tokyo),* 36, 609, 1988.

920. **Mwangi, R. W.,** Locust antifeedant activity in fruits of *Melia volkensii, Entomol. Exp. Appl.,* 32, 277, 1982.

921. **Mwangi, R. W. and Rembold, H.,** Growth-inhibiting and larvicidal effects of *Melia volkensii* extracts on *Aedes aegypti* larvae, *Entomol. Exp. Appl.,* 46, 103, 1988.

922. **Rajab, M. S., Bentley, M. D., Alford, A. R., and Mendel, M. J.,** A new limonoid insect antifeedant from the fruit of *Melia volkensii, J. Nat. Prod.,* 51, 168, 1988.

923. **Connolly, J. D. and Labbe, C.,** Tetranortriterpenoids and related compounds. Part 24. The interrelation of swietenine and swietenolide, the major tetranortriterpenoids from the seeds of *Swietenia macrophylla, J. Chem. Soc. Perkin Trans.,* 529, 1980.

924. **Daily, A., Seligmann, O., Lotter, H., and Wagner, H.,** 2-Hydroxy-swietenin, a new limonoid from *Swietenia mahagoni* DC, *Z. Naturforsch.,* 40C, 519, 1985.

925. **Martinez Nadal, N. C., Elba Montalvo, A., De La Torre, S., and Vega, G.,** Toxicological effects of active principles of West Indian caoba, *Swietenia mahogany, Caribbean J. Sci.,* 13, 131, 1973.

926. **Kipke, K.,** Active Insect Antifeedants from *Toona sureni* (Blume) Merrill (Meliaceae), Ph.D. thesis, University of Hohenheim, FRG, 1980, 117 pp.

927. **Kubo, I. and Klocke, J. A.,** An insect growth inhibitor from *Trichilia roka* (Meliaceae), *Experientia,* 38, 639, 1982.

928. Nakatani, M., James, J. G., and Nakanishi, K., Isolation and structures of trichilins, antifeedants against the southern armyworm, *J. Am. Chem. Soc.*, 103, 1228, 1981.

929. Nakatani, M., Iwashita, T., Naoki, H., and Hase, T., Structure of a limonoid antifeedant from *Trichilia roka*, *Phytochemistry*, 24, 195, 1985.

930. Lau, P. H. W. and Martin, J. C., Revision of the structure of xylomollin, *J. Am. Chem. Soc.*, 100, 7079, 1978.

931. Kubo, I. and Matsumoto, T., Potent insect antifeedants from the African medicinal plant *Bersama abyssinica*, Am. Chem. Soc. Symp. Ser. 1985, Bioregulation Pest Control, 1985, 183.

932. Kubo, I. and Matsumoto, T., Abyssinin, a potent insect antifeedant from an African medicinal plant, *Bersama abyssinica*, *Tetrahedron Lett.*, 25, 4601, 1984.

933. Amma, K. P., Chemical examination of plant products possessing insecticidal properties, *Travancore Univ. Res. Inst. Bull.*, Ser. A, 3, 129, 1954.

934. Porter, L. A., Picrotoxinin and related substances, *Chem. Ber.*, 67, 441, 1967.

935. Miller, T. A., Maynard, M., and Kennedy, J. M., Structure and insecticidal activity of picrotoxinin analogs, *Pesticide Biochem. Physiol.*, 10, 128, 1979.

936. Wada, K and Munakata, K., Naturally occurring insect control chemicals. Isoboldine, a feeding inhibitor and cocculolidine, an insecticide in the leaves of *Cocculus trilobus* DC, *J. Agr. Fd. Chem.*, 16, 471, 1968.

937. Engelhardt, R., The spirilla palms, *Moeller Garten Ztg.*, 4, 197, 1889 (in German).

938. Khare, B. P., *Insect Pests of Stored Grain and Their Control in Uttar Pradesh*, Patnagar University, India, 1972, 94.

939. Riley, C. V. and Howard, L. O., Hemp as a protection against weevils, *Insect Life*, 4, 223, 1891.

940. Wolcott, G. N., Organic termite repellents tested against *Cryptotermes brevis* Walker, *Puerto Rico Univ. J. Agr.*, 39, 115, 1955.

941. Von Arndt, U., Tests of the biological activity of wood as repellents for *Reticulotermes*, *Holzforschung*, 22, 104, 1968.

942. Wolcott, G. N., Stilbene and comparable materials for dry-wood termite control, *J. Econ. Entomol.*, 46, 374, 1953.

943. Gearien, J. E. and Klein, M., Isolation of 19α-H-lupeol from *Maclura pomifera*, *J. Pharm. Sci.*, 64, 104, 1975.

944. Uno, T., Isogai, A., and Suzuki, A., Evaluation of three phenolics from mulberry trees as feeding stimulant/deterrent for larvae of the silkworm, *Bombyx mori* L., *Appl. Entomol. Ecol.*, 17, 137, 1982.

945. Isogai, A., Murakoshi, S., Suzuki, A., and Tamura, S., Isolation from nutmeg of growth inhibitory substances to silkworm larvae, *Agr. Biol. Chem.*, 37, 889, 1973.

946. Miles, D. H., Ly, A. M., Randle, S. A., Hedin, P. A., and Burks, M. L., Alkaloidal insect antifeedants from *Virola calophylla* Warb., *J. Agr. Fd. Chem.*, 35, 794, 1987.

947. Katiyar, R. L. and Srivastava, K. P., An essential oil from Australian bottlebrush, *Callistemon lanceolatum* (Myrtaceae) with juvenoid properties against the red cotton bug, *Dysdercus koenigii* Fabr. (Heteroptera:Pyrrocoridae), *Entomon*, 7, 463, 1982.

948. Penfold, E. R. and Morrison, F. R., Uses of commercial eucalyptus oils, in *Bull. Museum Appl. Arts Sci. No. 17*, 3rd ed., Sydney, Australia, 1950.

949. Osmani, Z., Anees, I., and Naidu, M. B., Insect repellent creams from essential oils, *Pesticides (India)*, 6, (3), 19, 1972.

950. Mazanec, Z., Resistance of *Eucalyptus marginata* to *Perthida glyphopa* (Lepidoptera:Incurvariidae), *J. Austral. Entomol. Soc.*, 24, 209, 1985.

951. Rudman, P. and Gay, F. J., The causes of natural durability in timber. 6. Measurement of antitermitic properties of anthraquinones from *Tectona grandis* L. f. by a rapid seed-micromethod, *Holzforschung*, 15, 112, 1961.

952. Kambu, K., Di Phanzu, N., Coune, C., Wauters, J. N., and Angenot, L., Study of insecticidal and chemical properties of *Eucalyptus saligna* from Zaire, *Plant Med. Phytother.*, 16, 34, 1982.

953. Morrow, P. A. and Fox, L. R., Effects of variation in *Eucalyptus* essential oil yield on insect growth and grazing damage, *Oecologia*, 45, 209, 1980.

954. **Cobbinah, J. R., Morgan, F. D., and Douglas, T. J.,** Feeding responses of the gum leaf skeletonizer *Uraba lugens* Walker to sugars, amino acids, lipids, sterols, salts, vitamins and certain extracts of eucalypt leaves, *J. Austral. Entomol. Soc.,* 21, 225, 1982.

955. **French, J. R. J., Robinson, P. J., Yazaki, Y., and Hillis, W. E.,** Bioassays of extracts from white cypress pine (*Callitris columellaris* F. Muell.) against subterranean termites, *Holzforschung,* 33, 144, 1979.

956. **Yazaki, Y. and Hillis, W. E.,** Components of the extractives from *Callitris columellaris* F. Muell. heartwood which affect termites, *Holzforschung,* 31, 188, 1977.

957. **Erdtman, H.,** Heartwood extractives of conifers. Their fungicidal and insect-repellent properties and taxonomic interest, *Tappi,* 32, 305, 1949.

958. **Singh, D. and Agarwal, S. K.,** Himachalol and β-himachalene: insecticidal principles of Himalayan cedarwood oil, *J. Chem. Ecol.,* 14, 1145, 1988.

959. **Wagner, M. R., Ikeda, T., Benjamin, D. M., and Matsumura, F.,** Host derived chemicals: the basis for preferential feeding behavior of the larch sawfly, *Pristiphora ericksonii* (Hymenoptera:Tenthredinidae), on tamarack, *Larix laricina, Can. Entomol.,* 111, 165, 1979.

960. **Wagner, M. R., Benjamin, D. M., Clancy, K. M., and Schuh, B. A.,** Influence of diterpene resin acids on feeding and growth of larch sawfly, *Pristiphora ericksonii* (Hartig), *J. Chem. Ecol.,* 9, 119, 1983.

961. **Hesse, G., Kauth, H., and Wachter, R.,** Food attractants for *Hylobius abietis, Z. Angew. Entomol.,* 37, 239, 1955.

962. **Anderson, J. M. and Fisher, K. C.,** Repellency and host specificity in the white pine weevil, *Physiol. Zool.,* 29, 314, 1956.

963. **Christiansen, E., Waring, R. H., and Berryman, A. A.,** Resistance of conifers to bark beetle attack: searching for general relationships, *Forest Ecol. Management,* 22, 89, 1987.

964. **Nijholt, W. W., McMullen, L. H., and Safranyik, L.,** Pine oil protects living trees from attack by three bark beetle species, *Dendroctonus* spp. (Coleoptera:Scolytidae), *Can. Entomol.,* 113, 337, 1981.

965. **Ohnesorge, B. and Serafimovski, A.,** The suitability of omarika spruce for needle-eating and sucking arthropods, *Allgem. Forst Jagd.,* 132, 129, 1961.

966. **Harris, L. J., Borden, J. H., Pierce, H. D., Jr., and Oehlschlager, A. C.,** Cortical resin monoterpenes in Sitka spruce and resistance to the white pine weevil *Pissodes strobi* (Coleoptera:Curculionidae), *Can. J. Forest Res.,* 13, 350, 1983.

967. **Siegel, M.,** Thin-layer chromatography of *Pinus* alkaloids and their significance in the attack of pines by *Rhyacionia buoliana* (European pine shoot moth), *Biol. Zentralbl.,* 88, 629, 1969.

968. **All, J. N. and Benjamin, D. M.,** Olfactory response of *Neodiprion rugifrons* larvae to inhibitory extracts of juvenile jack pine, *Pinus banksiana,* foliage, *J. Georgia Entomol. Soc.,* 10, 300, 1975.

969. **All, J. N., Benjamin, D. M., and Matsumura, F.,** Influence of semi-purified constituents of juvenile jack pine *Pinus banksiana,* foliage and other pine-derived chemicals on feeding of *Neodiprion swainei* and *N. rugifrons* larvae, *Ann. Entomol. Soc. Am.,* 68, 1095, 1975.

970. **All, J. N. and Benjamin, D. M.,** Deterrent(s) in jack pine, *Pinus banksiana,* influencing larval feeding behavior and survival of *Neodiprion swainei* and *N. rugifrons, Ann. Entomol. Soc. Am.,* 68, 495, 1975.

971. **Schuh, B. A. and Benjamin, D. M.,** Evaluation of commercial resin acids as feeding deterrents against *Neodiprion dubiosus, N. lecontei,* and *N. rugifrons* (Hymenoptera:Diprionidae), *J. Econ. Entomol.,* 77, 802, 1984.

972. **Schuh, B. A. and Benjamin, D. M.,** The chemical feeding ecology of *Neodiprion dubiosus* Schedl., *N. rugifrons* Midd., and *N. lecontei* (Fitch) on jack pine (*Pinus banksiana* Lamb.), *J. Chem. Ecol.,* 10, 1071, 1984.

973. **All, J. N. and Benjamin, D. M.,** Potential of antifeedants to control larval feeding of selected *Neodiprion* sawflies (Hymenoptera:Diprionidae), *Can. Entomol.,* 108, 1137, 1976.

974. **Ikeda, T., Matsumura, F., and Benjamin, D. M.,** Chemical basis for feeding adaptation of pine sawflies *Neodiprion rugifrons* and *Neodiprion swainei, Science,* 197, 497, 1977.

975. **Ikeda, T., Matsumura, F., and Benjamin, D. M.,** Mechanism of feeding discrimination between matured and juvenile foliage by two species of pine sawflies, *J. Chem. Ecol.,* 3, 677, 1977.

976. **Richmond, C. E.,** Effectiveness of two pine oils for protecting lodgepole pine from attack by mountain pine beetle (Coleoptera:Scolytidae), *Can. Entomol.,* 117, 1445, 1985.

977. **Sumitomo, M., Shiraga, M., and Kondo, T.,** Ethane in pine needles preventing the feeding of the beetle, *Monochamus alternatus, J. Insect Physiol.,* 21, 713, 1975.

978. **Coyne, J. F. and Lott, L. H.,** Toxicity of substances in pine oleoresin to southern pine beetles, *J. Georgia Entomol. Soc.,* 11, 301, 1976.

979. **Carter, F. L., Stringer, C. A., and Taras, M. A.,** Termiticidal properties of slash pine wood related to position in the tree, *Wood Sci.,* 12, 46, 1979.

980. **Smith, R. H.,** The fumigant toxicity of three pine resins to *Dendroctonus brevicomis* and *D. jeffreyi, J. Econ. Entomol.,* 54, 365, 1961.

981. **Smith, R. H.,** Toxicity of pine resin vapors on the western pine beetle, *J. Econ. Entomol.,* 58, 509, 1965.

982. **Scheffrahn, R. H. and Rust, M. K.,** Drywood termite feeding deterrents in sugar pine and antitermitic activity of related compounds, *J. Chem. Ecol.,* 9, 39, 1983.

983. **Vandermar, T. J. D.,** Resistance of western white pine to feeding and oviposition by *Pissodes strobi* Peck in western Canada, *J. Chem. Ecol.,* 4, 641, 1978.

984. **Blanc, A. and Blanc, M.,** Study of the effects of extracts of *Pinus pinaster* on attraction and feeding by *Pissodes notatus* F. (Col., Curculionidae), *Ann. Zool. Ecol. Anim.,* 7, 525, 1975.

985. Sankei Chemicals Co., Cineole as an insect repellent, Jap. Patent 57,179,105, Apr. 27, 1981.

986. Sankei Chemicals Co., Jap. Patent 57,179,103, Apr. 27, 1981.

987. **Kangas, E., Perttunen, V., Oksanen, H., and Rinnea, M.,** Laboratory experiments on the olfactory orientation of *Blastophagus piniperda* L. (Coleoptera, Scolytidae) to substances isolated from pine rind, *Acta Entomol. Fenn.,* 22, 1967.

988. **Chararas, C.,** Attraction of scolytids to conifers and the role of terpenic substances extracted from the oleoresins, *Rev. Pathol. Veg. Entomol. Agr. France,* 38, 113, 1958 (In French).

989. **Khalsa, H. C., Nieam, B. S., and Agarwal, P. N.,** Resistance of coniferous timbers to lyctus attack, *Indian J. Entomol.,* 26, 113, 1965.

990. **Smelyanetz, V. P.,** Mechanisms of plant resistance in pine trees, *Pinus sylvestris.* I. Indicators of physiological state in interacting plant-insect populations, *Z. Angew. Entomol.,* 83, 225, 1977.

991. **Smelyanetz, V. P.,** Mechanisms of plant resistance in Scotch pine (*Pinus sylvestris*). V. Changes in pest population during feeding on pines having different degree of resistance (residual preferendum and phase of afteraction), *Z. Angew. Entomol.,* 84, 344, 1977.

992. **Steiner, K.,** Genetic differences in resistance of Scotch pine to Eastern pine-shoot borer, *Great Lakes Entomol.,* 7, 103, 1974.

993. **Smythe, R. V. and Carter, F. L.,** Feeding responses to sound wood by *Coptotermes formosanus, Reticulitermes flavipes,* and *R. virginicus* (Isoptera:Rhinotermitidae), *Ann. Entomol. Soc. Am.,* 63, 841, 1970.

994. **Goyer, R. A. and Williams, V. G.,** The effects of feeding by *Leptoglossus corculus* (Say) and *Tetyra bipunctata* (Herrich and Schaffer) on loblolly pine (*Pinus taeda* L.) conelets, *J. Georgia Entomol. Soc.,* 16, 16, 1981.

995. **Gollob, L.,** Monoterpene composition in bark beetle-resistant loblolly pine, *Naturwissenschaften,* 67, 409, 1980.

996. **Higbee, E. C.,** Insecticidal plants in the Americas, *Pan Amer. Union Bull. No. 76,* 252, 1942.

997. **Matsui, K. and Munakata, K.,** The structure of piperenone, a new insect antifeeding substance from *Piper futokadsura, Tetrahedron Lett.,* 1905, 1975.

998. **Su, H. C. F. and Sondengam, B. L.,** Laboratory evaluation of toxicity of two alkaloidal amides of *Piper guineense* to four species of stored product insects, *J. Georgia Entomol. Soc.,* 15, 47, 1980.

999. **Okogun, J. I., Sondengam, B. L., and Kimbu, S. F.,** New amides from the extracts of *Piper guineense, Phytochemistry,* 16, 1295, 1977.

1000. **Tackie, A. N., Dwuna-Badu, D., Ayim, J. S. K., El Sohly, H. N., Knapp, J. E., Slatkin, D. J., and Schiff, P. L., Jr.,** *N*-Isobutyloctadeca-*trans*-2,*trans*-4-dienamide: a new constituent of *Piper guineense, Phytochemistry,* 14, 1888, 1975.

1001. **Addae-Mensah, I., Torio, F. C., Dimonyeka, C. I., Baxter, I., and Sanders, J. K. M.,** Novel amide alkaloids from the roots of *Piper guineense, Phytochemistry,* 16, 757, 1977.

1002. **Vig, O. P., Ahuja, V. D., Sharma, M. L., and Sharma, S. D.,** Syntheses of some naturally occurring amides, *Indian J. Chem.,* 13, 1358, 1975.
1003. **Presborn, S. B. and Wymore, F. H.,** Attempts to protect sweet corn from infestations of the corn earworm, *Heliothis obsoleta* (Fabr.), *J. Econ. Entomol.,* 22, 666, 1929.
1004. **Lathrop, F. H. and Keirstead, L. G.,** Black pepper to control the bean weevil, *J. Econ. Entomol.,* 39, 534, 1946.
1005. **Su, H. C. F.,** Insecticidal properties of black pepper to rice weevils and cowpea weevils, *J. Econ. Entomol.,* 70, 18, 1977.
1006. **Su, H. C. F.,** Laboratory study on toxic effect of black pepper varieties to three species of stored-product insects, *J. Georgia Entomol. Soc.,* 13, 266, 1978.
1007. **Su, H. C. F. and Horvat, R.,** Isolation, identification, and insecticidal properties of *Piper nigrum* amides, *J. Agr. Fd. Chem.,* 29, 115, 1981.
1008. **Anonymous,** Peppery pesticide, *Agr. Res. (USDA),* 26, (11), 15, 1978.
1009. **Scott, W. P. and McKibben, G. H.,** Toxicity of black pepper extract to boll weevils, *J. Econ. Entomol.,* 71, 343, 1978.
1010. **Raina, M. L., Dhar, K. L., and Atal, C. K.,** Occurrence of *N*-isobutyl eicosa-*trans*-2-*trans*-4-dienamide in *Piper nigrum, Planta Med.,* 30, 198, 1976.
1011. **Miyakado, M., Nakayama, I., Yoshioka, H., and Nakatani, N.,** The *Piperaceae* amides. I. Structure of pipercide, a new insecticidal amide from *Piper nigrum* L., *Agr. Biol. Chem.,* 43, 1609, 1979.
1012. **Miyakado, M. and Yoshioka, M.,** The *Piperaceae* amides. II. Synthesis of pipercide, a new insecticidal amide from *Piper nigrum* L., *Agr. Biol. Chem.,* 43, 2413, 1979.
1013. **Harvill, E. K., Hartzell, A., and Arthur, J. M.,** Toxicity of piperine solutions to house flies, *Contrib. Boyce Thompson Inst.,* 13, 87, 1943.
1014. **Gupta, O. P., Dhar, K. L., and Atal, C. K.,** Structure of new amide from *Piper officinarum, Phytochemistry,* 15, 425, 1976.
1015. **Gupta, O. P., Gupta, S. C., Dhar, K. L., and Atal, C. K.,** A new amide from *Piper officinarum, Phytochemistry,* 16, 1436, 1977.
1016. **Gupta, O. P., Gupta, S. C., Dhar, K. L., and Atal, C. K.,** Structure of a new amide, filfiline from *Piper officinarum, Indian J. Chem.,* 148, 912, 1976.
1017. **Banerji, A., Rej, R. N., and Ghosh, P. C.,** Isolation of *N*-isobutyl-deca-*trans*-2-*trans*-4-dienamide from *Piper sylvaticum* Roxb., *Experientia,* 3, 223, 1974.
1018. **Chartol, A.,** *Piper umbellatum,* a plant insecticide, *Med. Trop.,* 24, 743, 1964 (in French).
1019. **Kubo, I., Taniguchi, M., Chapya, A., and Tsujimoto, K.,** An insect antifeedant and antimalarial agent from *Plumbago capensis, Planta Med (Suppl.),* 185, 1980.
1020. **Kubo, I., Uchida, M., and Klocke, J. A.,** An insect ecdysis inhibitor from the African medicinal plant, *Plumbago capensis* (Plumbaginaceae), a naturally occurring chitin synthetase inhibitor, *Agr. Biol. Chem.,* 47, 911, 1983.
1021. **Kubo, I., Klocke, J. A., Matsumoto, T., and Kamikawa, T.,** Plumbagin as a model for insect ecdysis inhibitory activity, in *IUPAC Pesticide Chemistry (Human Welfare and the Environment),* Miyamoto, J., et al., Eds., Pergamon, New York, 1983, 169.
1022. **Jonasson, T.,** Resistance of oat plants to larval attack by the fruit fly, *Oscinella fruit* L. (Dipt., Chloropidae), *Z. Angew. Entomol.,* 93, 508, 1982.
1023. **Mikolajczak, K. L., Freedman, B., Zilkowski, B. W., Smith, C. R., Jr., and Burkholder, W. E.,** Effect of oat constituents on aggregation behavior of *Oryzaephilus surinamensis, J. Agr. Fd. Chem.,* 31, 30, 1983.
1024. **Tiwari, B. K., Rajpai, V. N., and Agarwal, P. N.,** Evaluation of insecticidal, fumigant and repellent properties of lemongrass oil, *Indian J. Exp. Biol.,* 4, 128, 1966.
1025. **Kim, M., Koh, H.-S., Obata, T., Fukami, H., and Ishii, S.,** Isolation and identification of *trans*-aconitic acid as the antifeedant in barnyard grass against the brown planthopper, *Nilaparvata lugens* (Stal.) (Homoptera:Delphacidae), *Appl. Entomol. Ecol.,* 11, 53, 1976.
1026. **Starks, K. J. and Mirkes, K. A.,** Yellow sugarcane aphid: plant resistance in cereal crops, *J. Econ. Entomol.,* 72, 486, 1979.

1027. **Verma, G. D. and Mehto, D. N.,** Relative susceptibility of some important wheat and barley varieties to *Rhopalosiphum maidis* Fitch, *Natl. Acad. Sci. Lett. (India),* 1, 315, 1978.

1028. **Corcuera, L. J., Argandona, V. H., and Niemeyer, H. M.,** Proc. Int. Symp. on Chemistry and Biology of Hydroxamic Acids, 1982, 111.

1029. **Menendez Ramos, R.,** *Melinis minutiflora* and the tick, *Rev. Agr. Puerto Rico,* 12, 219, 1924 (in Spanish).

1030. **Smyth, E.,** Why not trap-crops that entrap?, *J. Econ. Entomol.,* 18, 550, 1925.

1031. **Rosenfeld, A. H.,** Why not trap-crops that entrap?, *J. Econ Entomol.,* 18, 639, 1925.

1032. **Finnemore, H.,** *The Essential Oils,* London, 1926, 880 pp.

1033. **Anonymous,** *Oil, Paint Drug Reptr.,* 144, 5, 1943.

1034. **Woodhead, S., Padgham, D. E., and Bernays, E. A.,** Insect feeding on different sorghum cultivars in relation to cyanide and phenolic acid content, *Ann. Appl. Biol.,* 95, 151, 1980.

1035. **Atkin, D. S. J. and Hamilton, R. J.,** The effects of plant waxes on insects, *J. Nat. Prod.,* 45, 694, 1982.

1036. **Atkin, D. S. J. and Hamilton, R. J.,** The change with age in the epicuticular wax of *Sorghum bicolor, J. Nat. Prod.,* 45, 697, 1982.

1037. **Adams, C. M. and Bernays, E. A.,** The effect of combinations of deterrents on the feeding behavior of *Locusta migratoria, Entomol. Exp. Appl.,* 23, 101, 1978.

1038. **Woodhead, S.,** *p*-Hydroxybenzaldehyde in the surface wax of sorghum: its importance in seedling resistance to acridids, *Entomol. Exp. Appl.,* 31, 296, 1982.

1039. **McGinnis, A. and Kasting, R.,** Comparison of tissues from solid and hollow stemmed spring wheats during growth. III. An ether-soluble substance toxic to larvae of the pale western cutworm, *Agrotis orthogonia* Merr. (Lepidoptera:Noctuidae), *Entomol. Exp. Appl.,* 5, 313, 1962.

1040. **Thorsteinson, A. J. and Nayar, J. K.,** Plant phospholipids as feeding stimulants for grasshoppers, *Can. J. Zool.,* 41, 931, 1963.

1041. **Carlson, S. K., Patterson, F. L., and Gallun, R. L.,** Inheritance of resistance to hessian fly derived from *Triticum turgidum, Crop. Sci.,* 18, 1011, 1978.

1042. **Jain, S. C., Nowicki, S., Eisner, T., and Meinwald, J.,** Insect repellents from vetiver oil. I. Zizanal and epizizanal, *Tetrahedron Lett.,* 23, 4639, 1982.

1043. **Beck, S. D.,** The European corn borer, *Pyrausta nubilalis* (Hubn.) and its principal host plant. VI. Host plant resistance to larval establishment, *J. Insect Physiol.,* 1, 158, 1957.

1044. **Beck, S. D.,** The European corn borer, *Pyrausta nubilalis* (Hubn.) and its principal host plant. VII. Larval feeding behavior and host plant resistance, *Ann. Entomol. Soc. Am.,* 53, 206, 1960.

1045. **Gahukar, R. T.,** Feeding behavior and food intake of *Ostrinia nubilalis* larvae (Lep. Pyraustidae) in the presence of DIMBOA, *Ann. Soc. Entomol. France,* [N.S.], 15, 649, 1979 (in French).

1046. **Klun, J. A. and Robinson, J. F.,** Concentration of two 1,4-benzoxazinones in dent corn at various stages of development of the plant and its relation to resistance of the host plant to the European corn borer, *J. Econ. Entomol.,* 62, 214, 1969.

1047. **Smissman, E. E., Lapidus, J. B., and Beck, S. D.,** Corn resistance factor, *J. Org. Chem.,* 22, 220, 1957.

1048. **Klun, J. A. and Brindley, T. A.,** Role of 6-methoxybenzoxazolinone in inbred resistance of host plant (maize) to first-brood larvae of European corn borer, *J. Econ. Entomol.,* 59, 711, 1966.

1049. **Klun, J. A., Tripton, C. L., and Brindley, T. A.,** 2,4-Dihydroxy-7-methoxy-1,4-benzoxazin-3-one (DIMBOA), an active agent in the resistance of maize to the European corn borer, *J. Econ. Entomol.,* 60, 1529, 1967.

1050. **Robinson, J. F., Klun, J. A., and Brindley, T. A.,** European corn borer: a nonpreference mechanism of leaf feeding resistance and its relationship to 1,4-benzoxazin-3-one concentration in dent corn tissue, *J. Econ. Entomol.,* 71, 461, 1978.

1051. **Robinson, J. F., Klun, J. A., Guthrie, W. D., and Brindley, T. A.,** European corn borer (Lepidoptera:Pyralidae) leaf feeding resistance:dimboa bioassays, *J. Kansas Entomol. Soc.,* 55, 357, 1982.

1052. **Straub, R. W. and Fairchild, M. L.,** Laboratory studies of resistance in corn to the corn earworm, *J. Econ. Entomol.,* 63, 1901, 1970.

1053. **Wiseman, B. R., McMillian, W., W., and Widstrom, N. W.,** Potential of resistant corn to reduce corn earworm production, *Florida Entomol.,* 61, 92, 1978.
1054. **Waiss, A. C., Jr., Chan, B. G., Elliger, C. A., Wiseman, B. R., McMillian, W. W., Widstrom, N. W., Zuber, M. S., and Keaster, A. J.,** Maysin, a flavone glycoside from corn silks with antibiotic activity toward corn earworm, *J. Econ. Entomol.,* 72, 256, 1979.
1055. **Elliger, C. A., Chan, B. G., Waiss, A. C., Jr., Lundin, R. E., and Haddon, W. F.,** C-Glycosylflavones from *Zea mays* that inhibit insect development, *Phytochemistry,* 19, 293, 1980.
1056. **Kubo, I. and Kamikawa, T.,** Identification and efficient synthesis of 6-methoxy-2-benzoxazoli-none (MBOA), an insect antifeedant, *Experientia,* 39, 355, 1983.
1057. **Fortier, G., Arnason, J. T., Lambert, J. D. H., McNeill, J., Nozzolillo, C., and Philogene, B. J. R.,** Local and improved corns (*Zea mays*) in small farm agriculture in Belize, C. A.; their taxonomy, productivity, and resistance to *Sitophilus zeamais, Phytoprotection,* 63, 68, 1982.
1058. **Qi, Y.-T. and Burkholder, W. E.,** Protection of stored wheat from the granary weevil by vegetable oils, *J. Econ. Entomol.,* 74, 502, 1981.
1059. **Serratos, A., Arnason, J. T., Nozzolillo, C., Lambert, J. D. H., Philogene, B. J. R., Fulcher, G., Davison, K.,, Peacock, L., Atkinson, J., and Morand, P.,** Factors contributing to resistance of exotic maize populations to maize weevil, *Sitophilus zeamais, J. Chem. Ecol.,* 13, 751, 1987.
1060. **Stockel, J. P.,** Effect of aromatic extracts of corn seeds on the reproductive activity of *Sitotroga cerealella* Oliv. (Lepidoptera, Gelechiidae) under laboratory conditions, *C. R. Acad. Sci. Paris,* 292, 653, 1981.
1061. **Anon.,** Insect repellents, *Discovery,* 9, 294, 1948.
1062. **Kubo, I., Matsumoto, T., and Klocke, J. A.,** Multichemical resistance of the conifer *Podocarpus gracilior* (Podocarpaceae) to insect attack, *J. Chem. Ecol.,* 10, 547, 1984.
1063. **Saeki, I., Sumimoto, M., and Kondo, T.,** The termiticidal substances from the wood of *Podocarpus macrophyllum* D. Don., *Holzforschung,* 24, 83, 1970.
1064. **Singh, P., Fenemore, P. G., and Russell, G. B.,** Insect-control chemicals from plants. II. Effects of five natural norditerpene dilactones on the development of the housefly, *Austral. J. Biol. Sci.,* 26, 911, 1973.
1065. **Russell, G. B., Fennemore, P. G., and Singh, P.,** Insect-control chemicals from plants. Nagilactone C, a toxic substance from the leaves of *Podocarpus nivalis* and *phallii, Austral. J. Biol. Sci.,* 25, 1025, 1972.
1066. **Russell, G. B., Fenemore, P. G., and Singh, P.,** Structures of hallactones A and B, insect toxins from *Podocarpus hallii, J. Chem. Soc., Chem. Commun.,* 166, 1973.
1067. **Nakanishi, K., Koreeda, M., Sasaki, S., Chang, M. L., and Hsu, H. Y.,** The structure of ponasterone A, and insect moulting hormone from the leaves of *Podocarpus nakai* Hay, *J. Chem. Soc. Chem. Commun.,* 24, 915, 1966.
1068. **Zimmerman, M.,** Facultative deposition of an oviposition-deterring pheromone by *Hylemya, Environ. Entomol.,* 11, 519, 1982.
1069. **Barnes, C. S., and Loder, J. W.,** The structure of polygodial: a new sesquiterpene dialdehyde from *Polygonum hydropiper* L., *Austral. J. Chem.,* 15, 322, 1962.
1070. **Howell, S. C., Ley, S. V., Mahon, M., and Worthington, P. A.,** Synthesis of cinnamolide and polygodial, *J. Chem. Soc. Chem. Commun.,* 507, 1981.
1071. **Jallali-Naini, M., Boussac, G., Lemsitre, P., Larcheveque, M., Guillerm, D., and Lallemand, J. Y.,** Efficient total syntheses of polygodial and drimenin, *Tetrahedron Lett.,* 22, 2995, 1981.
1072. **Fukuyama, Y., Sato, T., Asakawa, Y., and Takemoto, T.,** A potent cytotoxic warburganal and related drimane-type sesquiterpenoids from *Polygonum hydropiper, Phytochemistry,* 21, 2895, 1982.
1073. **Abivardi, C. and Benz, G.,** Oviposition-deterring activity of sixteen extracts of medicinal plants, extensively used in modern medicine, against *Cydia pomonella* L. (Lepidoptera:Tortricidae), *Mitteil. Schweiz. Entomol. Ges.,* 59, 31, 1986.
1074. **Frear, D. E. H.,** *A Catalogue of Insecticides and Fungicides. II. Chemical Fungicides and Plant Insecticides,* Chronica Botanica Co., Waltham, Massachusetts, 1948.
1075. **Saito, M.,** Insecticide, Jap. Patent 3,299, 1951.

1076. **Arnason, J. T., Philogene, B. J. R., Donskov, N., Muir, A., and Towers, G. H. N.,** Psilotin, an insect feeding deterrent and growth reducer from *Psilotum nudum, Biochem. System. Ecol.,* 14, 287, 1986.

1077. **McIndoo, N. E.,** *A Review of the Insecticide Uses of the Rotenone Bearing Plants, 1938/44,* USDA Agr. Res. Service, Bureau of Entomology and Plant Quarantine No. E-713, 1947.

1078. **Deshpande, R. S., Adhikar, P. R., and Tipnis, H. P.,** Stored grain pest control agents from *Nigella sative* L. and *Pogostemon heyneanus* Benth., *Bull. Grain Technol.,* 12, 232, 1974.

1079. **Honigberger, J. M.,** *Thirty-Five Years in the East. The Medical Part,* London, vol.2, 1852.

1080. **Dymock, W., Warden, C. J. H., and Hooper, D.,** *Pharmacographia Indica,* London, vol. 3, 1893.

1081. **Hassanali, A., Lwande, W., and Gebreyesus, T.,** Structure-activity studies of acridone feeding deterrents, in *Natural Pesticides From the Neem Tree and Other Tropical Plants,* Schmutterer, H. and Ascher, K. R. S., Eds., GTZ Press, Eschborn, FRG, 1984, 75.

1082. **Dahlman, D. L. and Johnson, V.,** *Heteromeles arbutifolia* (Rosaceae:Pomoideae) found toxic to insects, *Entomol. News,* 91, 141, 1980.

1083. **Free, D. J.,** Resistance to development of larvae of the apple maggot in crab apples, *J. Econ. Entomol.,* 70, 611, 1977.

1084. **Briggs, J. B.,** Insect resistance in fruit plants, *Ann. Appl. Biol.,* 56, 325, 1965.

1085. **Buttery, R. G., Soderstrom, E. L., Seifert, R. M., Ling, L. C., and Haddon, W. F.,** Components of almond hulls: possible navel orangeworm attractants and growth inhibitors, *J. Agr. Fd. Chem.,* 28, 353, 1980.

1086. **Kaethler, F., Free, D. J., and Bown, A. W.,** HCN: a feeding deterrent in peach to the oblique-banded leafroller, *Choristoneura rosaceana* (Lepidoptera: Tortricidae), *Ann. Entomol. Soc. Am.,* 75, 568, 1982.

1087. **Gurai, S.,** Insecticidal property of branches of *Padus racemosa, Veterinarya,* 27, (7), 39, 1950.

1088. **Wolcott, G. N.,** The insects of "almendron", *Prunus occidentalis* Sw., *Puerto Rico Univ. J. Agr.,* 40, 203, 1956.

1089. **Chang, J. F. and Philogene, B. J. R.,** Response of *Psylla pyricola* (Homoptera:Psyllidae) to, and characterization of polar and lipid fractions of *Pyrus* sp. leaves, *Phytoprotection,* 59, 28, 1978.

1090. **Minaeff, M. G. and Wright, J. H.,** Mothproofing, *Ind. Engin. Chem.,* 21, 1187, 1929.

1091. **Rizvi, S. J. H., Pandey, S. K., Mukerjee, D., and Mathur, S. N.,** 1,3,7-Trimethylxanthine, a new chemosterilant for stored grain pest, *Callosobruchus chinensis* (L.), *Z. Angew. Entomol.,* 90, 378, 1980.

1092. **Bry, R. E.,** personal communication, 1978.

1093. **Chatterjee, P. N., Singh, P., and Sivaramakrishnan, R.,** Efficacy of various insecticides including indigenous vegetable products. II. Preliminary note on testing the flea repellent properties of an essential oil from *Boenninghausenia albiflora* (Hook) Reichb., *Indian Forester,* 324, 1968.

1094. **Sood, V. K., Karnik, M. G., Sagan, V., and Sood, K. C.,** Preliminary investigation on the essential oil of *Boenninghausenia albiflora* (Hook) Reichb., *Indian Forester,* 92, 295, 1966.

1095. **Su, H. C. F., Speirs, R. D., and Mahany, P. G.,** Citrus oils as protectants of black-eyed peas against cowpea weevils: laboratory evaluation, *J. Econ. Entomol.,* 65, 1433, 1972.

1096. **Su, H. C. F., Speirs, R. D., and Mahany, P. G.,** Toxicity of citrus oils to several stored-product insects: laboratory evaluation, *J. Econ. Entomol.,* 65, 1438, 1972.

1097. **Taylor, W. E. and Vickery, B.,** Insecticidal properties of limonene, a constituent of citrus oil, *Ghana J. Agr. Sci.,* 7, 61, 1974.

1098. **Su, H. C. F.,** Toxicity of a chemical component of lemon oil to cowpea weevils, *J. Georgia Entomol. Soc.,* 11, 297, 1976.

1099. **Su, H. C. F. and Horvat, R.,** Isolation and characterization of four major components from insecticidally active lemon peel extract, *J. Agr. Fd. Chem.,* 35, 509, 1987.

1100. **Alford, A. R., Cullen, J. A., Storch, R. H., and Bentley, M. D.,** Antifeedant activity of limonin against the Colorado potato beetle (Coleoptera:Chrysomellidae), *J. Econ. Entomol.,* 80, 575, 1987.

1101. **Hassanali, A. and Bentley, M. D.,** Comparison of the insect antifeedant activities of some limonoids, in *Natural Pesticides From the Neem Tree and Other Tropical Plants,* Schmutterer, H. and Ascher, K. R. S., Eds., GTZ Press, Eschborn, FRG, 1987, 683.
1102. **Olenev, N. O.,** On the effect of the toxic principles of higher plants on Ixodid ticks, *Dokl. Akad. Nauk SSSR,* 71, 1119, 1950.
1103. **Bennett, R. D. and Hasegawa, S.,** Isolimonic acid, a new citrus limonoid, *Phytochemistry,* 19, 2417, 1980.
1104. **Rousseff, R. L.,** Preparative high performance liquid chromatography of bitter and nonbitter citrus limonoids, in *Semiochemistry: Flavors and Pheromones,* Acree, T. E. and Soderlund, D. M., Eds., De Gruyter, New York, 1985, 275.
1105. **Jones, V., Pollard, G. V., and Seafirth, C. E.,** Chemical deterrency of *Citrus reticulata* (Blanco) to the leaf-cutting ant, *Acromyrmex octospinosus* (Reich), *Insect Sci. Appl.,* 8, 99, 1987.
1106. **Bennett, R. D. and Hasegawa, S.,** Limoneids of calamondin seeds, *Tetrahedron,* 37, 17, 1981.
1107. **Anonymous,** Scientists peel an orange, discover a powerful insecticide, St. Petersburg (Florida) Times, Oct. 17, 1983, Bl.
1108. **Anon.,** Natural chemicals protect citrus from caribfly, *Citrus Ind.,* 58(12), 12, 1977.
1109. **Altieri, M. A., Lippmann, M., Schmidt, L. L., and Kubo, I.,** Antifeedant effects of nomilin on *Spodoptera frugiperda* (J. E. Smith) (Lepidoptera, Noctuidae) and *Trichoplusia ni* (Hübner) (Lepidoptera, Noctuidae) under laboratory and lathehouse conditions, *Protect. Ecol.,* 6, 91, 1984.
1110. **Hassanali, A., Bentley, M. D., Ole Sitayo, E. N., Njoroge, P. E. W., and Yatagai, M.,** Studies on limonoid insect antifeedants, *Insect Sci. Appl.,* 7, 495, 1986.
1111. **Mester, I., Szedrei, K., and Reisch, J.,** Constituents of *Clausena anisata* (Willd.) Oliv. (Rutaceae). 1. Coumarins from the root bark, *Planta Med.,* 32, 81, 1977 (in German).
1112. **Meyer, T. M.,** The essential oil from the leaves of *Clausena anisata* Hook, *Rec. Trav. Chim. Pays-Bas,* 66, 395, 1947.
1113. **Okunade, A. L. and Olaifa, J. I.,** Estragole, an acute toxic principle from the volatile oil of the leaves of *Clausena anisata, J. Nat. Prod.,* 50, 990, 1987.
1114. **Berenbaum, M.,** Toxicity of a furanocoumarin in armyworms: a case of biosynthetic escape from insect herbivores, *Science,* 201, 532, 1978.
1115. **Chou, F. Y., Hostettmann, K., Kubo, I., Nakanishi, K., and Taniguchi, M.,** Isolation of an insect antifeedant *N*-methylflindersine and several benz [C] - phenanthridine alkaloids from East African plants: a comment on Chelerythrine, *Heterocycles,* 7, 169, 1977.
1116. **Yajima, T., Kato, N., and Munakata, K.,** Isolation of insect anti-feeding principles in *Orixa japonica* Thunb., *Agr. Biol. Chem.,* 41, 1263, 1977.
1117. **Ayafor, J. F., Sondengam, B. L., Connolly, J. D., Rycroft, D. S., and Okogun, J. I.,** Tetranortriterpenoids and related compounds. 26. Tecleanin, a possible precursor of limonin, and other new tetranortriterpenoids from *Teclea grandifolia* Engl. (Rutaceae), *J. Chem. Soc. Perkin I,* 1750, 1981.
1118. **Lwande, W., Gebreyesus, T., Chapya, A., Macfoy, C., Hassanalli, A., and Okech, M.,** 9-Acridone insect antifeedant alkaloids from *Teclea trichocarpa* bark, *Insect Sci. Appl.,* 4, 393, 1983.
1119. **Jacobson, M.,** Herculin, a pungent insecticidal constituent of southern prickly ash bark, *J. Am. Chem. Soc.,* 70, 4234, 1948.
1120. **Stermitz, F. R. and Sharifi, I. A.,** Alkaloids of *Zanthoxylum monophyllum* and *Z. punctatum,* *Phytochemistry,* 2003, 1977.
1121. **Capimera, J. L., and Stermitz, F. R.,** Laboratory evaluation of zanthophylline as a feeding deterrent for range caterpillar, migratory grasshopper, alfalfa weevil, and greenbug, *J. Chem. Ecol.,* 5, 767, 1979.
1122. **Clark, J. V.,** Feeding deterrent receptors in the last instar African armyworm *Spodoptera exempta:* a study using salicin and caffein, *Entomol. Exp. Appl.,* 29, 189, 1981.
1123. **Kurir, A.,** On the occurrence of *Earias chlorana* with special regard to its feeding plants, *Centralbl. Gesamte Forstwirt.,* 84, 231, 1967.
1124. **Matsuda, K. and Senbo, S.,** Chlorogenic acid as a feeding deterrent for the salicaceae-feeding leaf beetle, *Lochmaeae capreae cribrato* (Coleoptera:Chrysomelidae) and other species of leaf beetles, *Appl. Entomol. Zool.,* 21, 411, 1986.

1125. Shankaranarayana, K. H., Shivaramakrishnan, V. R., Ayyar, K. S., and Sen Sarma, P. K., Isolation of a compound from the bark of sandal, *Santalum album* L. and its activity against some lepidopterous and coleopterous insects, *J. Entomol. Res.*, 3, 116, 1979.

1126. **Shankaranarayana, K. H., Ayyar, K. S., and Krishna Rao, C. S.**, Insect growth inhibitor from the bark of *Santalum album*, *Phytochemistry*, 19, 1239, 1980.

1127. **Hoover, S. L.**, Insecticide, U.S. Patent 1,619,258, Mar. 1, 1927.

1128. **Von Sandermann, W., Dietrichs, H. S., and Gottwald, A.**, Examination of ancient woods and their significance for wood protection, *Holz Roh-u. Werkstoff*, 16, 191, 1958.

1129. **Von Sandermann, W. and Funke, H.**, Termite resistance of old temple woods of the Mayan era by means of saponins, *Naturwissenschaften*, 57, 407, 1970.

1130. **Waldbauer, G. P., Yamamoto, R. T., and Bowers, W. S.**, Laboratory rearing of the tobacco hornworm, *Protoparce sexta* (Lepidoptera, Sphingidae), *J. Econ. Entomol.*, 57, 93, 1964.

1131. **Liu, H.-W., Kusumi, T., and Nakanishi, K.**, A hydroperoxychroman with insect antifeedant properties from an African shrub. Characterization of fully-substituted aromatic structures, *J. Chem. Soc. Chem. Commun.*, 1971, 1981.

1132. **Hassanali, A., Bentley, M. D., Slawin, A. M. Z., Williams, D. J., Shephard, R. N., and Chapya, A. W.**, Pedonin, a spiro tetranortriterpenoid insect antifeedant from *Harrisonia abyssinica, Phytochemistry*, 26, 573, 1987.

1133. **Klocke, J. A., Arisawa, M., Honda, S. S., Kinghorn, A. D., Cordell, C. A., and Farnsworth, N. R.**, Growth inhibitory insecticidal and antifeedant effects of some antileukemic and cytotoxic quassinoids on two species of agricultural pests, *Experientia*, 41, 379, 1985.

1134. **Leskinen, V., Polonsky, J., and Bhatnagar, S.**, Antifeedant activity of quassinoids, *J. Chem. Ecol.*, 10, 1497, 1984.

1135. **Lidert, Z., Wing, K., Polonsky, J., Imakura, Y., Okano, M., Tani, S., Lin, Y.-M., Kiyokawa, H., and Lee, K.-H.**, Insect antifeedant and growth inhibitory activity of forty-six quassinoids on two species of agricultural pests, *J. Nat. Prod.*, 50, 442, 1987.

1136. **Yamamoto, R. T. and Fraenkel, G.**, Assay of the principal gustatory stimulant for the tobacco hornworm *Protoparce sexta*, from solanaceous plants, *Ann. Entomol. Soc. Am.*, 36, 802, 1960.

1137. **Koch, U. A.**, Action of insecticides and plant alkaloids on the digestive tract of larvae and adults of the Colorado potato beetle (*Leptinotarsa decemlineata*), *Entomol. Exp. Appl.*, 3, 103, 1960.

1138. **Devitt, B. D., Philogene, B. J. R., and Hinks, G. F.**, Effects of veratrine, berberine, nicotine, and atropine on developmental characteristics and survival of the dark-sided cutworm, *Euxoa messoria* (Lepidoptera:Noctuidae), *Phytoprotection*, 61, 88, 1980.

1139. **Majule, F. O.**, An investigation into the insecticide effect of capsicums, *Tech. Rept. Nigerian Stored Prod. Res. Inst.*, 1974, (11), 89, 1977.

1140. **Chand, P.**, Host preference in *Diacrisia obliqua* Wlk. (Lepidoptera) in the field, *Sci. Culture (India)*, 41, 604, 1975.

1141. **Kirtaniya, S. D., Ghosh, M. R., Mitra, S. R., Adityachaudhury, N., and Chatterjee, S. R.**, Note on insecticidal properties of the fruits of chilli, *Indian J. Agr. Sci.*, 50, 510, 1980.

1142. **Levinson, H. Z.**, The defensive role of alkaloids in insects and plants, *Experientia*, 32, 408, 1976.

1143. **Corby, H. D. R.**, Report of a study of a pest (*Pachymerus longus* Pie.) causing damage to groundnuts in the Wurkum District of the Muri Division of Adamaura, Rept. Nigeria Dept. Agr., Samaru, Zaria, No. 9302/167, 1941.

1144. **Ascher, K. R. S., Eliyahu, M., Goldman, A., Kirson, I., Abraham, A., Jacobson, M., and Schmutterer, H.**, The antifeedant effect of some new withanolides on three insect species, *Spodoptera littoralis, Epilachna varivestis*, and *Tribolium castaneum, Phytoparasitica*, 15, 15, 1987.

1145. **Adam, C., Chien, N. Q., and Khoi, N. H.**, Dunawithanine A and B, first plant withanolide glycosides from *Dunalia australis, Naturwissenschaften*, 68, 425, 1981.

1146. **DeBoer, G., and Hanson, F. E.**, Feeding responses to solanaceous allelochemicals by larvae of the tobacco hornworm, *Manduca sexta, Entomol. Exp. Appl.*, 45, 123, 1987.

1147. **Elliger, G. A., Wong, Y., Chan, B. G., and Waiss, A. C., Jr.**, Growth inhibitors in tomato (*Lycopersicon*) to tomato fruitworm (*Heliothis zea*), *J. Chem. Ecol.* 7, 753, 1981.

1148. **Duffey, S. S. and Isman, M. B.**, Inhibition of insect larval growth by phenolics in glandular trichomes of tomato leaves, *Experientia*, 37, 574, 1981.

1149. **Campbell, B. C. and Duffey, S. S.,** Tomatine and parasitic wasps: potential incompatibility of plant antibiosis with biological control, *Science,* 205, 700, 1979.

1150. **De Wilde, J., Sloof, R., and Bongers, W.,** A comparative study of feeding and oviposition preference in the Colorado beetle (*Leptinotarsa decemlineata* Say), *Ghent Landbouw. Meded.,* 25, 1340, 1960.

1151. **Anon.,** Tomato product repels grain weevils, *Agr. Res. (USDA),* 34, (8), 5, 1986.

1152. **Kennedy, G. G. and Yamamoto, R. T.,** A toxic factor causing resistance in a wild tomato to the tobacco hornworm and some other insects, *Entomol. Exp. Appl.,* 26, 121, 1979.

1153. **Williams, W. G., Kennedy, G. G., Yamamoto, R. T., Thacker, J. D., and Bordner, J.,** 2-Tridecanone: a naturally occurring insecticide from the wild tomato *Lycopersicon hirsutum* f. *glabratum, Science,* 207, 888, 1980.

1154. **Kennedy, G. G.,** 2-Tridecanone, tomatoes and *Heliothis zea:* potential incompatibility of plant antibiotics with insecticidal control, *Entomol. Exp. Appl.,* 35, 305, 1984.

1155. **Dymock, M. B., Kennedy, G. G., and Williams, W. G.,** Toxicity studies of analogs of 2-tridecanone, a naturally occurring toxicant from a wild tomato, *J. Chem. Ecol.,* 8, 837, 1982.

1156. **Juvik, J. A., Berlinger, M. J., Ben-David, T., and Rudich, J.,** Resistance among accessions of the genera *Lycopersicon* and *Solanum* to four of the main insect pests in Israel, *Phytoparasitica,* 10, 145, 1982.

1157. **Nalbandov, O., Yamamoto, R. T., and Fraenkel, C. S.,** Nicandrenone, a new compound with insecticidal properties isolated from *Nicandra physalodes, J. Agr. Fd. Chem.,* 12, 55, 1964.

1158. **Bates, R. B. and Eckert, D. J.,** Nicadrenone, an insecticidal plant steroid derivative with ring D aromatic, *J. Am. Chem. Soc.,* 94, 8258, 1972.

1159. **Subramanian, S. S., Sethi, P. D., and Adam, G.,** Structure of nicandrenone from *Nicandra physalodes, Indian J. Pharm.,* 35, 123, 1973.

1160. **Altrichter, J.,** The mysterious shoo-fly plant, *Esplor. Sci. Res.,* 1, (2), 10, 1938.

1161. **Glotter, E., Ascher, K. R. S., and Kirson, I.,** Naturally occurring steroids from solanaceous plants as potential insect antifeedants, BARD Project Special Rept. 1-142-80, Rehovot, Israel, Dec. 1, 1986, 130 pp.

1162. **Ascher, K. R. S., Schmutterer, H., Glotter, E., and Kirson, I.,** Withanolides and related ergostane-type steroids as antifeedants for larvae of *Epilachna varivestis* (Coleoptera:Coccinellidae), *Phytoparasitica,* 9, 197, 1981.

1163. **Von Schreiber, K.,** Solanum alkaloids. I. Solacauline, a new glycoalkaloid from the leaves of *Solanum acaule, Berichte,* 87, 1007, 1954.

1164. **Gibson, R. W. and PPickett, J. A.,** Wild potato repels aphids by release of aphid alarm pheromone, *Nature,* 302, 608, 1983.

1165. **Kuhn, R. and Low, I.,** New alkaloidglycosides in the leaves of *Solanum chacoense, Angew. Chem.,* 69, 236, 1957.

1166. **Sinden, S. L.,** Potato plants make their own insect repellent, *Agr. Res. (USDA),* 36, (2), 15, 1988.

1167. **Stürkow, B. and Low, I.,** The action of several alkaloid glycosides on the Colorado potato beetle, *Leptinotarsa decemlineata, Entomol. Exp. Appl.,* 4, 133, 1961 (in German).

1168. **Dalman, D. L.,** Responses of *Empoasca fabae* (Harris) (Cicadellidae, Homoptera) to selected alkaloids and alkaloidal glycosides of *Solanum* species, *Diss. Abstr.,* 26, 6245, 1966.

1169. **Torka, M.,** On the feeding test with *Leptinotarsa*-resistant potatoes, *Entomol. Exp. Appl.,* 1, 3, 1958.

1170. **Van de Klashorst, G. and Tingey, W. M.,** Effect of seedling age, environmental temperature, and foliar total glycoalkaloids on resistance of five *Solanum* genotypes to the potato leafhopper, *Environ. Entomol.,* 8, 690, 1979.

1171. **Sen, A.,** Chemical composition and morphology of epicuticular waxes from leaves of *Solanum tuberosum, Z. Naturforsch.,* 42C, 1153, 1987.

1172. **Bentley, M. D., Leonard, D. E., and Bushway, R. J.,** *Solanum* alkaloids as larval feeding deterrents for spruce budworm, *Choristoneura fumiferana* (Lepidoptera:Tortricidae), *Ann. Entomol. Soc. Am.,* 77, 401, 1984.

1173. **Harrison, C. D. and Mitchell, B. K.,** Host-plant acceptance by geographic populations of the Colorado potato beetle, *Leptinotarsa decemlineata.* Role of solanaceous alkaloids as sensory deterrents, *J. Chem. Ecol.,* 14, 777, 1988.

1174. **Pelletier, S. V., Mody, N. V., Nowacki, J., and Bhattacharyya, J.,** Carbon-13 nuclear magnetic resonance spectral analysis of withanolides and their derivatives, *J. Nat. Prod.,* 42, 512, 1979.

1175. **Hirayama, M., Gamoh, K., and Idekawa, N.,** Synthesis of the steroidal lactone moiety of withanolides, *Chem. Lett. (Tokyo),* 491, 1982.

1176. **Ascher, K. R. S., Nemny, N. E., Eliyahu, M., Kirson, I., Abraham, A., and Glotter, E.,** Insect antifeedant properties of withanolides and related steroids from Solanaceae, *Experientia,* 36, 62, 1980.

1177. **Ascher, K. R. S., Eliyahu, M., Glotter, E., Kirson, I., and Abraham, A.,** Distribution of the chemotypes of *Withania somnifera* in some areas of Israel: feeding studies with *Spodoptera littoralis* larvae and chemical examination of withanolide content, *Phytoparasitica,* 12, 148, 1984.

1178. **Ascher, K. R. S., Nemny, N. E., Eliyahu, M., Kirson, I., Abraham, A., and Glotter, E.,** Insect antifeedant ptoperties of withanolides and related steroids from Solanaceae, *Experientia,* 36, 998, 1980.

1179. **Sakata, K., Aoki, K., Chang, C.-F., Sakurai, A., Tamura, S., and Murakoshi, S.,** Stemospironine, a new insecticidal alkaloid of *Stemona japonica* Miq. Isolation, structural determination and activity, *Agr. Biol. Chem.,* 42, 457, 1978.

1180. **Beroza, M. and LaBresque, G. C.,** Chemosterilant activity of oils, especially oil of *Sterculia foetida,* in the house fly, *J. Econ. Entomol.,* 60, 196, 1967.

1181. **Walton, G. P. and Gardiner, R. F.,** Cocoa byproducts and their utilization as fertilizer materials, U.S. Dept. Agr. Bull. 1413, 1926, 44 pp.

1182. **Schanne, C.,** Isolation of a substance from *Taxus baccata* L. and *Taxus baccata* cv. *fastigiata* Loud. with insecticidal and metamorphosis-disrupting property on *Epilachna varivestis* Muls. (Col., Coccinellidae), *J. Appl. Entomol.,* 105, 303, 1988.

1183. **Perusse, F., Duval, A., Chawla, S. S., and Perron, J. M.,** Effects of *Taxus canadensis* on two insects (Homoptera:Aphididae; Coleoptera:Tenebrionidae), *Can. Entomol.,* 109, 268, 1977.

1184. **Scheffrahn, R. H., Hou, R.-C., Su, N.-Y., Huffman, J. B., Midland, S. L., and Sims, J. J.,** Allelochemical resistance of bald cypress, *Taxodium distichum,* heartwood to the subterranean termite, *Coptotermes formosanus, J. Chem. Ecol.,* 14, 765, 1988.

1185. **Jones, C. G., Aldrich, J. R., and Blum, M. S.,** 2-Furaldehyde from bald cypress. A chemical rationale for the demise of the Georgia silkworm industry, *J. Chem. Ecol.,* 7, 89, 1981.

1186. **Numata, A., Kitajima, A., Katsuno, T., Yamamoto, K., Nagakama, N., Takahashi, C., Fujiki, R., and Nabae, M.,** An antifeedant for the yellow butterfly larvae in *Camellia japonica:* a revised structure of camellidin II, *Chem. Pharm. Bull.,* 35, 3948, 1987.

1187. **Sivapalan, P. and Shivanandarajah, V.,** Inhibitory effects of extracts from seeds and roots of tea, seeds of *Sapindus emarginatus,* azasterols and steroidal amines on the development of *Xyleborus fornicatus* in vitro, *Entomol. Exp. Appl.,* 22, 274, 1977.

1188. **Dixon, E. R.,** Attack response of the smaller European elm bark beetle, *Scolytus multistriatus* in confinement, *J. Econ. Entomol.,* 57, 170, 1964.

1189. **Munakata, K.,** Insect antifeeding substances in plant leaves, *Agr. Biol. Chem.,* 40, 57, 1976.

1190. **Hosozawa, S., Kato, N., and Munakata, K.,** Diterpenes from *Caryopteris divaricata, Phytochemistry,* 13, 1019, 1974.

1191. **Hosozawa, S., Kato, N., and Munakata, K.,** Absolute configuration of caryoptin and 3-epicaryoptin—an exception in the exciton chirality method, *Tetrahedron Lett.,* 3753, 1974.

1192. **Hosozawa, S., Kato, N., and Munakata, K.,** Antifeeding active substances for insect in *Caryopteris divaricata* Maxim., *Agr. Biol. Chem.,* 38, 823, 1974.

1193. **Hosozawa, S., Kato, N., and Munakata, K.,** Diterpenoids from *Clerodendron calamitosum, Phytochemistry,* 13, 308, 1974.

1194. **Geuskens, R. B. M., Luteijn, J. M., and Schoonhoven, L. M.,** Antifeedant activity of some ajugarin derivatives in three lepidopterous species, *Experientia,* 39, 403, 1983.

1195. **Cooper, R., Solomon, P. H., Kubo, I., Nakanishi, K., Shoolery, J. N., and Occolowitz, J. L.,** Myricoside, an African armyworm antifeedant: separation by droplet countercurrent chromatography, *J. Am. Chem. Soc.,* 102, 7953, 1980.

1196. **Antonius, A. G. and Saito, T.,** Mode of action of antifeeding compounds in the larvae of the tobacco cutworm, *Spodoptera litura* (F.) (Lepidoptera:Noctuidae). I. Antifeeding activities of chlordimeform and some plant diterpenes, *Appl. Entomol. Zool.,* 16, 328, 1981.

1197. **Antonius, A. G. and Saito, T.,** Antifeeding activities of chlordimeform and some plant diterpenes when injected into tobacco cutworm, *J. Pesticide Sci.,* 7, 385, 1982.

1198. **Antonius, A. G. and Saito, T.,** Mode of action of antifeeding compounds in the larvae of the tobacco cutworm, *Spodoptera litura* (F.) (Lepidoptera:Noctuidae). III. Sensory responses of the larval chemoreceptors to chlordimeform and clerodin, *Appl. Entomol. Zool.,* 18, 40, 1983.

1199. **Kato, N., Takahashi, M., Shibayama, M., and Munakata, K.,** Antifeeding active substances for insects in *Clerodendron trichotomum* Thunb., *Agr. Biol. Chem.,* 36, 2579, 1972.

1200. **Kato, N., Shibayama, S., Munakata, K., and Katayama, C.,** Structure of the diterpene clerodendrin A, *Chem. Commun.,* 1632, 1971.

1201. **Kato, N., Munakata, K., and Katayama, C.,** Crystal and molecular structure of the *p*-bromobenzoate chlorohydrin of clerodendrin A, *J. Chem. Soc. Perkin Trans. II,* 69, 1973.

1202. **Kato, N., Shibayama, M., and Munakata, A.,** Structure of the diterpene clerodendrin A, *J. Chem. Soc. Perkin Trans. I,* 712, 1973.

1203. **Attri, B. S. and Singh, R. P.,** A note on the biological activity of *Lantana camara* L., *Indian J. Entomol.,* 39, 384, 1977.

1204. **Chogo, J.,** unpublished results.

1205. **Chogo, J. and Crank, G.,** Essential oil and leaf constituents of *Lippia ukambensis* from Tanzania, *J. Nat. Prod.,* 45, 186, 1982.

1206. **Bernays, E. and De Luca, C.,** Insect antifeedant properties of an iridoid glycoside: ipolamide, *Experientia,* 37, 1289, 1981.

1207. **Rimpler, H. and Christiansen, I.,** Tectograndinol, a new diterpene from *Tectona grandis* L. fil, *Z. Naturforsch.,* 32C, 724, 1977 (in German).

1208. **Kubo, I., Matsumoto, A., and Ayafor, J. F.,** Efficient isolation of a large amount of 20-hydroxyecdysone from *Vitex madiensis* (Verbenaceae) by droplet countercurrent chromatography, *Agr. Biol. Chem.,* 48, 1683, 1984.

1209. **Sander, L.,** The tsetse flies (Glossinae Wiedemann), *Arch. Schiff. Tropen. Hyg.,* 9, 355, 1905.

1210. **Jilani, G., Saxena, R. C., and Rueda, B. P.,** Repellent and growth-inhibiting effects of turmeric oil, sweetflag oil, neem oil, and "Margosan-O" on red flour beetle (Coleoptera:Tenebrionidae), *J. Econ. Entomol.,* 81, 1226, 1988.

1211. **Su, H. C. F., Horvat, R., and Jilani, C.,** Isolation, purification, and characterization of insect repellents from *Curcuma longa* L., *J. Agr. Fd. Chem.,* 30, 290, 1982.

1212. **Gundu Rao, H. R. and Majumder, S. K.,** *Sci. Culture (India),* 32, 461, 1963.

1213. **Dixit, R. S. and Perti, S. L.,** Indigenous insecticides. III. Insecticidal properties of some medicinal and aromatic plants, *Bull. Jammu Regional Res. Lab.,* 1, 169, 1963.

1214. **Jain, D. C.,** Antifeedant active saponin from *Balanites roxburghii* stem bark, *Phytochemistry,* 26, 2223, 1987.

Index

INDEX